COMMUNICATION SATELLITES IN THE GEOSTATIONARY ORBIT

DONALD M. JANSKY
MICHEL C. JERUCHIM

COMMUNICATION SATELLITES IN THE GEOSTATIONARY ORBIT

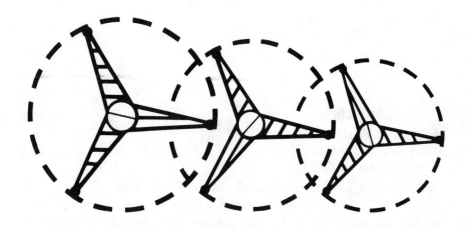

Artech House Inc.

Contents

Preface

In the remarkably short span of two decades, satellite communication has progressed from what was (in the public perception, at least) the fictional creation of imaginative writers to an established, major industry providing services as diverse as telephone interconnection, television distribution, data transfer, meteorological data, and maritime communication. Unquestionably, the pace of technological advance has been a major force in this rapid evolution. However, a probably equal contributor is the existence of a natural phenomenon called the "geostationary satellite orbit," which we shall henceforth refer to as the geostationary orbit or GSO. The GSO can be abstractly described as an imaginary circle in the plane of the equator, 22,300 miles above the earth, where the gravitational forces are such that an object (e.g., a satellite) properly placed there will appear from the earth to be stationary. The operational consequences of this property are what makes the GSO significant. While there are many important satellite applications using non-geostationary orbits, there is little doubt that the growth of satellite communications is due in large measure to the existence of the GSO.

Satellites cannot be placed arbitrarily closely along the GSO, not because of the possibility of collision, but because of harmful electromagnetic interference. Allowable interference considerations determine required satellite spacings in terms of degrees of angular separation for satellites using the same frequency band. Since the dimension of the GSO is 360°, there is therefore a finite number of such satellites that can be accommodated. It is thus recognized that the GSO is a limited natural resource, although it is often overlooked that it is not an expendable one. The radio frequency spectrum is also a limited natural resource of recognized value for which there is much demand. Communications, of course, cannot take place without using spectrum. Thus, the recognition of the joint and inextricable nature of these two aspects is frequently acknowledged by the term orbit/spectrum resource. The effective utilization of this resource is the subject of this book.

ix

Internationally, the orbit/spectrum resource is regulated by the International Telecommunication Union (ITU). Inevitably, regulations are shaped not only by technical factors, but at least equally by domestic and international policy. These two aspects intertwine and affect each other in complex ways. The subject of orbit/spectrum utilization, as we define it here, is concerned with the regulatory and technical factors that affect the use of that resource.

There has been much written on this subject, particularly on the technical aspects, in documents produced by the International Radio Consultative Committee (CCIR) of the ITU. However, most of this material is scattered throughout the literature or in CCIR reports. No systematic book-length exposition of orbit/spectrum utilization exists. Such a book now appears appropriate in view of the growing recognition of the subject's fundamental importance, a fact underscored by the forthcoming World Administrative Radio Conference (to be held in two sessions in 1985/1988) to plan the use of the GSO.

It is not possible to treat the subject exhaustively in a single volume. However, we hope that this book will provide the reader with a solid basis for independently pursuing further research. Our intended audience consists of practicing spectrum managers in government and industry, policy planning personnel, and communications engineers and system designers who, until recently, have too often been unaware of the regulatory and "sharing" constraints within which their systems would operate. This book would also be suitable as a college or graduate level text by selecting a suitable combination of the topics treated.

Although the regulatory/policy aspects and the technical factors are mutually interactive in practice, for the purpose of exposition it is convenient to separate them. Accordingly, the book is divided initially into two parts: Part I on policy and regulation, prepared by Donald M. Jansky, and Part II on technical considerations, prepared by Michel C. Jeruchim. A third part, also prepared by Jeruchim, examines the critical question of interference.

Chapter 1 of Part I serves as a general introduction through a brief discussion of the physical properties of the GSO, followed by a description of the types of services using geostationary satellites that have evolved. Then, a short historical overview is provided, indicating the major developments, both of a technical as well as institutional nature, that have led to the current intensive utilization of the GSO.

As we mentioned, the fundamental physical constraint in orbit utilization is the inevitable presence of interference. Interference between different geostationary satellite networks is what sets lower limits in angular separation between satellites. However, interference also takes place between satellite systems and terrestrial systems, which does not have a direct bearing on satellite spacing *per se,* but which does have implications on the parameters of both types of systems. When more than one system (space or terrestrial) operates simultaneously in a common frequency band, this situation is referred to as "sharing"; i.e., the resource (spectrum or orbit/spectrum) is shared by both systems. Sharing implies mutual interference. Thus, a major regulatory problem is the establishment of rules of sharing, namely the constraints under which systems must operate in order to sustain acceptably low levels of interference. Chapter 2 addresses the sharing question and the rules that have been incorporated into the international regulatory framework. The current procedures by which interference compatibility between existing and planned satellite networks is assured, include a simplified calculation which permits these often cumbersome procedures to be curtailed. The details of the calculation method are provided in Chapter 2, along with the question of sharing between terrestrial and space systems. In order to facilitate such sharing the ITU's *Radio Regulations* impose certain constraints on the power or power density of both types of services so that adherence to these "interface" conditions obviates the need for detailed interference calculations for all possible interference paths between terrestrial and space stations. These constraints are also listed in this chapter.

Chapter 3 takes up domestic and international policy considerations. A sketch of the evolution of U.S. policy is given, leading to a description of the principles dominating current policy, and followed by a similar treatment of the forces that have shaped world views on the use of the geostationary orbit. The discussion centers on the events of the last two decades and the chapter concludes with a look ahead to the third decade of GSO exploitation, and beyond.

Part II of the book addresses technical issues. Chapter 4 is devoted to the examination of technical factors influencing orbit/spectrum utilization. In particular, the relationship between orbit/spectrum measures (e.g., intersatellite spacing) and a variety of technical parameters is developed. Because the relationship is exceedingly complex when all the interacting networks can have a range of different characteristics from one another, a simplified but nevertheless very useful model of orbit occupancy has been adopted. This is the "homogeneous" model, wherein all networks are assumed to be identical, whatever their characteristics may be. Chapter 4 examines the homogeneous case exclusively. One of the main objectives of this chapter is to provide an idea of the trade-offs available between orbit utilization and the technical characteristics. To this end, many trade-off curves are given, capturing the equations in graphical form.

Chapter 5 continues the study of technical factors affecting orbit utilization, but generally without the homogeneous restriction. The topics taken up are not easily or meaningfully formulated within the homogeneous model. Of course, the real world is non-homogeneous, and even though this situation is less amenable to analytic treatment, this chapter attempts to bring some insight into this more general case. Included in this chapter is a discussion of operational considerations, which in practice may have a substantial effect on orbit utilization because of the constraints they impose.

We have already mentioned that interference is the fundamental limitation in orbit/spectrum utilization. In the actual design and operation of a network, it is necessary to quantify the effect of a given level of a given type of interference on the performance of any communication system receiving that interference. This topic is the subject of Part III, Chapters 6 and 7. Chapter 6 is concerned with the case where the wanted (interfered-with) signal is analog, and Chapter 7 treats the case where the wanted signal is digital in nature. There is a virtually limitless number of combinations of wanted and unwanted signal types, but in these chapters we confine ourselves to a comparatively few cases that are of particular importance in satellite communication.

Generally, in the past, systems have been designed in more or less conventional ways, that is without taking explicit measures to minimize interference. Once the system was designed, one could compute its sensitivity to interference, and locate the space and earth stations such that only acceptable interference was received (or caused to other systems). However, as use of the geostationary orbit becomes more intensive, especially in certain arcs of the orbit or in certain frequency bands, it may become necessary in the future to take more positive measures to reduce interference. Chapter 8 provides a brief survey of such measures, which include active interference cancellation techniques, shaped satellite antenna beams, and adaptive depolarization compensation.

The appendices at the end of the book support the discussion in various sections of Chapters 4-8, with the exception of Appendix L, which is referred to in Chapter 3.

In writing the book it has become clear that we owe a great deal of our understanding of the subject matter to the existing literature and to our interaction with many workers in the field over the last fifteen years or so. To all these sources and colleagues we acknowledge our debt. We would like to thank a number of individuals who were helpful during the preparation of the book. Michel Jeruchim extends his appreciation in particular to: Dr. M. Afifi, for carefully reviewing Chapter 8, suggesting some improvements thereto, and for discussing antenna properties; Dr. J.H. Moore, for useful discussions on a number of topics, particularly polarization; Dr. B.A. Pontano, for reviewing his work on interference cancellation; E.E. Reinhart, for incisive and detailed comments on organization of the material; and M.R. Wachs for making available the results of his extensive experimental work. Donald Jansky would like to acknowledge both direct and indirect contributions from members of the FCC staff, in particular R. Lepkowski and R. Herbstreit. Both authors would like to publicly acknowledge the forbearance shown by their families during the preparation of the book.

This book is the personal endeavor of both authors and any opinions or statements expressed herein do not necessarily reflect the positions or concurrence of the organizations with which they are affiliated.

PART ONE

Chapter One

OVERVIEW AND HISTORICAL PERSPECTIVE OF THE GSO

1.0 Introduction

The purpose of this text is to provide a comprehensive treatment of how during the last two decades the geostationary orbit has been developed, utilized, and managed, as well as a description of the domestic and international policies which have brought it to its present state of exploitation. This will take the form of chapters on the factors affecting orbit utilization, interference, sharing, and policy and regulation.

This chapter provides the context for the subsequent parts of the book. It includes sections on the physical properties of the GSO, the nature of the telecommunication services provided by satellites in the GSO, and a general historical development of its use.

The geostationary satellite orbit (GSO) is a phenomenon which derives its peculiar characteristics from the basic laws of physics, notably Newton's laws of gravitational attraction.

When the forces of gravity balance for a finite but small mass compared to the mass of the earth, the distance of this small mass from the earth is approximately 22,300 miles (36,000 km). This occurs at a point directly over the equator. Thus, a circular boundary plane, extending through the earth's equator at 22,300 miles, constitutes the locus of the geostationary orbit. If the gravitational field at this altitude were perfectly uniform, an earth satellite placed into it would rotate about the earth at the same rate that the earth rotates on its axis. To an observer on the earth this gives the appearance of such a body being stationary.

This orbit at said altitude has a circumference of 165,000 miles. The aforementioned geometry is indicated at Figure 1-1.

The reality of exploiting this unique orbital placement of satellites was made possible through the development of rocket launch capability and the dependability of sufficient power to carry sufficient payload to this height to eventually enable viable utilization. (Figure 1-2 shows how satellite insertion into this orbit is achieved.)

1.1 Physical Properties [1]

The GSO belongs to a broader family of orbits called "geosynchronous"; these orbits are generally inclined with respect to the equatorial plane, they may be circular or elliptic, but satellites move on them with the same period, and the same sense of rotation, as the Earth. Geosynchronous satellites are seen from the ground to describe 24-hour periodic figures with varying shapes.

As seen from the satellite, the Earth subtends an angle of 17.4 degrees; in other words, it is seen as a circle which, if measured from the center of the Earth (so that angles can be expressed in longitude and latitude) subtends an angle of 162.6 degrees. Thus a GSO satellite can be seen from any point at the surface of the Earth which lies within a circle distance of 81.3 degrees from the sub-satellite point. On the circumference of this circle, the GSO satellite is seen on the local horizon (elevation, or "site angle" is equal to zero). In practice, because of atmospheric attenuation, the site angle of the satellite must be more than zero, and its minimum value is considered here to be equal to 10 degrees, which corresponds to a circle of 71.45 degrees radius, within which the satellite is visible; or, conversely, the GSO will be visible from any point within this circle. All the points within this circle constitute the *service zone* for that particular GSO satellite (whether this zone will actually be serviced will, of course,

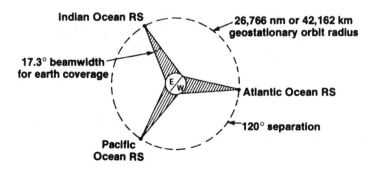

Figure 1-1 Geometry of the GSO
Source: International Aeronautical Federation

depend on the satellite's antenna coverage). There is a mathematical relation between the geographical coordinates (longitude, latitude) of the points from which a GSO satellite can be seen with a given site angle which defines the potential service area. As an example, a satellite located at 63 degrees east longitude (i.e., hovering above the Indian Ocean) can be seen (with a site angle of 10 degrees) from the following points, and, *a fortiori*, all regions between them:

Longitude	Latitude	Notes
8 W	5 N	Africa — Liberia
134 E	2 N	Western part of New Guinea
131 E	32 S	Western Australia included
131 E	32 N	Japan — Southern tip of Kyushu
56 E	71 N	USSR — southern tip of Nova Zemlia
5 E	53 N	Western Europe — Netherlands

Source: International Aeronautical Federation.

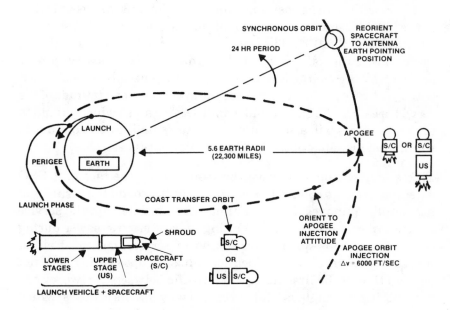

Figure 1-2 Spacecraft Placement in Earth Synchronous Orbit
Source: Aerospace Division, Hughes Aircraft Co.

The lesson to be drawn from this is the following:

If one wishes to cover a maximal service zone, as for instance in the above example, the satellite's position is highly constrained; satellites intended for intercontinental, or generally, global service, must by necessity be in certain longitude slots, once the service zone has been defined. This comment applies not only to telecommunication services, but also to the observation zone in cse of meteorological or Earth observation missions.

If, instead of a singular point of the GSO, one considers a sizeable portion, or arc of the GSO (with extreme points q and p), from which a given zone of the Earth can be serviced, this zone coincides with the part that is common to the two individual zones defined by the two points q and p. If the arc is large (i.e., much flexibility is afforded to position the satellite), the zone which can be serviced from any point of that arc is comparatively small. This is the case for regional or domestic telecommunication services. An extreme case is provided by a service zone consisting of one single point (indeed, some scientific GSO satellites require only one ground station): if this is so, the satellite can be positioned anywhere on a 130 degrees portion of the GSO (assuming latitude of ground station equal to $\pm 40°$ and site angle equal to $10°$). The lesson to be learned from this notion of service arc is the following: satellites that service or observe a relatively small area of the Earth's surface can generally be positioned with greater flexibility, the more so the lower the mean latitude of the served area. On the other hand, small service areas which extend to higher latitudes will also be limited in their satellite positions due to the limited visibility of the GSO from high latitudes.

This statement must however be tempered by the fact that GSO satellites undergo solar eclipses twice yearly, around the vernal and autumnal equinoxes. If one cannot afford to interrupt the service and in order to obviate the resulting lack of electric power during eclipses, and consequent interruption of service, one must either place another satellite at a location whose eclipse period does not overlap that of the first, and switch traffic between the two, or one must position a single satellite in such a way that its eclipses (which are accurately predictable) occur at a time when traffic interruptions are acceptable; e.g., the single satellite should be placed at a position such that eclipses would occur after midnight in the service area; this is the case for a satellite located westward of the service area's westernmost meridian. Another type of limitation in a satellite's position concerns scientific satellites with special requirements such as positioning along particular magnetic field lines. These considerations, then, restrict the positioning flexibility mentioned above.

Altogether, as a result of these various constraints imposed on the positions of GSO satellites (and not counting radio-frequency effects), the geostationary satellite orbit is not, and will most probably not in future be populated with a uniform density of satellites, and this obviously tends to aggravate the congestion problems in the GSO (in other words, the positioning constraints imply that the total useful service arc is less than 360°, and the GSO's total capacity is correspondingly decreased).

1.2 Geostationary Satellite Services

The popularity of the GSO is described below. This popularity is both a benefit and a challenge. It is a benefit because of the broad scale of services which can be provided, and a challenge to ensure an equitable availability of these services.

As Table 1-1 indicates there are now over 100 satellites proposed or operating in the geostationary orbit. The investment in these systems is several billions of dollars. An entire industry has developed through the exploitation of this unique geophysical realization of nature and man's technological ingenuity. Additionally, the telecommunication services made possible by the satellites placed in this orbit have contributed to significant social and economic change in the areas of worldwide voice, TV, and data communications; weather protection, and navigation.

The types of services which use the GSO are detailed below. The material is taken from a study of the International Aeronautical Federation (IAF) done for the UN.

1.2.1 Satellite Communications Services

Fixed Satellite Service

Traditionally, this has consisted of telephone, telegram and telex trunk services, as well as television (TV) distribution. The former category may be routinely transmitted by cable, even across oceans. But TV transmissions have wide-band requirements and special cables are needed in this case if RF transmissions (either by terrestrial means or by satellites) are unavailable.

(1) Business Services — In addition to the established communication services satellites are being used to provide mail, telephone, telex, telegram, and other services on a limited scale (e.g., facsimile text and picture transmission, banking, and reservation data networks). These "new" or "business" services include the following:

extensive, high-speed transmission of documents (text, graphs, pictures), including newspaper print;

high-speed data transmission among widely separated computers (at rates in the megabit per second range);

audio conference with visual aids; and

video teleconferences (whereby groups of people at different sites can confer with each other through live television transmissions).

All the above services can be usefully carried out by satellites, but the last application is of particular relevance to the satellite business because of its likely rapid expansion, and of its high requirements in bandwidth (i.e., large capacity transmission systems are necessary). While satellites offer the technical means to satisfy these new needs, they will be in competition with terrestrial means (either RF microwave beams or fibre optics cables). As a consequence of uncertainties about economic and other factors, the relative share of this new traffic among the various available techniques is hard to predict; this is compounded by the difficulty of forecasting the overall demand for entirely new kinds of services.

(2) Intercontinental Service — Since the introduction of satellite communications, the ability of these systems to cope with larger traffic volume has stimulated demand, which has been ever-increasing, both in total traffic and in the fraction of traffic carried by satellites. Two international organizations (INTELSAT and INTERSPUTNIK) have until now provided this service. In the case of INTELSAT, the traffic increases year by year at a rate of 20%, and is expected to become five times larger in the next ten years. To cope with this situation, INTELSAT will be developing higher capacity

satellites with improved RF technologies and will utilize several of these in each of its three "ocean" regions. TV transmissions across oceans have been on an occasional basis. It should also be noted that satellite communications systems can operate over very large geographical scales and yet remain domestic and be categorized as such, in that they operate within national borders (e.g., USSR, USA, and, in the near future, the French Telecom-1 satellite). From the point of view of technical constraints (service area, elevation angle at edge of service area, etc.) such domestic systems are similar to global, or intercontinental, services.

(3) Regional and Domestic Services — A number of domestic satellite communication systems have been put into operation up to now, and more are planned in the future. The trend lines show that the overall growth of domestic communication satellites, which is more conveniently expressed in terms of number of transponders, is equal to some 50 transponders per year. Many countries do not have very large needs, and thus, instead of acquiring for themselves entire satellite systems (which either have the required small capacity but are uneconomical, or which are oversized and thus equally uneconomical), several countries have resorted to such schemes as leasing transponder capacity on the INTELSAT system (16 transponders were leased in 1979, on average about one per country) or by developing cooperative systems (e.g., European Communication Satellites or ARABSAT).

For traditional communication services, the trend towards using satellites will depend to a large extent on the available terrestrial systems. Thus, in countries with very little existing infrastructure, it may be cheaper, and probably faster, to have recourse to satellites for both telephone and TV distribution. It is, in this case, of prime importance that the system adopted be capable of handling traffic not only between large capacity centers with subsequent redistribution by terrestrial routes. This is the case for international traffic (e.g., INTELSAT) which usually requires large ground equipment; instead the system must be able to distribute the traffic among a large number of small ground stations (community ground stations with 3 to 7 meter antenna diameter) in small towns and villages, in particular, in remote rural areas (e.g., the recently proposed Global Domestic Satellite System).

The situation is different in countries with already well-developed communication systems. The question here is not so much to foresee the total needs for the traditional (low-capacity) services, but rather to estimate whether it will be economical to cater to this increased traffic by means of satellites. It can be shown that whereas satellites are more advantageous over very large distances, terrestrial means are less expensive over small distances. The average distance at which the costs are about equal (called the cross-over point) depends on many factors, and is the subject of much controversy; the cross-over distance could be as low as 500 km, or as large as 3,000 km, according to various authors, and to the assumptions made (i.e., depending on types of services, and kind of technical solutions, satellite systems can also save costs of ground equipment, as compared to terrestrial services).

The video teleconference service is of particular importance since it is likely to expand rapidly and is extremely demanding in terms of satellite capacity. Not only are terrestrial means inadequate, but also present-day satellites have too small a channel capacity (typically 20 to 50 two-way video circuits) to make the usage of this service affordable (i.e., the cost per video circuit is too high). It is thus necessary to design satellites with very high capacity, and develop techniques to achieve this. Also, an affordable system will require very low cost, ground based equipment. (Some demand projections for these services are shown in Figure 1-3.)*

Broadcast Satellite Services

This service provides television and radio programs for direct reception by individual home or community receivers and further cable distribution inside buildings. Among several pilot programs carried out in the USA, Canada, India, and Japan, it is particularly noteworthy to mention the experiment in direct TV broadcasting to community sets, which was conducted in India in 1975-1976, using the USA's ATS-6 satellite and its high power UHF equipment, as well as 2400 community reception sets. From the pilot programs carried out so far, it appears that the necessary technology is available, the aim being to bring the costs of the individual ground receivers down to levels which are affordable by many users in order to make

---

*The figure shows additional capacity as a function of time.

the system worthwhile. The demand for this service depends firstly on the extent to which ground television networks (microwave beams and cable TV) cover a given country or region, and secondly, if this criterion is well satisfied, on the demand for more TV programs (including educational TV, for instance). The conditions of course vary considerably from country to country, but some trends in favor of satellite TV can be seen as shown by the following cases: remote areas, in particular in mountainous regions; and inside large cities where it appears that installation of new cable or microwave networks is cumbersome and expensive. The U.S. has recently authorized several systems for commercial operation.

(1) Direct Sound Broadcasting (UHF) — UHF sound broadcasting via satellite could improve the quality and the coverage of this service as compared with present terrestrial means. Although part of the

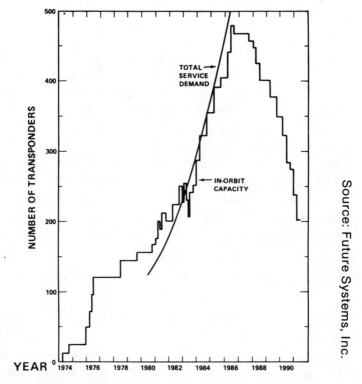

Figure 1-3 In-Orbit Capacity and Demand for Operating, Planned and Projected Satellites (excluding TV distribution)

equipment necessary for this new service has been developed, no satellite experiments have been conducted so far. The aim of this service is to broadcast radio sound programs to receivers of sizes and quality standards comparable to the current FM-VHF ones (for instance receivers in motor vehicles or for high-fidelity sound reception). This in turn would entail rather large dimensions for the satellite's antennas (10-20 m diameter). Similar to TV broadcasting via satellites, it is thought that this new service should be tested on an experimental basis, and that perhaps this could happen in the second half of this decade. It is not possible at this stage to make any further forecasts, but from the point of view of the GSO congestion, it is likely that this service may have a frequency band allocated to it (WARC 79 encouraged experiments at about 1.5 GHz) and that orbital positions may be assigned in a scheme similar to the one for TV broadcast by satellite.

Mobile Satellite Services

This category includes the RF communications of all types of mobiles (meaning all moving objects) at the surface of the earth; be it at sea, in the air, or on the ground. Long distance communications with mobiles by conventional (i.e., terrestrial) means are hampered by a number of unfavorable factors, such as lack of reliability due to fading, spectrum congestion, and very long access times in some cases. Satellite systems could solve most of these problems by providing near global coverage, immediate access for the user, the possibility of interconnection with terrestrial networks, high reliability coupled with adequate capacity, and a communication cost almost independent of distance.

A mobile satellite communications system is different from a fixed service system in several respects:

it is a low capacity system (channel capacity not exceeding a few hundred voice channels in the next decade);

access to the satellite is needed from many parts of the globe by many thousands of users; and

a large number of users implies that the cost to each user for investments, in particular, must be low; this in turn means that the mobile terminals need to be small, which leads to the requirement of efficient transmitting and receiving systems on the satellite (i.e., both high electric power and large antennas are needed). In this latter respect, the mobile satellite service is quite similar to the broadcast service; it is, on the other hand, very different because by definition the mobile service requires 2-way communication.

(1) Maritime Satellite Services — Following the privately owned MARISAT system, INMARSAT, an inter-governmental organization created in 1979, will operate a *global* maritime service by leasing the satellite system. The INMARSAT system could consist of either or both of the following components: two MARECS spacecraft provided by the European Space Agency, and Maritime Communication Service (MCS) packages on three INTELSAT V satellites.

Apart from the global INMARSAT system, one may notice other systems like the US Navy's Fleetsatcom, the USSR Volna system, and the Japanese Aeronautical and Maritime Engineering Satellite (AMES) which is planned for launch in 1985. Among various future prospects, one category concerns ships equipped with significantly smaller terminals than the current INMARSAT standards. Instead of aiming at ships with tonnages of thousands, one addresses here a market consisting of ships as small as one hundred tons. However, to start this new type of service, which would require more satellites, one might need to set up first a demonstration program (probably on a non-reimbursable basis).

(2) Aeronautical Satellite Services — In addition to the general services listed above, aeronautical traffic could benefit from the following functions: (a) position determinations; (b) positions of other aircraft; and (c) data channels.

(3) Land Mobile Satellite Services — Specific services in this category include:

communication with vehicles in remote and underdeveloped areas;

provision of distress communication channels for vehicles in remote areas; and

emergency communications for areas suffering disasters which disable or overload the normal communications system.

Limited improvements achievable with conventional terrestrial means constitute a determining factor in the demand for this type of service by means of satellites. Similar arguments hold as in the case of communications with small ships.

It appears that Canada will embark on its "M-Sat" program which is intended to operate with 4 MHz up, and 4 down, in the 806-890 MHz band.

1.2.2 Meteorological Satellite Services

The objectives of geostationary meteorological satellites are two-fold:

> (a) imaging of the earth's surface and cloud cover in order to produce very short-term weather forecasts, and to obtain a number of basic meteorological parameters; and

> (b) meteorological and hydrological data collection and distribution.

Several agencies (in the USA, Japan; and the European Space Agency) have contributed to setting up a pre-operational network of meteorological satellites which, in addition to some low-altitude polar orbiting satellites, consists of five satellites located at 70° longitude intervals from each other. This network serves the purposes of two programs instituted by the World Meteorological Organization (WMO), namely: the permanent World Weather Watch (WWW) program, which is of an operational nature, and the experimental Global Atmospheric Research Program (GARP). The latter program is witness to the fact that meteorological satellites contribute largely to our understanding of the earth's weather and climate system, and, due to its scientific character, it is natural that the GARP was undertaken jointly with the International Council of Scientific Unions.

1.2.3 Space Research and Exploration

The GSO offers specific advantages to space research in the following areas:

(a) The GSO is an ideal location to carry out measurements of the particles, the electromagnetic fields, and the plasma properties in the very center of the magnetosphere, in order to advance our knowledge of its properties; particularly in relation to the effects produced by various solar and interplanetary conditions.

(b) The GSO allows making images of the same portion of the earth at frequent intervals. Images obtained by meteorological satellites are useful for scientific purposes; e.g., for the study of the earth's atmosphere, or of its climate system.

(c) For economical reasons, it is desirable here, as well as in other areas, to use the GSO for a satellite's communication function. In the case of a scientific satellite placed in the GSO, only one ground station will be required. Apart from the cases mentioned above, one should mention astronomy satellites (except when the particle environment in the GSO disturbs the measurements, as in high-energy astrophysics).

(d) GSO satellites can be used to synchronize, by means of laser pulses, atomic clocks located at great distances from each other (e.g., LASSO experiment on the European SIRIO-2 spacecraft). The objective of such experiments is to provide a highly accurate tool in the area of geodynamics and geodesy. Note that a possible application of such a system could be to assist in the calibration of global navigation systems.

(e) Geostationary satellites can support radio astronomers and geophysicists alike in the area of "very long baseline interferometry," namely by transmission of wideband noise-like signals from various ground based radiotelescopes to a central processing facility. Terrestrial distances involved are of the order of several thousand kilometers. Experiments in this area have been made (Canada's Hermes project) and are planned in future (experiment on the European L-SAT spacecraft).

1.3 Historical Development

While Arthur Clark* may have conceptualized the use of the geosynchronous orbit, it took a great engineering feat to achieve its practicability. As a result of a great accident of technological breakthrough in nuclear weaponry† the USSR put its Sputnik into space in 1957. There ensued a great technological burst in space.

---

*Arthur Clark, *Wireless World,* (1945).

†The U.S. made a breakthrough in technological weaponry about seven years in advance of the USSR, thereby negating the need for large launch vehicles and as a result did not have an inventory of expendable large capacity vehicles to use for satellite launch purposes.

1.3.1 Domestic History

The United States launched the AT&T-built Telstar, which was in fact the world's first in-orbit communication satellite. This was a low orbit satellite with very limited capacity. It required very expensive, large antenna, earth tracking stations to utilize it. Despite these limitations, this and other communication satellite technology offered sufficient promise for Congress to hold hearings beginning in 1960 and 1961, culminating in the Communications Satellite Act of 1962. The purpose of this legislation was to create a corporation which would exploit space technology to create a "single" global communications satellite system. This constituted the first statement of communications satellite (COMSAT) policy on the part of the U.S. (For further discussion, see Chapter 3.)

The enactment of this legislation was controversial, both domestically and internationally, but laid the foundation for the exploitation of the unique properties of the geostationary orbit. Domestically, the principal focus of controversy was the structure of the organization to be created; i.e., should it be (1) commercial/private sector, (2) AT&T dominated, (3) a separate entity, or (4) Federal Government controlled. The compromise choice was the COMSAT Corporation in a form not quite as it is today. At the outset commercial communication carriers played a dominant role on the Board of Directors of COMSAT. (See Chapter 3.)

The legislation allowed for considerable amounts of capital because it was anticipated that the expense of constructing a global satellite system on the model of Telstar, and associated Andover-like earth stations, would be very great.

Even as late as 1965 the debate was still not settled with regard to whether the "global" system would be based on medium altitude satellite technology or geostationary satellite technology. Syncom II was successfully placed into orbit in 1963. In 1965 Early Bird, later to be known as INTELSAT I, was launched with the capability of 240 voice circuits or one television channel. The success of Early Bird effectively demonstrated the superiority of the geostationary approach. Figure 1-4 shows the evolution of communication satellite types.

In 1965, the American Broadcasting Company and Hughes Aircraft

Corporation petitioned the Federal Communications Commission to build and operate a US domestic satellite system. There ensued a series of developments (which will be discussed further in Chapter 3) which eventually led to a policy which would permit the development of not one but several US domestic satellite systems, and thereby contribute to the destabilization of the AT&T monopoly over US domestic communications.

1.3.2 International History

Internationally, the USSR opposed the concept of a "commercially" based global system on the grounds that it was in violation of United Nations principles. This debate occured in the UN Subcommittee on the Peaceful Uses of Outer Space. USSR was supported by two Eastern European countries, and gained considerable support from many quarters.

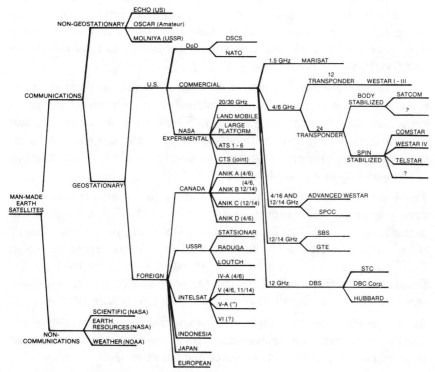

Figure 1-4 Man-Made Earth Satellites
Source: International Aeronautical Federation

However, in 1965, 15 countries agreed to an Interim Communication Satellite Organization (ICSO) and were soon joined by a number of others. The initial concept of an International Satellite Organization was based on the assumption that 15 to 20 developed countries would be candidates for participation, and a slow evolution over a period of one decade was envisioned because only developed countries would be able to make the investment in the technology to support a medium altitude system.

Within five years of the formation of COMSAT: 60 countries were members of ICSO; there were 2,129 geostationary satellite circuits and 50 earth stations, and many more underway. A revolution had been accomplished in international telecommunications, due in large measure to the successful implementation of geostationary satellite technology.

Many developing countries, for a small investment, could become part of a profit-making organization, and, more importantly, could communicate directly with many other countries without the need of first passing through an expensive communication hub, such as London, Paris, or New York.

As countries became members of this international consortium, there grew an increased desire to control their own telecommunications destiny. Whereas for decades ITT, or GT&E, for example, had run the communications infrastructure of many developing countries, this relationship no longer remained tenable with the coming of the geostationary satellite.

The international organization providing for international treaty agreements on telecommunications, the ITU, first took steps to accommodate satellite telecommunications at an Extraordinary Administrative Radio Conference in 1963. Administrative Radio Conferences are called under ITU procedures to deal with the international apportionment of electromagnetic spectrum; i.e., what airwaves will be used by what countries, under what conditions.

Such a conference was necessary, of course, because satellites are totally dependent on transmitted and received electromagnetic emissions to accomplish their intended functions.

The EARC set aside several hundreds of megahertz of spectrum to

adopted a coordination procedure for satellites in the GSO, and recognized interference levels for cumulative and single entry interference.

In 1977, the ITU's requirements for regulating the GSO were enhanced through a WARC dealing with broadcasting satellites in the band 11.7-12.2 GHz, a plan for the use of such satellites in Regions 1 and 3 (Europe, Asia, and Africa). The plan pre-assigned orbit location and frequency to respective countries, predicated on certain technical characteristics.

The 1979 WARC modified but did not drastically change the basic structure of GSO utilization. The station keeping standards were improved, and greater interference allowances were recognized. The most far-reaching agreement at this conference regarding the GSO was *Resolution 3* [9]. This resolution, adopted under pressure from developing countries, called for planning of all services in reference to the GSO and a future space conference. The reference conference is to be held in two sessions in 1985 and 1988. A formal agenda has yet to be adopted, but already its prospect is being felt by developed nations.

The 1960s were a decade of experimentation; the 1970s one of development of this unique capability. The decade of the 1980s is a period of implementation and, along with it, one of enhanced policy need. Chapter 3 deals with what is already an accellerated period of policy-making activity in both domestic and international forums.

1.4 Summary [10]

The recently completed UN Conference on the Peaceful Uses of Outer Space devoted an entire section to the GSO. It sets forth the characteristics of this orbit which have made it sought after by all mankind, and added impetus to the need to manage its use in an equitable fashion. The following are paragraphs taken from the section of this report dealing with the GSO.*

---

*Report of the Second United Nations Conference on the Peaceful Uses of Outer Space, Section G, Vienna, 1982, para. 277-288.

277. GSO is a unique natural resource of vital importance to a variety of space applications, including communications, meteorology, broadcasting, data relay from and tracking of orbital satellites, etc. It could also be used for such possible future applications as solar power satellites. Though not depletable, GSO is a limited natural resource. Therefore, as with any limited resource, its optimal utilization requires planning and/or arrangements.

278. Utilization of GSO cannot be considered in isolation: the associated issue of use of the radio frequency (RF) spectrum must be simultaneously looked at. The RF spectrum is also — like GSO — a limited (in practice) though non-depleting resource. While in theory it does extend indefinitely, practical constraints limit its present use to a comparatively small band. Hence, its optimal use also requires planning and/or arrangements.

279. It is in the light of this data that the members of the ITU have been making concerted efforts to evolve systems for planning and regulating the use of the GSO and the RF spectrum since 1963. It is noted that the forthcoming ITU Conferences in 1985 and 1987[8] which will continue this process in the light of technical progress and in the light of the broad considerations outlined here, will decide, in accordance with resolution 3 of WARC 1979, which space services and frequency bands should be planned ...

288. In conclusion, considering the long-term implications of the growing activities in GSO, any solution on the use of the GSO should be both equitable and flexible and take into consideration the economic, technical and legal aspects.

TABLE 1-1
GEOSTATIONARY SATELLITES TO BE IN OPERATION IN 1983-1984

Longitude	Name	Country/Organisation	Service
W 171	TDRS-West	USA	I
W 170	Statsionar-10	USSR	F
W 170	Volna-7	USSR	M
W 170	Gals-4	USSR	F
W 170	Loutch-P4	USSR	F
W 140	CBSS	Canada	R
W 140	SATCOM F5	USA	F
W 135	DSCS III	USA	F
W 135	NATO-3B	NATO	F
W 135	GOFS-West	USA	MET
W 132*	SATCOM III	USA	F
W 130	DSCS-11 F9	USA	F
W 128	COMSTAR D-1	USA	F
W 125	SBS-C	USA	F
W 122	SBS-A	USA	F
W 119	SBC-B	USA	F
W 119	SATCOM II	USA	F
W 116	Anik-C2	Canada	F
W 114	Anik-D	Canada	F
W 112.5	Anik-C1	USA	F
W 109	Anik-B1	Canada	F
W 109	Anik-C3	Canada	F
W 106	SBS	USA	F
W 103	GTE	USA	F
W 100	Fltsatcom-1	USA	M
W 100	GTE	USA	F
W 99	WESTAR-4	USA	F
W 95	COMSTAR D-2	USA	F
W 92	CBSS	Canada	B
W 91	WESTAR	USA	F
W 87	COMSTAR D-3	USA	F
E 0	Meteosat-2	ESA	MET
E 5	OTS	ESA	F, E
E 10	Eutelsat 1	ESA/Eutelsat	F
E 13	Eutelsat 1-2	ESA/Eutelsat	F
E 19	Arabsat-1	ASCO	F
E 20	Sirio-2	ESA	MET, S
E 26	Arabsat-2	ASCO	F
E 35	Statsionar-2	USSR	F
E 35	Loutch-P1	USSR	F
E 45	Statsionar-9	USSR	F
E 45	Volna-3	USSR	M
E 45	Gals-2	USSR	F
E 45	Loutch-P2	USSR	F
E 53	Statsionar-5	USSR	F
E 53	Loutch-2	USSR	F
E 53	Volna-4	USSR	M
E 57	IOR, Leases	Intelsat	F
E 60	IOR, Spare, MP	Intelsat	F, M
E 60	USGCSS, IO	USA	F
E 63	IOR, Primary	Intelsat	F, M
E 66	IOR, Spare	Intelsat	F, M
E 70	STW-2	China	F - Exp
E 74	GOMSS	USSR	MET
E 74	Insat-IA	India	F,B,MET
E 75	Fltsatcom-2	USA	M
E 80	Statsionar-1	USSR	F
E 85	Statsionar-3	USSR	F
E 85	Volna-5	USSR	M
E 85	Gals-3	USSR	F
E 85	Loutch-P3	USSR	F

COM	Communications (General)	I	Intersatellite link
B	Broadcast satellite service	M	Mobile satellite service
F	Fixed satellite service	MET	Meteorological satellite service

Source: ITU, International Frequency Registration Board

Chapter Two
COMSAT SHARING OF THE GSO

2.0 Introduction

When the first major ITU WARC dealing with satellite systems and the geostationary orbit was held in 1971 [8], the electromagnetic spectrum was already allocated to various terrestrial, mobile, broadcasting services up to 40 GHz. The 1971 conference recognized 13 satellite services, and made available over 15,000 MHz of spectrum to support these services in this already allocated spectrum. Only a few MHz of this spectrum were made available on an exclusive basis for a particular satellite service. These new allocations were made possible because of sharing with other services.

The charts in Table 2-1 show the allocations as they now stand after WARC 79 [9]. They are in the primary "window" for space services, 1-14.5 GHz. This window is that band of radio spectrum where propagation and satellite system transmission characteristics become optimum for satellite system technology.

There are 10 different sharing pairings involving the fixed satellite service including sharing with itself. This chapter describes the technical considerations involved in sharing. The basic sharing considerations involving communication satellites fall into two categories:

Sharing With Other Satellite Systems (Section 2.1); and
Sharing With Terrestrial Systems (Section 2.2).
Each category is treated below.

2.1 Sharing With Other Satellite Systems

2.1.1 Fixed Satellites Sharing With Other Fixed Satellites

The most important service with which communication satellites must share spectrum is itself. The complete list of bands for the primary window made available to this service is in Table 2-1. The table also shows the bandwidth of the allocation and the direction which it is used. The chart also shows those other services, and associated bands, where there is sharing with the fixed satellite services. There are 24 separate bands below 35 GHz, and 14 above. Special aspects of these bands are treated in the appropriate sharing section.

Report 455-2 [5] of the CCIR gives an intensive treatment of the sharing constraints between co-channel communication satellites. Basically, the constraints focus on a trade-off between insuring economically feasible communication satellite system design and efficient orbit/spectrum utilization.

The nature of sharing between communication satellites in a given frequency band is generally dependent on:

number of co-channel satellites;

radiation pattern of earth and space station antennas;

polarization differences between wanted and unwanted signals;

relative power-flux density levels of wanted and unwanted signals; and

portion of total satellite system noise allowance allocated to interference from other satellite networks.

The fundamental element which limits sharing is interference and is treated extensively elsewhere in this text.

The nature of sharing between several satellite systems has been incorporated into international rules and regulations. The nature of this consideration is predicated on the configurations in Figures 2-1 and 2-2.

The sharing constraints operative at present in the ITU are the following:

station keeping capability $\pm 0.1°$

satellite pointing accuracy $0.3°$

use of a reference antenna pattern of $32-25 \log\theta$

use of Appendix 29 of the Radio Regulations as a coordination mechanism.

The use of Appendix 29 of the Radio Regulations contains several additional implicit sharing constraints. This procedure was incorporated in the International Radio Regulations to provide a basis for communication satellite systems to ascertain whether or not any two systems might have a sharing problem. The procedure, contained in paragraphs 1060 and 1067, reads as follows [9]:

1059 *Requirement for Coordination*

1060 6. (1) Before an administration (or, in the case of a space station, one acting on behalf of a group of named administrations) notifies to the Board or brings into use any frequency assignment to a space station on a geostationary satellite or to an earth station that is to communicate with a space station on a geostrationary satellite, it shall, except in the cases described in Nos. **1066** to **1071**, effect coordination of the assignment with any other administration whose assignment, for a space station on a geostationary satellite or for an earth station that communicates with a space station or a geostationary satellite, might be affected . . .

1066 (3) No coordination under No. 1060 is required:

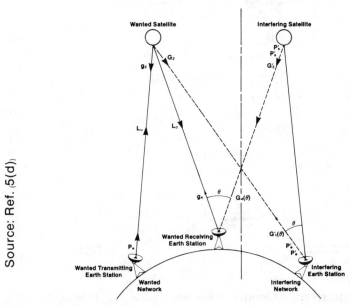

Source: Ref. 5(d)).

Figure 2-1 Interference Geometry Between Satellite Networks (Case I)

1067 a) when the use of a new frequency assignment will cause, to any service of another administration, an increase in the noise temperature of any space station receiver or earth station receiver, or an increase in the equivalent satellite link noise temperature, as appropriate, calculated in accordance with the method given in Appendix 29, which does not exceed the threshold value defined therein.

In carrying out these calculations the interference levels for cumulative and single entry interference levels are those cited in CCIR *Recommendation 466* (Table 2-2) [11] below. It is evident that these levels are different for different types of system configurations.

If this calculation results in a 4% increase in noise to one or another network involved, then further analysis is required. This procedure is intended to be only a gross indicator of possible sharing problems.

Having determined that "coordination is necessary" then the parties involved may engage in discussions concerning interference occurring between two systems at a more detailed level. This examination is carried out on the basis of mutually agreed interference

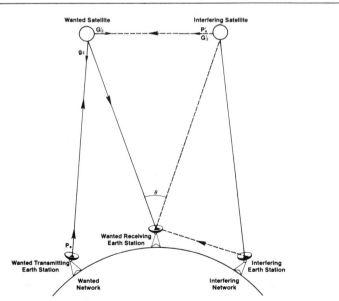

Source: Ref. [5(d)]

Figure 2-2 Interference Geometry Between Satellite Networks (Case II)

calculation methods, and carrier-to-interference ratio requirements of each system.

2.1.2 Fixed Satellites Sharing With Broadcasting Satellites

There are several different aspects to fixed satellites sharing with direct broadcast satellites. They include the following:

Sharing on a co-regional and primary basis in the band 2655-2690 MHz (Regions 2 and 3), and 12.1-12.3 GHz (Region 2); and

Sharing on an interregional basis: 11.7 to 12.1/12.3 for DBS (Regions 1 and 3); FSS (space-to-earth) for Region 2; and 12.3 to 12.7/12.75 GHz, FSS (space-to-earth), Regions 1 and 3; and FSS (earth-to-space), Region 1 and DBS (Regions 2 and 3).

(1.) 2,500-2,690 MHz

While the band is allocated to both the DBS and FSS services, actual systems have not been proposed for implementation. The principal sharing constraint is the power-flux density limit which varies from –137 to –152 dBW/m^2, depending on the angle of elevation. This limit is intended to protect the terrestrial environment, given that the bandwidth puts significant limits on the capacity that could be achieved in a practical satellite system. The CCIR has not studied the co-service sharing situation in any detail.

(2.) 12 GHz

The complex satellite sharing situation at this frequency is shown in Table 2-3. As evident, it is primarily an interregional sharing situation.

The problem of sharing between BSS and FSS systems, particularly in the space-to-earth direction, is one of non-homogeneous system characteristics:

sharing between BSS in Regions 1 and 3 and FSS in Region 2;

the areas served by the two services are separated generally by large bodies of water with boundaries running north to south facilitating reduction of interference through satellite antenna side-lobe discriminiation;

a detailed BSS plan exists in Regions 1 and 3 thereby putting the burden of sharing on Region 2.

Sharing criteria between the services, in principle, can be established in terms of PFD limits as minimum space station separation.

Similar considerations apply to the FSS in Regions 1 and 3 sharing with BSS in Region 2. The specifics of this sharing will be determined at the forthcoming Regional Administrative Radio Conference to be held in 1983 to plan the BSS service in Region 2 [12].

2.1.3 Sharing Between the FSS, EESS and Meteorological Satellites

The FSS shares with the meteorogical satellite service (METSAT) in the bands 7450-7550 MHz (space-to-earth) and 8175-8215 MHz (earth-to-space). The earth exploration satellite service and the FSS share in the bands:

8025-8175 MHz,
8175-8215 MHz,
8215-8400 MHz (Region 2), and
18.6-18.8 GHz (Region 2).

The METSAT service is utilized under a worldwide plan in which there are certain limited fixed locations of satellites in the GSO. Since the position and characteristics of these satellites are well known, the FSS should have little trouble establishing the necessary sharing criteria.

In the EESS service the principal use of GSO satellites will be for the relay of data gathered by low orbiting EESS satellites to central radiation earth stations. In light of the limited number of expected satellites, the sharing constraints should not be significant.

The 18.6-18.8 GHz band is shared between FSS and passive sensors of the EESS service. The principal sharing concern is to allow satisfactory operation of the sensors without adversely affecting the FSS.

2.1.4. Mobile Satellites

This service shares with the FSS at 5000-5250 and 15.4-15.7 GHz; this sharing is in conjunction with the aeronautical mobile satellite service. The envisioned purpose of the FSS in these circumstances is to provide feeder links to an aeronautical satellite; therefore, it is expected that the necessary sharing constraints can be worked out in the worldwide aeronautical community.

2.2 Sharing With Terrestrial Systems

As noted earlier, at the time that satellite services came into existence the radio spectrum was already allocated through 40 GHz. Therefore, it became imperative to establish sharing criteria which would allow coexistence of terrestrial services and space services.

The term terrestrial service usually refers to the fixed service. However, in this section it also refers to the mobile service, and the radio determination services which share with the FSS.

2.2.1 Fixed and Mobile Sharing With FSS

There are a number of frequency bands shared on a primary basis between the fixed satellite service (FSS) and the fixed and mobile services. Some of these bands are heavily used by the fixed satellite and fixed services and it can be expected that such growth will eventually develop in the other shared bands, particularly below 40 GHz. Therefore, the sharing and interference criteria have been extensively developed.

As for sharing between the fixed satellite service and the mobile services, no specific interference criteria have been developed due to the limited usage of the mobile services in these bands. In general, provisions of Articles 27 and 28 of the *Radio Regulations* would also apply. Therefore, the discussion in this section pertains primarily to sharing between the fixed satellite service and the fixed service (FS); although, by extension, it may be applied to the mobile service.

(1.) The Radio Regulations [9]

Considering the extensive amount of sharing between space and terrestrial services that exists in the *Table of Frequency Allocations,* the number of technical regulations is relatively small. The regulations dealing with sharing between these two groups of services, space and terrestrial, are formulated in such a fashion as to be quite general, instead of addressing the sharing between a specific space service and a terrestrial service. The emphasis is on the rational and efficient use of the radio frequency spectrum resource.

Frequency sharing between space and terrestrial services began in earnest with the allocation at the 1963 EARC of bands shared between the fixed satellite service and terrestrial fixed (and certain mobile) services, notably at 4 and 6 GHz. The widespread successful application of sharing led to the extension of the principle to other frequency bands by WARC-ST, 1971.*

(2.) Performance and Interference Criteria

Both fixed satellite systems and fixed service systems are designed to satisfy certain overall service requirements in conformance with particular performance objectives, e.g., Hypothetical Reference Circuits (HRC) and Hypothetical Reference Digital Path (HRDP). These criteria are related to the type of traffic (e.g., multichannel telephony, television, etc.) and the type of modulation (e.g., analog, digital) in use within the system. Several CCIR recommendations have been developed — these are: for the fixed satellite service, *Recommendation 353-3* (analog) and *Recommendation 522* (8-bit PCM) [13]; and for the fixed service, *Recommendation 378-3* (analog), *Recommendation AA/9.*

The performance criteria are relevant to the interference criteria insofar as the latter must be compatible with the former; i.e., interference should have a limited effect on performance so that systems designers retain reasonable control over their systems' performance in the presence of interference. Numerically this often means that certain parameters associated with an interference criterion should be small fractions of the corresponding parameter of the performance criterion.

The interference criteria have been established specifically for interference from either service into the other; thus it must be remembered that there may be interference present which does not originate from the other service (e.g., intraservice interference, intrasystem interference). Moreover, the interference criteria, like the performance criteria, constitute the total cumulative maximum permissible interference into the hypothetical reference circuit from all possible systems in the interfering service. This has caused a need for the concept of a single-entry interference criterion which is defined as the maximum permissible interference from a single sta-

---

*These sharing constraints are established through limits on earth station, space station, and terrestrial stations powers. These are set forth in Tables 2-4 and 2-5 which are excerpts of Articles 27 & 28, respectively, of the *Radio Regulations.*

tion in the interfering service to a station of the service interfered with. This single-entry interference criterion is in some cases the level of interference which is applied to the assessment of interference between two specific systems in the two services.

Criteria of maximum permissible interference have been set forth in the following documents [14]:

for the fixed satellite service
Analog telephony: Rec. 356-4
8-Bit PCM telephony: Rec. 558

for the fixed service
Analog telephony: Rec. 357-3

At the Interim Meetings, *Report AA/4-9* was adopted relating to interference criteria for digital radio relay systems in bands shared with the Fixed Satellite Service.

(3.) Sharing Criteria — General

It has already been stated that the interference criteria are not directly applied in interference cases between specific systems or stations in the two services, but that in those cases single-entry criteria may be used. However, this is not practical when considering mutual interference between terrestrial stations and space stations in the fixed satellite service.

For example, while it would be possible to compute the interference effects from the emissions of a given space station on a single radio relay system, the calculation of cumulative interference effects from many space stations upon each of the large number of radio relay systems in existence but yet to be implemented, is an impractical task. In view of the comparative uniformity of the characteristics of line-of-sight radio relay systems, it has been found possible to provide protection for terrestrial radio relay systems by placing general restrictions on the emissions from space stations. These restrictions are given in Article 28 of the *Radio Regulations* and *Recommendation 358-2 (MOD I)*, and are discussed in *Report 387-3 (MOD I)*.

The restrictions are expressed in terms of values of maximum permissible power-flux density in a reference bandwidth, produced at the surface of the earth by the emissions of any one space station under assumed free space conditions.

Furthermore, in order to bound interference from earth stations into terrestrial stations, minimum elevation angles are specified in Article 28.

A similar situation prevails when assessing interference from transmitting stations in the fixed device to space stations of the fixed satellite service. In this case it was decided that the geostationary satellite orbit in its entirety should be protected, and this has been accomplished by limiting the maximum e.i.r.p. which any transmitting station of the fixed service would be permitted to radiate towards the geostationary satellite orbit as well as by limiting the power which may be applied to a terrestrial transmitting antenna. The latter provision would tend to bound the cumulative unwanted power at the geostationary satellite orbit which may reach it through the side-lobes of the many individual terrestrial transmitting stations. These restrictions are given in Article 27 and *Report 393* [15]. The restrictions in Article 27 also tend to bound interference in receiving earth stations.

The sharing criteria which were adopted in the currently allocated shared frequency bands are best categorized in terms of the earlier defined interference interfaces, and are discussed in the following sections.

(4.) Sharing Criteria — Transmitting Stations in the Fixed Service Into Receiving Earth Stations

This interference situation is resolved by detailed coordination on a case-by-case basis. To identify specific cases requiring such detailed coordination, WARC 71 adopted a method for drawing a geographical area round the receiving earth station such that fixed stations outside that area are not likely to cause interference. This method was subsequently refined by the CCIR in *Report 382-3* [16] and WARC 79 for frequencies below 40 GHz and appears in the *Final Acts* of WARC 79 as Appendix 28. WARC 79 in *Resolution AJ* requested the CCIR to continue to study this matter for subsequent modifications to Appendix 28. This method was devised so that earth station planners would then be able, through calculations and mutual consultation, to ascertain that interference from terrestrial transmitting stations located within the coordination area would be acceptable or could be made acceptable. Conversely, the planner of a terrestrial station to be located witin the coordination area of an existing earth station would have to ascertain by calculations and consultation that interference caused to the earth station would be acceptable. The mechanism of the coordination process constituted part of the substance of Article 11 and of Appendix 28 to the *Radio Regulations* [17].

For frequencies above 40 GHz, new draft *Report AB/4-9* discussed separation distances between earth stations and fixed stations. In this report, based upon certain assumed parameters, it is concluded that due to high propagation losses and atmospheric absorption, the minimum distance from terrestrial stations which are *outside* the main beam of the earth station is very small (less than 2 km), and is on the order of 50-150 km for those *within* the main beam, depending upon the elevation angle.

Documentation relevant to the assessment of interference from a terrestrial transmitting station to a receiving earth station is contained in [18]:

Report 388-3
Report 449-1
Report 448-2

The antenna patterns used are important in determining both the coordination area and the assessment of interference. The CCIR has adopted a reference pattern for the side-lobes and these are given in *Recommendation 465-1* and *Report 391-2* [19].

In addition to the above considerations, it should be noted that the emission constraints imposed on the fixed service as given in the preceding discussion are generally too stringent to allow satisfactory operation of transhorizon radio relay systems.

(5.) Sharing Criteria — Space Stations To Terrestrial Receiving Stations

For this interference interface, protection of receiving stations in the fixed service is provided by the adoption of maximum permissible power-flux densities (pfd) at the earth's surface due to a space station in the fixed satellite service.

This matter constitutes part of the substance of Article 28 of the *Radio Regulations*. The relevant CCIR text is *Recommendation*

358-2 which recommends the maximum pfd at the surface of the earth by emissions of a satellite for all conditions and methods of modulation. While this recommendation covers specific frequency bands between 1 and 23 GHz, WARC 79 decided to extend the pfd values specified for the 17.7-19.7 GHz band to cover specific bands between 31 and 40.5 GHz, pending further study by CCIR. For frequency bands above 40 GHz, new draft *Report AB/4-9* derives permissible pfd limits using assumed parameters for likely systems. The values are on the order of 10-20 dB higher than those for bands below 40 GHz.

Additional material relevant to the protection of systems in the fixed service against interference from space station in the fixed satellite service is contained in the texts [20, 21]:

> *Report 387-3,* which deals with the derivation of power flux densities between 1 and 23 GHz, and

> *Report 792,* which gives a method for the calculation of the maximum power density in a 4 kHz band for angle-modulated carriers.

New Draft *Report AA/4-9* discusses the factors involved in determining C/I criteria for sharing between digital radio relay systems and fixed satellite systems.

(6.) Sharing Criteria — Terrestrial Transmitting Stations To Receiving Geostationary Space Stations

This matter constitutes part of the substance of Article 27 of the *Radio Regulations*. The relevant text for this interference interface is *Recommendation 406-4* [22] which recommends certain power (e.i.r.p.) and pointing angle restrictions on radio relay systems operating in shared bands above 1 GHz. Again, WARC 79 decided to modify Article 27 to account for pointing angle restrictions above 15 GHz pertinent to terrestrial systems introduced after 1 January 1982. CCIR is requested to make a recommendation as to the need for such restrictions.

Additional pertinent material is found in *Report 790* [23].

A calculation method by which it is possible to determine the angle between the main beam axis of a terrestrial transmitting antenna and the closest point on the geostationary orbit as seen from the location of the terrestrial station is given in *Report 393-3* [15].

(7.) *Sharing Criteria — Transmitting Earth Stations To Terrestrial Receiving Stations*

Interference through this interface is dealt with in case-by-case coordination, with the aid of a coordination area.

Relevant to this matter is *Recommendation 524* which bounds the e.i.r.p. density (e.i.r.p. per 4 kHz) of transmitting earth stations in their antenna side-lobes and in the direction of the geostationary satellite orbit. To the extent that compliance with *Recommendation 524* [24] establishes an actual maximum power density applied to the transmit earth station's antenna, use of the CCIR reference antenna pattern for earth stations (*Recommendation 465-1*) would yield a maximum e.i.r.p. density towards the physical horizon as required in connection with the coordination procedure.

2.2.2 Radio Determination Services Sharing With FSS

(1.) *Radionavigation Service Sharing with the Fixed Satellite Service Near 14 GHz*

WARC 79 adopted "footnote 856" which stipulates that the use of the band 14-14.3 GHz by the radionavigation service shall be such as to provide sufficient protection to space stations in the fixed satellite service. [9].

Recommendation 496-2 gives the following provisional limits [25]:

where there are few simultaneous interference sources, the maximum value of the peak power flux-density produced at any point in the geostationary satellite orbit by any radionavigation transmitter in the band 14-14.3 GHz should not exceed -150 dB (W/m^2) in any 1 MHz band;

where the value of D, as defined below, exceeds 2×10^{-4} the maximum value of peak power-flux density produced at the geostationary satellite orbit by any radionavigation transmitter should not exceed

$$-187 - 10 \log_{10} D \ dB(W/m^2)$$

where D is the estimated geographical density of radionavigation transmitters per km^2 simultaneously active in any 1 MHz band, taking into account future needs and averaged over the territory

of the administration concerned or over an area of 10^6 km^2, whichever is less.*

(2.) Radiolocation Service Sharing With the Fixed Satellite Service

The only case involving sharing between the radiolocation service
and the fixed satellite service on a primary basis, that will exist after
1985, is in the 5725-5850 MHz band. Reports *AC/1* and *AE/1* were
adopted at the Interim Meetings. Until 1985, sharing occurs with
airborne radars in the 3400-3600 MHz band; and primary status of
the service will be dropped after 1985 as per "footnote 784."

---

*See also *Report 560-1*, Reference 26.

TABLE 2-1
SPECTRUM ALLOCATIONS FOR COMMUNICATIONS SATELLITES

MHz
1 215-1 530

Source: Ref. [9]

Allocation to Services		
Region 1	Region 2	Region 3
1 215 — 1 240 RADIOLOCATION RADIONAVIGATION-SATELLITE (space-to-Earth) 710 711 712 713		
1 240 — 1 260 RADIOLOCATION RADIONAVIGATION-SATELLITE (space-to-Earth) 710 Amateur 711 712 713 714		
1 400 — 1 427 EARTH EXPLORATION-SATELLITE (passive) RADIO ASTRONOMY SPACE RESEARCH (passive) 721 722		
1 427 — 1 429 SPACE OPERATION (Earth-to-space) FIXED MOBILE except aeronautical mobile 722		
1 525 — 1 530 SPACE OPERATION (space-to-Earth) FIXED Earth Exploration-Satellite Mobile except aeronautical mobile 724 722 725	**1 525 — 1 530** SPACE OPERATION (space-to-Earth) Earth Exploration-Satellite Fixed Mobile 723 722	**1 525 — 1 530** SPACE OPERATION (space-to-Earth) FIXED Earth Exploration-Satellite Mobile 723 724 722

TABLE 2-1, *cont.*

MHz
1 530-1 700

Allocation to Services		
Region 1	Region 2	Region 3
1 530 — 1 535 SPACE OPERATION (space-to-Earth) MARITIME MOBILE- SATELLITE (space-to-Earth) Earth Exploration-Satellite Fixed Mobile except aeronautical mobile 722 726	1 530 — 1 535 SPACE OPERATION (space-to-Earth) MARITIME MOBILE-SATELLITE (space-to-Earth) Earth Exploration-Satellite Fixed Mobile 723 722 726	
1 535 — 1 544	MARITIME MOBILE-SATELLITE (space-to-Earth) 722 727	
1 544 — 1 545	MOBILE-SATELLITE (space-to-Earth) 722 727 728	
1 545 — 1 559	AERONAUTICAL MOBILE-SATELLITE (R) (space-to-Earth) 722 727 729 730	
1 690 — 1 700 METEOROLOGICAL AIDS METEOROLOGICAL- SATELLITE (space-to-Earth) Fixed Mobile except aeronautical mobile 671 722 741	1 690 — 1 700 METEOROLOGICAL AIDS METEOROLOGICAL-SATELLITE (space-to-Earth) 671 722 740 742	

TABLE 2-1, *cont.*

MHz
1 700-2 655

Allocation to Services		
Region 1	Region 2	Region 3
1 700 — 1 710 FIXED METEOROLOGICAL- SATELLITE (space-to-Earth) Mobile except aeronautical mobile 671 722	1 700 — 1 710 FIXED METEOROLOGICAL-SATELLITE (space-to-Earth) MOBILE except aeronautical mobile 671 722 743	
2 290 — 2 300 FIXED SPACE RESEARCH (deep space) (space-to-Earth) Mobile except aeronautical mobile	2 290 — 2 300 FIXED MOBILE except aeronautical mobile SPACE RESEARCH (deep space) (space-to-Earth)	
2 500 — 2 655 FIXED 762 763 764 MOBILE except aeronautical mobile BROADCASTING- SATELLITE 757 760	2 500 — 2 655 FIXED 762 764 FIXED-SATELLITE (space-to-Earth) 761 MOBILE except aeronautical mobile BROADCASTING- SATELLITE 757 760	2 500 — 2 535 FIXED 762 764 FIXED-SATELLITE (space-to-Earth) 761 MOBILE except aeronautical mobile BROADCASTING- SATELLITE 757 760 754
		2 535 — 2 655 FIXED 762 764 MOBILE except aeronautical mobile BROADCASTING- SATELLITE 757 760

TABLE 2-1, *cont.*

MHz
2 655-3 500

Allocation to Services		
Region 1	Region 2	Region 3
2 655 — 2 690	**2 655 — 2 690**	**2 655 — 2 690**
FIXED 762 763 764	FIXED 762 764	FIXED 762 764
MOBILE except aeronautical mobile	FIXED-SATELLITE (Earth-to-space) (space-to-Earth) 761	FIXED-SATELLITE (Earth-to-space) 761
BROADCASTING-SATELLITE 757 760	MOBILE except aeronautical mobile	MOBILE except aeronautical mobile
Earth Exploration-Satellite (passive)	BROADCASTING-SATELLITE 757 760	BROADCASTING-SATELLITE 757 760
Radio Astronomy	Earth Exploration-Satellite (passive)	Earth Exploration-Satellite (passive)
Space Research (passive)	Radio Astronomy	Radio Astronomy
	Space Research (passive)	Space Research (passive)
758 759 765	765	765 766

2 690 — 2 700	EARTH EXPLORATION-SATELLITE (passive)
	RADIO ASTRONOMY
	SPACE RESEARCH (passive)
	767 768 769

3 400 — 3 600	**3 400 — 3 500**
FIXED	FIXED
FIXED-SATELLITE (space-to-Earth)	FIXED-SATELLITE (space-to-Earth)
Mobile	Amateur
Radiolocation	Mobile
	Radiolocation 784
	664 783

MHz
7 550-8 215

TABLE 2-1, *cont.*

Allocation to Services		
Region 1	Region 2	Region 3
7 750 — 7 900 FIXED MOBILE except aeronautical mobile		
7 900 — 7 975 FIXED FIXED-SATELLITE (Earth-to-space) MOBILE 812		
7 975 — 8 025 FIXED FIXED-SATELLITE (Earth-to-space) MOBILE 812		
8 025 — 8 175 FIXED FIXED-SATELLITE (Earth-to-space) MOBILE Earth Exploration-Satellite (space-to-Earth) 813 815	**8 025 — 8 175** EARTH EXPLORATION- SATELLITE (space-to-Earth) FIXED FIXED-SATELLITE (Earth-to-space) MOBILE 814	**8 025 — 8 175** FIXED FIXED-SATELLITE (Earth-to-space) MOBILE Earth Exploration-Satellite (space-to-Earth) 813 815
8 175 — 8 215 FIXED FIXED-SATELLITE (Earth-to-space) METEOROLOGICAL- SATELLITE (Earth-to-space) MOBILE Earth Exploration-Satellite (space-to-Earth) 813 815	**8 175 — 8 215** EARTH EXPLORATION- SATELLITE (space-to-Earth) FIXED FIXED-SATELLITE (Earth-to-space) METEOROLOGICAL- SATELLITE (Earth-to-space) MOBILE 814	**8 175 — 8 215** FIXED FIXED-SATELLITE (Earth-to-space) METEOROLOGICAL- SATELLITE (Earth-to-space) MOBILE Earth Exploration-Satellite (space-to-Earth) 813 815

TABLE 2-1, *cont.*

MHz
8 215-11 700

Allocation to Services		
Region 1	Region 2	Region 3
8 215 — 8 400 **FIXED** FIXED-SATELLITE (Earth-to-space) MOBILE Earth Exploration-Satellite (space-to-Earth) 813 815	**8 215 — 8 400** EARTH EXPLORATION- SATELLITE (space-to-Earth) FIXED FIXED-SATELLITE (Earth-to-space) MOBILE 814	**8 215 — 8 400** FIXED FIXED-SATELLITE (Earth-to-space) MOBILE Earth Exploration-Satellite (space-to-Earth) 813 815
10.6 — 10.68	EARTH EXPLORATION-SATELLITE (passive) FIXED MOBILE except aeronautical mobile RADIO ASTRONOMY SPACE RESEARCH (passive) Radiolocation 831 832	
10.68 — 10.7	EARTH EXPLORATION-SATELLITE (passive) RADIO ASTRONOMY SPACE RESEARCH (passive) 833 834	
10.7 — 11.7 FIXED FIXED-SATELLITE (space-to-Earth) (Earth-to-space) 835 MOBILE except aeronautical mobile	**10.7 — 11.7** 　　　FIXED 　　FIXED-SATELLITE (space-to-Earth) 　　MOBILE except aeronautical mobile	

MHz
3 600-5 925

TABLE 2-1, *cont.*

Allocation to Services		
Region 1	Region 2	Region 3
3 600 — 4 200 FIXED FIXED-SATELLITE (space-to-Earth) Mobile	**3 500 — 3 700** FIXED FIXED-SATELLITE (space-to-Earth) MOBILE except aeronautical mobile Radiolocation 784 786	
	3 700 — 4 200 FIXED FIXED-SATELLITE (space-to-Earth) MOBILE except aeronautical mobile 787	
5 725 — 5 850 FIXED-SATELLITE (Earth-to-space) RADIOLOCATION Amateur 801 803 805 806 807 808	**5 725 — 5 850** RADIOLOCATION Amateur 803 805 806 808	
5 850 — 5 925 FIXED FIXED-SATELLITE (Earth-to-space) MOBILE 806	**5 850 — 5 925** FIXED FIXED-SATELLITE (Earth-to-space) MOBILE Amateur Radiolocation 806	**5 850 — 5 925** FIXED FIXED-SATELLITE (Earth-to-space) MOBILE Radiolocation 806

TABLE 2-1, *cont.*   **MHz**
                                        5 925-7 750

Allocation to Services		
Region 1	Region 2	Region 3
5 925 — 7 075	FIXED FIXED-SATELLITE (Earth-to-space) MOBILE 791 809	
7 250 — 7 300	FIXED FIXED-SATELLITE (space-to-Earth) MOBILE 812	
7 300 — 7 450	FIXED FIXED-SATELLITE (space-to-Earth) MOBILE except aeronautical mobile 812	
7 450 — 7 550	FIXED FIXED-SATELLITE (space-to-Earth) METEOROLOGICAL-SATELLITE (space-to-Earth) MOBILE except aeronautical mobile	
7 550 — 7 750	FIXED FIXED-SATELLITE (space-to-Earth) MOBILE except aeronautical mobile	

GHz
11.7-12.75

TABLE 2-1, *cont.*

Allocation to Services		
Region 1	Region 2	Region 3

11.7 — 12.5
 	
FIXED
 	
BROADCASTING
 	
BROADCASTING-
  SATELLITE
 	
Mobile except
  aeronautical mobile
	11.7 — 12.1
	
	FIXED 837
	
	FIXED-SATELLITE
	  (space-to-Earth)
	
	Mobile except
	  aeronautical mobile
	
	836 839 840
		11.7 — 12.2
		
		FIXED
		
		MOBILE except
		  aeronautical mobile
		
		BROADCASTING
		
		BROADCASTING-
		  SATELLITE
		
		838 840

	12.1 — 12.3
	
	FIXED 837
	
	FIXED-SATELLITE
	  (space-to-Earth)
	
	MOBILE except
	  aeronautical mobile
	
	BROADCASTING
	
	BROADCASTING-
	  SATELLITE
	
	839 840 841
	842 843 844
		12.2 — 12.5
		
		FIXED
		
		MOBILE except
		  aeronautical mobile
		
		BROADCASTING

838 840	**12.3 — 12.7**
	
	FIXED
	
	MOBILE except
	  aeronautical mobile
	
	BROADCASTING
	
	BROADCASTING-
	  SATELLITE
	
	839 840 843 844 846
		838 840 845

12.5 — 12.75
 	
FIXED-SATELLITE
  (space-to-Earth)
  (Earth-to-space)
		12.5 — 12.75
		
		FIXED
		
		FIXED-SATELLITE
		  (space-to-Earth)
		
		MOBILE except
		  aeronautical mobile
		
		BROADCASTING-
		  SATELLITE 847

	12.7 — 12.75
	
	FIXED
	
	FIXED-SATELLITE
	  (Earth-to-space)
	
	MOBILE except
	  aeronautical mobile	

840 848 849 850	840	840

TABLE 2-1, *cont.*

GHz
12.75-13.25

Allocation to Services		
Region 1	Region 2	Region 3
12.75 — 13.25 FIXED FIXED-SATELLITE (Earth-to-space) MOBILE Space Research (deep space) (space-to-Earth)		
14 — 14.25 FIXED-SATELLITE (Earth-to-space) 858 RADIONAVIGATION 856 Space Research 857 859		
14.3 — 14.4 FIXED FIXED-SATELLITE (Earth-to-space) 858 MOBILE except aeronautical mobile Radionavigation- Satellite 859	**14.3 — 14.4** FIXED-SATELLITE (Earth-to-space) 858 Radionavigation- Satellite 859	**14.3 — 14.4** FIXED FIXED-SATELLITE (Earth-to-space) 858 MOBILE except aeronautical mobile Radionavigation- Satellite 859
14.4 — 14.47 FIXED FIXED-SATELLITE (Earth-to-space) 858 MOBILE except aeronautical mobile Space Research (space-to-Earth) 859		
14.47 — 14.5 FIXED FIXED-SATELLITE (Earth-to-space) 858 MOBILE except aeronautical mobile Radio Astronomy		

TABLE 2-2
CCIR Recommendation 466 and Appendix 29
Source: Ref. [9,11]

I. Appendix 29

This procedure is known as the "increase in equivalent noise temperature" *[ΔT] method. It consists in simulating the effects of interference to an increase in the thermal noise on the wanted link, assuming that the power density of the interfering signal is uniformly distributed over the frequency with a value equal to its maximum value. A pessimistic assumption is also adopted for the interfering signal spectrum. The ratio $\Delta T/T$ is expressed as a percentage.*

Two cases are possible: either the wanted and interfering networks share a frequency band in the same transmission direction, or they share a frequency band in opposite transmission directions (two-way use).

(a) *Case I: Frequency band shared in the same transmission direction*

Generally speaking, the interference caused by one satellite network to another is a combination of two types of interference: one on the earth-to-space links and the other on the space-to-earth links (*see Figure 1*).

The increase in noise temperature results from interference affecting both the satellite receiver of the wanted link and the earth station receiver of the wanted system link. (*The interfering paths are shown as dotted lines.*)

Interference on the earth-to-space path is calculated using the following parameters:

P'_e = transmitting spectral power density of the interfering earth station (dB (W/Hz)) (*);

$G'_1(\theta)$ = transmitting antenna gain of the interfering earth station (dB) in the direction of the wanted satellite, where θ is the topocentric angular separation between satellites;

G_2 = receiving antenna gain of the wanted satellite (dB) in the direction of the interfering earth station;

L_u = free space transmission loss (dB) on the earth-to-space link.

The power flux-density of interference at the wanted satellite receiver input is then given by:

$$I_u = P'_e + G'_1(\theta) + G_2 - L_u$$

(*) To be averaged over the reference bandwidth (4 kHz for frequencies below 15 GHz, and 1 MHz for frequencies above 15 GHz) for all kinds of emissions and necessary bandwidths. The most recent version of Report 792 should be used to the extent applicable.

The conversion of this density into increase in noise temperature yields the following expression for the increase in the wanted satellite receiving noise temperature ΔT_s:

$$10 \log_{10} (k \, \Delta T_s) = P'_e + G'_1 (\theta) + G_2 - L_u$$

where k is Boltzmann's constant, equal to 1.38×10^{-23} J/K.

Interference on the space-to-earth path is calculated on the basis of the following parameters:

P'_s = transmitting spectral power density of the interfering satellite (dB (W/Hz)) (*);

G'_3 = transmitting antenna gain of the interfering satellite (dB) in the direction of the wanted earth station;

$G_4 (\theta)$ = receiving antenna gain of the wanted receiving earth station (dB) in the direction of the interfering satellite;

L_d = free space transmission loss (dB) on the space-to-earth link.

The power flux-density of interference at the wanted earth station receiver input is given by:

$$I_d = P'_s + G'_3 + G_4 (\theta) - L_d$$

The conversion of this density into increase in noise temperature yields the following expression for the increase in the wanted receiving earth station's noise temperature ΔT_e:

$$10 \log_{10} (k \, \Delta T_e) + P'_s + G'_3 + G_4 (\theta) - L_d$$

Let ΔT be the apparent increase - due to the total interference caused on the wanted link - in the noise temperature for the complete satellite link at the input of the wanted receiving earth station. This increase is the sum of the previous increases referred to the wanted station receiver input. If y is the transmission gain of the satellite link, evaluated from the wanted satellite receiving antenna output to the wanted receiving earth station receiving antenna output (the numerical power ratio is usually less than 1), we may write:

$$\Delta T + y \, \Delta T_s + \Delta T_e$$

From this we calculate the value of $\Delta T/T$ (as a percentage):

$$\frac{\Delta T}{T} \text{ (as } \% \text{)} = \frac{100}{T} (y \, \Delta T_s + \Delta T_e)$$

(b) *Case II: Frequency band shared in opposite transmission directions*

In this case, interference occurs between satellites, on the one hand, and earth stations, on the other. This procedure concerns only interference between

satellites. The increase in the wanted receiving satellite earth temperature ΔT_s
is given by:

$$10 \log (k \, \Delta T_s) = P'_s + G'_3 + G_2 - L_{ss}$$

where P'_s has the same meaning as in (a) and where G'_3 and G_2 have the same
definition as before, except that they should be taken in the direction of the
other satellite. Lastly, L_{ss} is the free space transmission loss (dB) on the
inter-satellite link.

The increase in the equivalent noise temperature of the link ΔT is calculated
from:

$$\Delta T = \gamma \, \Delta T_s$$

where γ has the same meaning as above. Hence, finally:

$$\frac{\Delta T}{T} (\text{in } \%) = \frac{100}{T} \gamma \, \Delta T_s$$

(c) *Possible consideration of polarization isolation*

If Administrations agree to take account of any polarization isolation factor
due to the fact that the networks involved use cross-polarization, the above
formulae become:

$$\text{Case I} \quad \frac{\Delta T}{T} (\text{in } \%) = \frac{100}{T} \left(\frac{\gamma^{\Delta T_S}}{Y_u} + \frac{\Delta T_e}{Y_d} \right)$$

$$\text{Case II} \quad \frac{\Delta T}{T} (\text{in } \%) = \frac{100}{T} \frac{\gamma^{\Delta T_S}}{Y_{ss}}$$

where Y_u, Y_d, Y_{ss} are polarizaton isolation factors for up-link, down-link and
inter-satellite link, respectively. They are equal to 4 if the two polarizations are
circular and opposite, to 1.4 if one polarization is circular and the other linear,
and to 1 in other cases.

Examination of results (Appendix 29)

The value of $\Delta T/T$ obtained should be compared with the threshold value fixed
at 4%. If the value obtained is less than 4%, the interference will be regarded as
acceptable and no coordination is necessary between the networks involved.
Otherwise, coordination must be effected, which usually calls for detalied
interference calculations taking into account the types of carrier modulation of
the two networks.

In certain cases, the earth-to-space links should be dealt with separately
(where a single link has common frequencies or where modulation is changed
on board the satellite). In these cases, ΔT_e or ΔT_s are calculated and the

relative increase $\Delta T_s / T_c$ is compared to the threshold value of 4% (T_s and T_c being the receiving temperatures of the satellite and of an earth station). In these cases, the condition for the avoidance of coordination is that each of the calculated values should be less than 4%.

In the particular case of narrow-band carriers being interfered by slow-swept television carriers, the 4% criteria may not afford enough protection to the interfered network. This should be taken into account by Administrations when using this method.

II. Recommendation 466

Recommendation 466-2 (MOD I)

MAXIMUM PERMISSIBLE LEVEL OF INTERFERENCE IN A TELEPHONE CHANNEL OF A GEOSTATIONARY SATELLITE NETWORK IN THE FIXED-SATELLITE SERVICE EMPLOYING FREQUENCY MODULATION WITH FREQUENCY-DIVISION MULTIPLEX, CAUSED BY OTHER NETWORKS OF THIS SERVICE

(Study Programme 2J-2/4)   **(1970 - 1974 - 1978)**

The CCIR,

CONSIDERING

[a]   that geostationary satellite networks in the Fixed Satellite Service operate in the same frequency bands;

[b]   that interference between networks in the Fixed Satellite Service contributes to the noise in the network;

[c]   that it is desirable that the interference noise in the telephone channels of networks in the Fixed Satellite Service caused by transmitters of different networks in that service should be such as to give a reasonable orbit utilization efficiency;

[d]   that the overall performance of a network should essentially be under the control of the system designer;

[e]   that it is necessary to protect a network in the Fixed Satellite Service from interference by other such networks;

[f]   that it is necesasry to specify the maximum permissible interference power in a telephone channel, to determine space station and earth station characteristics;

[g]   that the value selected for the maximum permissible single entry of interference noise should not result in too wide a minimum orbital spacing;

[h]   that for co-ordination between two satellite networks, it is necessary to

know the maximum permissible level of interference in a telephone channel of the wanted network caused by transmitters of the other network;

[*j*]   that networks in the Fixed Satellite Service may receive interference both into the space station receiver and into the earth station receiver;

[*k*]   that the mean interference noise power should be an appropriate fraction of the total noise power permitted in the hypothetical reference circuit;

[*l*]   that in many cases the largest interference contributions to a geostationary satellite network will be from the networks using geostationary satellites in orbit nearby and serving an overlapping coverage area. The value of interference from any other network will generally be less;

[*m*]   that the accumulation of interference from many entries from other satellite networks and from terrestrial stations is not likely to be equal to the arithmetic sum of the individual interference entries at their maximum value across the band, and the total interference noise level in any single channel may be significantly less;

[*n*]   that the levels of interference between geostationary satellite networks in the Fixed Satellite Service operating in frequency bands below 10 GHz are not expected to undergo a large variation with time,

RECOMMENDS

1.   that different geostationary satellite networks in the Fixed Satellite Service operating in the same frequency bands, below 10 GHz, be designed in such a manner that the interference noise power at a point of zero relative level in any telephone channel of a hypothetical reference circuit of a network in the Fixed Satellite Service, employing frequency modulation, caused by the aggregate of the earth station and space station transmitters of other fixed satellit networks, should not exceed:

1.1   in frequency bands in which the network does not practise frequency re-use; 2000 pW0p, psophometrically weighted one minute mean power for 20% of any month . . .

6.   *Examples of small transportable earth stations*

In the 6/4 GHz band, a number of transportable earth stations are operating now with various antennae diameters. In the 14/13 GHz band, most of the transportable stations have antennae with around 3 m diameters.

6.1   *An example of a small transportable earth station for operation at 6/4 GHz.*

An air-transportable earth station, which may also be carried by an 8-ton truck, has been manufactured . . . and satisfactory performance has been achieved [*Okamoto et al., 1977*].

The station has a 3 meter diameter antenna, a peak e.i.r.p. of about 66 dBW and a *G/T* of about 20 dBK. The total weight is 5.5 tons and the power requirement, including air conditioning, is 7.5 kVA. The reflector is divided into seven sections each weighing about 20 kg, the total setting up time being about 2 hours by three persons. The station uses FDM-FM modulation and can support 6 two-way channels using a transponder similar to an INTELSAT-IVA global beam transponder, or 60 two-way channels using a shaped-beam transponder similar to Japanese CSE (Communication Satellite for Experimental Purpose) transponder, with a channel signal-to-noise ratio of about 43 dB.

6.2   *Examples of a small transportable earth station for operation at 30/20 GHz*

Two types of 30/20 GHz small transportable earth stations, which can be transported by a truck or a helicopter, have been manufactured and operated satisfactorily (Saruwatari *et al.*, 1978; and Egami *et al.*, 1980.

One of these stations has a 2.7-meter Cassegrain antenna. The total weight of the station is 5.8 tons and the power requirement is 12 kVA. The maximum e.i.r.p. is about 76 dBW and *G/T* is about 27 dBK in clear sky conditions. The station uses FDM-FM modulation and can transmit 132 two-way telephone channels. Setting-up time is about one hour after arrival.

The other station is an SCPC station which has a 2-meter Cassegrain antenna. The maximum e.i.r.p. is about 56.3 dBW and *G/T* is about 20.4 dBK with a clear sky condition. The station uses an adaptive delta-modulation/4-phase PSK/SCPC system and can transmit one two-way telephone channel.

REFERENCES

Egami, S., T. Okamoto, and H. Fuketa, (February 1980) K-band mobile earth station for domestic satellite communication system, *IEEE Trans. Comm.*, Vol. COM-28, No. 2.

Okamoto, T., S. Egami, and M. Oguchi, (1977) Small transportable earth station for satellite communication, *Japan Telecommunication Review*, Vol. 19, No. 3, pp. 230.

Saruwatari, T., K. Tsukamoto, M. Yokoyama, and S. Sasaoka, (1978) Digital transmission experiments with the CS satellite, *4th Int. Conf. on Digital Satellite Communications*, pp. 283-290.

TABLE 2-3

FSS & BSS SHARING IN THE 12 GHz BAND
Source: ITU, IWP 4/1 [**34**]

FSS and BSS sharing situations in the 12 GHz band

Frequency Band (GHz)	Region 1		Region 2		Region 3	
11.7 - 12.1	BSS	(S-E)	FSS BSS (FN 3787A)	(S-E) (S-E)	BSS	(S-FM)
12.1 - 12.2	BSS	(S-E)	FSS or BSS (FN3787B)	(S-E)	BSS	(S-E)
12.2 - 13.2	BSS	(S-E)	FSS or BSS (FM3787B)	(S-E)	FSS (FN3785B)	(S-E)
12.3 - 12.5	BSS	(S-E)	BSS FSS (FN3787F)	(S-E) (S-E)	FXX (FN3785B)	(S-E)
12.5 - 12.7	FSS FSS	(S-E) (E-S)	BSS FSS (FN3787F)	(S-E) (S-E)	FSS BSS (FM3785A)	(S-E) (S-E)
12.7 - 12.75	FSS FSS	(S-E) (E-S)	FSS FSS	(E-S) (E-S)	FSS BSS (FM3785A)	(S-E) (S-E)

TABLE 2-4

ARTICLES 27 & 28 of the RADIO REGULATIONS
Source: ITU, CCIR, *Radio Regulations*, 1981 [35]

CHAPTER VIII
Provisions Relating to Groups of Services and to Specific Services and Stations*

ARTICLE 27
Terrestrial Radiocommunication Services Sharing Frequency Bands with Space Radiocommunication Services Above 1 GHz

Section I. Choice of Sites and Frequencies

2501 §1. Sites and frequencies for terrestrial stations, operating in frequency bands shared with equal rights between terrestrial radiocommunication and space radiocommunication services, shall be selected having regard to the relevant CCIR Recommendations with respect to geographical separation from earth stations.

2502 §2. (1) As far as practicable, sites for transmitting[1] stations, in the fixed or mobile service, employing maximum values of equivalent isotropically radiated power (e.i.r.p.) exceeding +35 dBW in the frequency bands between 1 GHz and 10 GHz, should be selected so that the direction of maximum radiation of any antenna will be at least 2° away from the geostationary-satellite orbit, taking into account the effect of atmospheric refraction .

2503 (2) As far as practicable, sites for transmitting[1] stations, in the fixed or mobile service, employing maximum values of equivalent isotropically radiated power (e.i.r.p.) exceeding +45 dBW in the frequency bands between 10 GHz and 15 GHz, should be selected so that the direction of maximum radiation of any antenna will be at least 1.5° away from the geostationary-satellite orbit, taking into account the effect of atmospheric refraction .

2504 (3) In the frequency bands above 15 GHz there shall be no restriction[3] as to the direction of maximum radiation for stations in the fixed or mobile service.

Section II. Power Limits

2505 §3. (1) The maximum equivalent isotropically radiated power (e.i.r.p.) of a station in the fixed or mobile service shall not exceed +55 dBW.

2506 (2) Where compliance with No. **2502** is impracticable the maximum equivalent isotropically radiated power (e.i.r.p.) of a station in the fixed or mobile service shall not exceed:

+47 dBW in any direction within 0.5° of the geostationary-satellite orbit; or

+47 dBW to +55 dBW, on a linear decibel scale (8 dB per degree), in any direction between 0.5° and 1.5° of the geostationary-satellite orbit, taking into account the effect of atmospheric refraction .

2507   (3) The power delivered by a transmitter to the antenna of a
station in the fixed or mobile service in frequency bands between 1
GHz and 10 GHz shall not exceed +13 dBW.

(4) The power delivered by a transmitter to the antenna of a
station in the fixed or mobile service in frequency bands above 10
GHz shall not exceed +10 dBW.

2509   (5) The limits given in Nos. **2502, 2505, 2506** and **2507** apply
in the following frequency bands allocated to the fixed-satellite service, the meteorological-satellite service and the mobile-satellite
service for reception by space stations, where these bands are shared
with equal rights with the fixed or mobile service:

1 626.5 - 1 645.5 MHz	(for countries mentioned in No. **730**)
1 646.5 - 1 660 MHz	(for countries mentioned in No. **730**)
2 655 - 2 690 MHz	(for Regions 2 and 3)
5 725 - 5 755 MHz	(for countries of Region 1 mentioned in Nos. **803** and **805**)
5 755 - 5 850 MHz	(for countries of Region 1 mentioned in Nos. **803, 805** and **807**)
5 850 - 7 075 MHz	
7 900 - 8 400 MHz	

2510   (6) The limits given in Nos. **2503, 2505** and **2508** apply in the
following frequency bands allocated to the fixed-satellite service for
reception by space stations, where these bands are shared with equal
rights with the fixed or mobile service:

10.7 - 11.7 GHz	(for Region 1)
12.5 - 12.75 GHz	(for countries mentioned in Nos. **848** and **850**)
12.7 - 12.75 GHz	(for Region 2)
12.75 - 13.25 GHz	
14.0 - 14.25 GHz	(for countries mentioned in No. **857**)
14.25 - 14.3 GHz	(for countries mentioned in Nos. **857, 860** and **861**)
14.3 - 14.4 GHz	(for Regions 1 and 3)
14.4 - 14.5 GHz	
14.5 - 14.8 GHz	

2511   (7) The limits given in Nos. **2505** and **2508** apply in the following frequency bands allocated to the fixed-satellite service for reception by space stations, where these bands are shared with equal
rights with the fixed or mobile service:

17.7 - 18. 1 GHz	
27.0 - 27.5 GHz	(for Regions 2 and 3)
27.5 - 29.5 GHz	

TABLE 2-5

ARTICLE 28

Space Radiocommunication Services Sharing Frequency Bands with Terrestrial Radiocommunication Services Above 1 GHz

Section I. Choice of Sites and Frequencies

2539 §1. Sites and frequencies for earth station, operating in frequency bands shared with equal rights between terrestrial radiocommunication and space radiocommunication services, shall be selected having regard to the relevant CCIR Recommendations with respect to geographical separation from terrestrial stations.

Section II. Power Limits

2540 §2. (1) Earth stations.

2541 (2) The equivalent isotropically radiated power (e.i.r.p.) transmitted in any direction towards the horizon by an earth station operating in frequency bands between 1 GHz and 15 GHz shall not exceed the following limits except as provided in No. **2544** or **2546**:

+40 dBW in any 4 kHz band for $\theta \le 0°$

+40 + 3 θ dBW in any 4 kHz band for $0° < \theta \le 5$§

where θ is the angle of elevation of the horizon viewed from the centre of radiation of the antenna of the earth station and measured in degrees as positive above the horizontal plane and negative below it.

2542 (3) The equivalent isotropically radiated power (e.i.r.p.) transmitted in any direction towards the horizon by an earth station operating in frequency bands above 15 GHz shall not exceed the following limits except as provided in No. **2545** or **2546**:

+64 dBW in any 1 MHz band for $\theta \le 0°$

+64 + 3 θ dBW in any 1 MHz band for $0° < 0 \le 5°$

where θ is as defined in No. **2541**.

2543 (4) For angles of elevation of the horizon greater than 5° there shall be no restriction as to the equivalent isotropically radiated power (e.i.r.p.) transmitted by an earth station towards the horizon.

2544 (5) As an exception to the limits given in No. **2541**, the equivalent isotropically radiated power (e.i.r.p.) towards the horizon for an earth station in the space research service (deep space) shall not exceed +55 dBW in any 4 kHz band.

2545 (6) As an exception to the limits given in No. **2542**, the equivalent isotropically radiated power (e.i.r.p.) towards the horizon for an earth station in the space research service (deep-space) shall not exceed +79 dBW in any 1 MHz band.

2546   (7) The limits given in Nos. **2541, 2542, 2544** and **2545,** as applicable, may be exceeded by not more than 10 dB. However, when the resulting coordination area extends into the territory of another country, such increase shall be subject to agreement by the administration of that country.

2547   (8) The limits given in No. **2541** apply in the following frequency bands allocated to the fixed-satellite service, the earth exploration-satellite service, and in particular the meteorlogical-satellite service, the mobile-satellite service and the space research service for transmission by earth stations where these bands are shared with equal rights with the fixed or mobile service:

5 670	- 5 725	MHz	(for the countries mentioned in No. **804** with respect to the countries mentioned in Nos. **803** and **805**)
5 725	- 5 755	MHz	(for Region 1 with respect to the countries mentioned in Nos. **803** and **805**)
5 755	- 5 850	MHz	(for Region I with respect to the countries mentioned in Nos. **803, 805** and **807**)
5 850	- 8 400	MHz	
7 900	- 8 400	MHz	
10.7 -	11.7	GHz	(for Region 1)
12.5 -	12.75	GHz	(for Region 1 with respect to the countries mentioned in No. **848**)
12.7 -	12.75	GHz	(for Region 2)
12.75 -	13.25	GHz	
14.0 -	14.25	GHz	(with respect to the countries mentioned in No. **857**)
1 814.25 -	14.3	GHz	(with respect to the countries mentioned in Nos. **857, 860** and **861**)
14.3 -	14.4	GHz	(for Regions 1 and 3)
14.4 -	14.8	GHz	

2548   (9) The limits given in No. **2542** apply in the following frequency bands allocated to the fixedsatellite service, the earth exploration-satellite service, the mobile-satellite service and the space research service for transmission by earth stations where shared with equal rights with the fixed or mobile service:

17.7 - 18.1 GHz	
27.0 - 27.5 GHz	(for Regions 2 and 3)
27.5 - 29.5 GHz	
31.0 - 31.3 GHz	(for the countries mentioned in No. **885**)
34.2 - 35.2 GHz	(for the countries mentioned in Nos. **895** and **896** with respect to the countries mentioned in No. **894**)

Section III. Minimum Angle of Elevation

2549 §3. (1) Earth stations.

2550 (2) Earth station antennae shall not be employed for transmission at elevation angles of less than 3§ measured from the horizontal plane to the direction of maximum radiation, except when agreed to by administrations concerned and those whose services may be affected. In case of reception by an earth station, the above value shall be used for coordination purposes if the operating angle of elevation is less than that value.

2552 (3) As an exception to No. **2550**, earth station antennae in the space research service (near Earth) shall not be employed for transmission at elevation angles of less than 5§, and earth station antennae in the space research service (deep space) shall not be employed for transmission at elevation angles of less than 10§ both angles being those measured from the horizontal plane to the direction of maximum radiation. In the case of reception by an earth station, the above values shall be used for coordination purposes if the operating angle of elevation is less than those values.

Section IV. Limits of Power Flux-Density from Space Stations

2561 (3) Power flux-density limits between 2 500 MHz and 2 690 MHz.

2562 a) The power flux-density at the Earth's surface produced by emissions from a space station in the broadcasting-satellite service or the fixed-satellite service for all conditions and for all methods of modulation shall not exceed the following values:

— 152 d/B(W/m) in any 4 kHz band for angles of arraival between 0 and 5 degrees above the horizontal plane;

— 152 + 0.75(δ — 5) dB(W/m) in any 4 kHz band for angles of arrival δ (in degrees) between 5 and 25 degrees above the horizontal plane;

— 137 dB(W/m) in any 4 kHz band for angles of arrival between 25 and 90 degrees above the horizontal plane.

These limits relate to the power flux-density which would be obtained under assumed free-space propagation conditions.

2563 b) The limits given in No. **2562** apply in the frequency band:

2 500 - 2 690 MHz

which is shared by the broadcasting-satellite service or the fixed-satellite service with the fixed or mobile service.

2564 c) The power flux-density values given in No. **2562** are derived on the basis of protecting the fixed service using line-of-sight techniques. Where a fixed service using tropospheric scatter operates in the band mentioned in No. **2563** and where there is insufficient frequency separation, there must be sufficient angular separation between the direction to the space station and the direction of maximum radiation of the antenna of the receiving station of the fixed service using tropospheric scatter to ensure that the interference power at the receiver input of the station of the fixed service does not exceed — 168 dBW in any 4 kHz band.

2565   (4) Power flux-density limits between 3 400 MHz and 7 750 MHz.

2566   a) The power flux-density at the Earth's surface produced by emissions from a space station, including emissions from a reflecting satellite, for all conditions and for all methods of modulation, shall not exceed the following values:

> — 152 dB(W/m) in any 4 kHz band for angles of arrival between 0 and 5 degrees above the horizontal plane;

> — 142 dB(W/m) in any 4 kHz band for angles of arrival between 25 and 90 degrees above the horizontal plane.

These limits relate to the power flux-density which would be obtained under assumed free-space propagation conditions.

2567   b) The limits given in No. **2566** apply in the frequency bands listed in No. **2568** which are allocated to the following space radio-communication services:

> - — fixed-satellite service (space-to-Earth)

> — meteorological-satellite service (space-to-Earth)

> — mobile-satellite service

> — space research service

for transmission by space stations where these bands are shared with equal rights with the fixed or mobile service.

2568   3 400 - 4 200 MHz
4 500 - 4 800 MHz
5 670 - 5 725 MHz   (on the territory of countries mentioned
in Nos. **803** and **805**)

7 250 - 7 750 MHz

2569   (5) Power flux-density limits between 8 025 MHz and 11.7 GHz.

2570   a) The power flux-density at the Earth's surface produced by emissions from a space station, including emissions from a reflecting satellite, for all conditions and for all methods of modulation, shall not exceed the following values:

> — 150 dB(W/m) in any 4 kHz band for angles of arrival between 0 and 5 degrees above the horizontal plane;

> — $150 + 0.5(\delta - 5)$ dB(W/m) in any 4 kHz band for angles of arrival δ (in degrees) between 5 and 25 degrees above the horizontal plane;

— 140 dB(W/m) in any 4 kHz band for angles of arrival between 25 and 90 degrees above the horizontal plane.

These limits related to the power flux-density which would be obtained under assumed free-space propagation conditions.

2571   b) The limits given in No. **2570** apply in the frequency bands listed in No. **2572** which are allocated to the following space radio-communication services:

— earth exploration-satellite service (space-to-Earth)

— space research service (space-to-Earth)

— fixed-satellite service (space-to-Earth)

for transmission by space stations where these bands are shared with equal rights with the fixed or mobile service.

2572   8 025 - 8 500   MHz
              10.7 -   11.7 GHz

2573   (6) Power flux-density limits between 12.2 GHz and 12.75 GHz.

2574   a) The power flux-density at the Earth's surface produced by emissions from a space station, including emissions from a reflecting satellite, for all conditions and for all methods of modulation, shall not exceed the following values:

— 148 dB(W/m) in any 4 kHz band for angles of arrival between 0 and 5 degrees above the horizontal plane;

— $148 + 0.5(\delta - 5)$ dB(W/m) in any 4 kHz band for angles of arrival δ (in degrees) between 5 and 25 degrees above the horizontal plane;

— 138 dB(W/m) in any 4 kHz band for angles of arrival between 25 and 90 degrees above the horizontal plane.

These limits relate to the power flux-density which would be obtained under assumed free-space propagation conditions.

2575   The limits given in No. 2574 apply in the frequency bands indicated in No. **2576** which are allocated to the fixed-satellite service for transmission by space stations where these bands are shared with equal rights with the fixed or mobile service.

2576   12.2 - 12.5 GHz   (for Region 3)
              12.5 - 12.75 GHz   (for Region 3 and for Region 1 on the territory of countries mentioned in Nos. **848** and **850**).

2577   (7) Power flux-density limits between 17.7 GHz and 19.7 GHz.

2578   a) The power flux-density at the Earth's surface produced by emissions from a space station, including emissions from a reflecting satellite, for all conditions and for all methods of modulation, shall not exceed the following values:

— 115 dB(W/m) in any 1 MHz band for angles of arrival between 0 and 5 degrees above the horizontal plane;

— 115 + 0.5(δ —5) dB(W/m) in any 1 MHz band for angles of arrival δ (in degrees) between 5 and 25 degrees above the horizontal plane;

— 105 (dB(W/m) in any 1 MHz band for angles of arrival between 25 and 90 degrees above the horizontal plane.

Thees limits relate to the power flux-density which would be obtained under assumed free-space propagation conditions.

2579   b) The limits given in No. **2578** apply in the frequency band listed in No. **2580** which is allocated to the following space radio-communication services:

— fixed-satellite service (space-to-Earth)

— earth exploration-satellite including meteorological-satellite service (space-to-Earth)

Chapter Three

INTERNATIONAL AND DOMESTIC ORBIT/SPECTRUM POLICY

3.0 Introduction

The development of policy and regulation concerning use of communication satellites in the geostationary orbit has been characterized by two themes:

1. the interaction between policy institutions and technological development; and

2. the contrast between the international and domestic aspects of policy and regulation.

Fundamentally, the development and use of the geostationary orbit has been a classic struggle between technology and policy at both the international and domestic levels.

For the past two decades the United States has been the leader in both the technology and policy development. Today, on the threshold of the third decade of use of communication satellites in the GSO other influences are beginning to be felt, such as demands on the part of the "Third" or "developing" world which could significantly alter the regulations and technological development of the GSO.

3.1 The First Decade

The years 1962-1963 saw the marriage of a policy on communication satellites with the breakthrough in technology that would lead to the rapid exploitation of GSO communication satellite technology which in only two decades would revolutionize worldwide communications.

Prior to 1962 the American Telegraph and Telephone Company (AT&T), and particularly its subsidiary the Bell Telephone Laboratories, had demonstrated the feasibility of using satellites for communication purposes. This was accomplished with the launch of Telstar, a low orbiting satellite, working with a multi-million dollar horn antenna at Andover, Maine.

With the demonstration of this technology, the United States Congress proceeded to hold earnings on how it might be brought to fruition on an international basis. Hearings were held in the context of the Kennedy Administration which proposed that a government-owned entity be established to exploit the technology. On the other hand, the common carrier industry, led by AT&T, wanted to own and operate such an organization.

The result of these debates was the Communications Satellite Act of 1962. This legislation created the Communications Satellite Corporation (COMSAT) with a capitalization of $150 million. (See Chapter 1.) This unique, legislatively-established corporation represented a compromise between those advocating complete private sector control and those advocating a completely government-owned entity. COMSAT's Board of Directors was to be made up of one-third each of government, common carrier, and company-appointed directors.

This act stated the first communications satellite policy, which, unbeknownst to the legislators fashioning it, was to lead to the subsequent dramatic exploitation of the GSO. The policy was that the COMSAT Corporation was to bring into existence "a single global communications satellite system." As noted in Chapter 1, the concept behind this policy, and the basic rationale for the capitalization, was that of a communication satellite system utilizing Telstar/Andover earth station technology. To achieve continuity of service in such a system would require tons of low orbit/low capacity satellite which would be used with large, expensive complex tracking earth stations.

When asked about the prospects of communication satellites in the GSO, the Director of the Bell Telephone Laboratories testified that he did not envision their application for another ten years.

Thus, a legislative/policy framework was conceived in 1962 based on uncertain technology. Nonetheless, it proved to be the right framework to achieve, in only ten years time, the policy objective of a global communications satellite system, at a pace far in excess of anyone's expectations. In ten years, COMSAT and what came to be called INTELSAT had created an organization with 100 member nations, almost 300 earth stations, and a total system capacity of 16,000 voice circuits and eight television channels.

3.2 The Second Decade

3.2.1 U.S. Domestic Policy

The success of COMSAT and the geostationary satellite technology fueled U.S. domestic policy making. After launch of the Hughes Aircraft Company's Early Bird in 1965, Hughes teamed with the American Broadcasting Corporation to petition the Federal Communications Commission to establish a U.S. Domestic Communication Satellite System, for the purpose of providing television distribution.

ABC was motivated to take this step as a means of escaping the increasingly high tariffs being charged by AT&T for the distribution of TV programming. The three major networks were paying over $45 million annually in distance-based tariffs for TV program distribution via the terrestrial system.

It soon became evident that geostationary satellites had several characteristics which would have significant impact on the infrastructure of communications. These included:

1. *Connectivity:* the ability to instantly, and simultaneously, connect a single point to a multiplicity of other points; and

2. *Distance Independence:* the cost of providing service became relatively independent of distance. This dramatically improved the economics of long-distance communication.

The FCC accordingly initiated a regulatory procedure, which ultimately lead to the establishment of U.S. domestic policy on communications satellites and the use of the GSO.

Whereas the first operational use of the GSO was to establish a global communications network taking advantage of the unique properties of the GSO, the U.S. domestic communications policy deliberations were the first to consider questions concerning efficiency of use.

The chronology of U.S. domestic satellite policy making is outlined in Table 3-1. As will become evident, U.S. DOMSAT policy has been the laboratory for development of policies related to use of the GSO.

The FCC opened its inquiry into domestic satellite policy in 1965, and had, under the influence of an FCC controlled by Democrats, nearly concluded that a single entity should be given the domestic monopoly on communication satellites. However, such was the state of uncertainty with respect to policy in a number of telecommunication areas, including not only DOMSAT, but cable and common carrier competition, that President Johnson established a task force in 1967 to study these and other matters. By the Fall of 1968 the task force had drafted a report recommending a single entity. However, Richard Nixon had been elected President, and a change of party in the executive and regulatory (FCC) institutions was imminent.

The Nixon White House requested a delay in the FCC's implementation of its DOMSAT policy; and proceeded to review if for a year and a half. During this period an assessment was conducted to determine available orbit capacity, and to find out whether it could support a multiplicity of DOMSAT systems.

Based on studies done by NASA, the FCC, and the newly created Office of Telecommunications Policy (OTP), it was determined that there was sufficient capacity to support a number of satellite systems. As a result of these orbit use analyses, the Nixon Administration advocated on "open skies" policy for U.S. domestic satellites, i.e., *any* commercial entity which had the resources could put a satellite system in orbit, and seek to offer services for profit.

As the FCC was sorting out which policy to pursue, Canada decided to move forward with implementation of its own domestic satellite system. During the process of coordinating its systems with those of the U.S., it sought and obtained assurance that U.S. satellites would remain 5° away from Canadian satellites in the 6/4 GHz bands; the first pair of bands to be developed for commercial satellite use. This constituted the first North American orbit use policy to be adopted. Subsequent to an exchange of letters, the FCC adopted "open skies" as a matter of policy, and established a 4° spacing policy between U.S. domestic satellites. These policies have remained in force for approximately a decade.

3.2.2 International Regulation and Policy

During the past decade, the principal forum for debate on use of the geostationary orbit has been the International Telecommunication

Union (ITU). GSO policy and regulation has been manifested in almost every ITU conference in the last decade dealing with satellites.*

1971 — World Administrative Radio Conference on Space Telecommunications

This WARC adopted the first regulations and procedures in the form of a treaty governing the use of the GSO. These included:

a method for coordination among GSO satellite systems in the same frequency band;

recognition of a single entry and cumulative interference limits for FDM/FM systems;

language encouraging the accomodation of all users in the event of crowding; and

adoption of GSO station keeping and beam pointing tolerances.

1973 — ITU Plenipotentiary Conference [27]

Language concerning use of the GSO was explicitly included in the basic functions of the ITU. Specifically, Article 33 reads:**

Rational Use of the Radio Frequency Spectrum and of the Geostationary Satellite Orbit.

1. Members shall endeavor to limit the number of frequencies and the spectrum space used to the minimum essential to provide in a satisfactory manner the necessary service. To that end they shall endeavor to apply the latest technical advances as soon as possible.

2. In using frequency bands for space radio services, members shall bear in mind that radio frequencies and the geostationary orbit are limited natural resources, that they must be used efficiently and economically so that countries or groups of countries may have equitable access to both in conformity with the provisions of the Radio Regulations according to their needs and the technical facilities at their disposal.

---

*See Chapter 1 for discussion of the role of the ITU in the evolution of GSO policy and regulation.
**ITU, Plenipotentiary Conference, Malaga-Torremolinos, 1973, *International Telecommunication Convention*, Art. 33.

The 1973 Plenipotentiary Conference also passed a resolution calling for an international radio conference to plan the use of the 12 GHz band which had been allocated by the WARC-ST '71 to the Broadcasting Satellite Service.

1972-1978

CCIR IWP 4/1 met many times to consider reports and recommendations leading to improved efficiency of the GSO. The subjects treated have included: antenna radiation patterns, modulation, interference allowance, polarization, satellite station keeping, and pointing accuracy. The significance of these factors to improved orbit use has been included in CCIR *Report 453* [28].

1977 — World Administrative Radio Conference on Broadcasting Satellites [29]

This treaty conference, for the first time, adopted a detailed plan involving use of the GSO. The plan involved assigning specific frequencies and GSO orbit locations in the band 11.7 to 12.2 GHz, to each of the over 100 administrations of ITU Regions 1 and 3. Each frequency assignment was associated with a particular orbit location and service area. The service area could be either an entire country or a part of a country. Under this arrangement each service area received a minimum of four TV channels.

This unprecedented plan broke new ground in GSO policy. For the first time orbit/spectrum space was reserved *a priori* on the basis of other than real requirements. There were several motivations behind this plan:

Western European countries were anxious to implement point-to-point terrestrial systems on a shared basis in this band. This could not be done until a plan had been established;

Developing countries were fearful that countries such as the United States, which had both the technology and the launch capability, would place DBS satellites in the GSO to broadcast to their countries;

Some countries were concerned that by the time they wanted to use the GSO for DBS, there might not be any orbit or spectrum left. In other words, an *a priori* plan was one way of assuring *"guaranteed access to the GSO."*

Region 2, under the leadership of the United States, Canada, and Brazil, did not adopt this approach to using the GSO in this band. Instead, because of the shared use of this band at that time by both the BSS and FSS, Region 2 agreed to an orbit segmentation approach. In this scheme, part of the GSO would be set aside for FSS and part for BSS. Subsequent analysis indicated that capacity would not be sufficient to support FSS growth and this scheme was abandoned in preparation for WARC 79 [30].

1979 — WARC '79

More new ground was broken concerning the GSO at this conference. In addition to modifying the existing regulations with respect to use of the GSO, it also proceeded to be a forum in which significant differences were revealed between developed and developing world philosophies on the use of the GSO. These were manifested in *Resolution 3* which reads in part as follows:*

resolves

1. that a world space administrative radio conference shall be convened not later than 1984 to guaranteed practice for all countries equitable access to the geostationary-satellite orbit and the frequency bands allocated space services;

2. that this conference shall be held in two sessions;

3. that the first session shall:

3.1 decide which space services and frequency bands should be planned;

3.2 establish the principles, technical parameters and criteria for the planning, including those for orbit and frequency assignment of the space services and frequency bands identified as per paragraph 3.1, taking into account the relevant technical aspects concerning the special geographical situation of particular countries; and provide guidelines for associated regulatory procedures;

3.3 establish guidelines for regulatory procedures in respect of services and frequency bands not covered by paragraph 3.2;

---

*ITU, World Administrative Radio Conference, Geneva, 1979, *Resolution 3* [9].

3.4 consider other possible approaches that could meet the objective of resolves 1;

4. *that the second session shall be held not sooner than twelve months and not later than eighteen months after the first session and implement the decisions taken at the first session ...*

The resolution calls for the holding of Space Planning Conference(s) to "plan" all satellite bands. This resolution was drafted in a special Ad Hoc Group of WARC 79 and was hammered out over a period of two weeks. The participants included: Algeria, India, United States, United Kingdom, Australia, Columbia, and Venezuela.

This confrontation was brought about in part as a result of difficulties encounted during the application of *Appendix 29* coordination procedures by such countries as India, Indonesia, and Brazil. These were caused by the experimental introduction of an interference "scaling law" by INTELSAT for the purpose of coordinating between INTELSAT satellites and other satellite systems. This interference law tended to favor satellite networks which were similar to those of INTELSAT, but were a problem for those which were not. This was the case for many domestic types of systems which were interested in high power and less coverage area. Eventually this coordination approach was dropped.

Nevertheless, the die had been cast, and there was now a clear focus on the GSO as a unique resource for satellite communications, and soon to be the subject of extensive domestic and internationally policy debate.

3.3 The Third Decade

The third decade of the development and use of the GSO will probably see the most dramatic change in the policies governing its use. This is due in large measure to the success of its development. Figure 3-1*shows the present and foreseen development. The implication is that unless improvements in orbit capacity are realized, e.g., through reduced spacing, demand will exceed available capacity. The forums in which policy formulation and regulation will be carried out include: the ITU, INTELSAT, UN/UNISPACE, and various national and international telecommunication regulatory institutions.

---

*The capacity of this graph is additive.

3.3.1 International Policy

Reference is again made to the last sentence of Article 33 of the ITU charter.

> ... In using frequency bands for space radio services Members shall bear in mind that radio frequencies and the GSO are limited natural resources, that they must be used efficiently and economically so that countries or groups of countries may have equitable access to both in conformity with the provisions of the RR *according to their needs* and the technical facilities at their disposal.*

This phrase is critical regarding policy on the GSO. Some developing countries view this language as undermining the concept of "Equitable Access to the GSO," and therefore are interested in seeing it dropped. It goes to the heart of the fears of developing countries that developed countries in the not too distant future, under the guise of fulfilling their needs, will "use up" all available orbit positions. It is this sort of fear that led to *Resolution 3* at WARC 79.

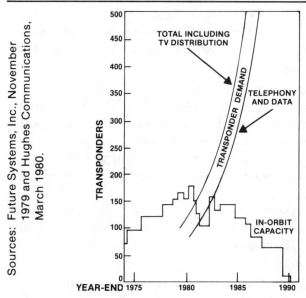

Figure 3-1 In-Orbit Capacity and Service Demand for Operating or Planned Satellites, 1980-1990

*The Nairobi Plenipotentiary Conference of the ITU modified this language to be consistent with UNISPACE '82 language.

The U.S. Policy on the matter of planning was stated at WARC 79:*
There has been extensive discussion over the nature of the planning to be undertaken at the next Space Conference, and various explanations have been given as to what is intended. For its part, the United States views the planning mandate of the next Space Conference as being very wide in scope, admitting of a broad range of possibilities ranging from detailed orbit/frequency assignment plans to more dynamic planning approaches that will provide access to the orbit/spectrum in an equitable manner as the real requirements of administrations arise. While we are of the view that a flexible and dynamic planning approach will be more responsive to user needs, we believe that the Conference will be free to decide for itself the best approach for ensuring equitable access to the orbit/spectrum based upon careful and comprehensive technical preparations and analysis. In this regard, the full and active support of the CCIR will be indispensable.

IWP 4/1 [32]

For a decade this interim CCIR working party of Study Group 4 has devoted its efforts exclusively to analyzing methods and techniques which could be used to make more efficient use of the GSO in the Fixed Satellite Service.

In February 1980, several months after the 1979 WARC, the CCIR, in response to *Resolution 3,* determined to give the IWP a special responsibility to provide the technical basis for the Space WARCs referenced in this resolution. Two meetings of this new group were held culminating in a 500-page report describing the various technical aspects of all the space services which would need to be taken into account in any planning activities.

Chapter 6 of this report was the most directly applicable to the question of GSO planning. It identified five planning methods and established a set of 12 criteria by which to evaluate these methods. The planning methods are:

A Priori Planning (20-year plan)   Orbit Segmentation

Five-Year Plan   Multilateral Coordination

Bilateral Coordination

*Summary Record of Part Two of the Tenth Meeting of Committee 6 "Regulatory Procedure," WARC-79, Doc. No. 846 (26 Nov. 1979) at Annex 1, pp. 6-7 [31].

In addition, there are a number of variations on these same themes. A chart with a coded evaluation of these methods is available in Figure 3-2. A description of these planning methods, and an interpretation of the criteria for evaluation of each is found in Apppendix L.

Traditionally, the ITU has had only one type of "plan," the *a priori* type. However, until the 1977 BSS WARC, it had never been applied explicitly to a space service. Experience to date with that plan has indicated its relative inflexibility in the face of real requirements. However, the method still has great appeal as a result of providing a *guarantee* of access to the orbit.

As is apparent the method with the greatest guarantee of access is also the least flexible and least responsive to real requirements. This is the dilemma faced in future international negotiations.

METHOD	A	B	C	D	E
CRITERIA 1	E	G	G	F	F
2	F	G	F	F	F
3	P	G	G	G	G
4	P	G	E	E	E
5	P	G	E	E	E
6	P	G	G	G	G
7	G	G	F	G	G
8	F	G	F	F	F
9	F	G	G	G	G
10	F	G	G	E	E
11	E	G	P	P	P

CODE	E = Excellent	1. Equitable Access	7. Terrestrial Share
A = A Priori	G = Good	2. Requirements	8. Space Share
B = 5-Year Plan	F = Fair	3. Changes	9. Efficient Orbit
C = Segmentation	P = Poor	4. Existing	10. Satellite System Cost
D = Multi-Lateral		5. Multi-Administration	11. Administrative Cost
E = Bilateral		6. Modern Technology	

Figure 3-2 Coded Evaluation Criteria for Planning Methods

As a result of actions taken at the XVth Plenary Assembly of the CCIR the work of preparing the technical inputs to the future WARC 85 is to be carried out by Conference Preparatory Meetings (CPMs). The first for the Space WARC will be held in 1984. Meanwhile, despite the fact that an agenda has not yet been set, preparatory activities are already underway [33]. The Executive Branch under NTIA and the Interdepartment Radio Advisory Committee, has established a special ad hoc committee; the FCC has established a public advisory committee. Both of these activities will be carried out over the next several years culminating in U.S. proposals to the Space WARC 85.

UNISPACE '82 [10]

In the Summer of 1982 the United Nations held its Second Conference on the Peaceful Uses of Outer Space. This conference, UNISPACE '82 served as another forum in which to debate the GSO, and influence world thinking on the development of policy. The report to the conference has over ten paragraphs devoted to the GSO; the issues, U.S., and developing country positions are given in Table 3-2.

While this is not a treaty conference, its report and recommendations will influence other conferences, notably the ITU Plenipotentiary Conference in the Fall of 1982.

WARC-BS '83 (Region 2)

The next radio conference which will influence policy on use of the GSO is the postponed BSS Planning Conference for Region 2 [12]. Subsequent examination of the results of WARC 77 revealed that the "arc segmentation" concept adopted for Region 2 would not yield sufficient capacity for Fixed Satellite System such as those being built by Satellite Business Systems. As a result, a successful reallocation was undertaken at WARC 79 providing a new allocation structure in the bands 11.7-12.7 GHz.

The net outcome was to provide separate 500 MHz bands, for both the fixed satellite and broadcasting satellite services. The requirement to "plan" the BSS in 1983 in Region 2 was retained.

In this interim period the technology has changed; the U.S. is on the verge of launching a number of real DBS systems, and thinking on the nature of GSO planning has evolved. In all likelihood, the U.S.

will be proposing an orbit planning scheme for Region 2 in the 12.2-12.7 GHz bands which allocates "blocks" of spectrum to particular service areas. The conceptual representation of this orbit planning approach is illustrated in Figure 3-3. It constitutes a different and unique form of planning, not heretofore witnessed in the ITU. It has the desirable features of, at the same time:

providing guaranteed access to the GSO for every administration; and providing sufficient flexibility so as not to inhibit the implementation of practical systems serving real requirements. In addition, it can accommodate changes in technology.

This planning approach is probably unique to the Direct Broadcast Satellite Service. This is due to the required homogeneity of program delivery. What makes this planning approach exceptional is that individual orbit location to service area couplings may decide, whatever appropriate parameters, and in whatever timeframe, what is suitable for the administrations concerned. At the same time, the planning approach takes advantage of natural geographic isolations.

This planning approach may not be the most orbit/spectrum efficient. If this were the only criterion, the spectrum would probably never be used for lack of an economically viable system.

The Region 2 Conference on Broadcasting Satellites is demonstrating how effective use of the GSO can be accomplished and at the same time meet the concerns of both developing and developed countries. In addition, it demonstrates that each band of frequencies in which the GSO may be utilized might be open to a unique solution.

3.3.2 Domestic Policy and Regulation

In many ways the use of the GSO by the U.S. has been the laboratory for policy and regulatory experimentation with respect to the GSO. The U.S. has had the technology, launch capability, and markets to support the development of communication satellites in the GSO.

As noted earlier, after an eight year assessment, including Presidential task forces, orbit use studies, numerous notices of inquiry by the FCC, an "open skies" policy was adopted by the FCC for domestic commercial COMSAT access to the GSO. Any organization with

Adopted: October 1, 1981
Released: November 18, 1981

Comments: March 8, 1982
Reply Comments: April 26, 1982

RESPONDANT	4/6 GHz 3°	4/6 GHz 2°	12/14 GHz 2°	ANTENNA STANDARD	REMARKS
ALASKCOM	NO	NO	–	WAIVER	ALASKA SHOULD BE EXEMPT
API	YES	MAYBE		G'FATHER	SUPPORTS INC. CAPACITY/CONCERN FOR STRINGENT STDS
ASC	NOW	COSTLY		6DB-6GHZ	SUGGESTS RANGE OF 2.5-3.0°
AT&T		VIABLE		NEEDED	FAVORS 2.5°
AP				ACTUAL	10° CAN OPERATE AT 2° AT ACCEPT. COSTS/USE PATTERN
CMI				NO	REGULATE INTERFERENCE & SPECTRAL CONTENT
CBS	YES?	YES?	YES?	YES?	QUITE OPTIMISTIC/DBS INTO 12 GHZ FSS
NTIA	YES	YES	YES	YES	AGREE IN PRINCIPLE/NARC 85/PROCEED IN ORDERLY FASHION
COMSAT	YES	AVERAGE	SAME-4/6	LIM. EIRP	REG. INTERFERENCE/REF. PATTERN/GRANDFATHER
COMSAT GEN.	YES	WAIT	POSSIBLE	WHY	USE 55-70° ARC/SATURATE BEFORE FURTHER REDUCTION
EIA	YES	2.5°?	NO	MODIFY	ORBIT PLANE ONLY/EIRP SPECTRAL DENSITIES INSTEAD
EQUATORIAL	YES	YES	DEFER	MODIFY	LARGE SPACE PLATFORMS
FSI	YES	NO	NO		1.5 M TVROS AT 12 GHZ
GSAT	YES	NO	NO		2° BY REVERSE FREQUENCY USEAGE
HBO		NOT ALL	YES		DON'T FAVOR HYBRIDS
HUGHES	NOW		WAIT		HAVE SIDELOBE CANCELER
LOCKHEED		YES	YES	GUIDELINE	REG. INTERFERENCE/TVRO STD A (18 DB) & STD B (14 DB)
M/A-COM		PLAN			TIME-PHASE-CAREFUL PLAN/6':4-3°/10':INC AT 3'/12':2. OK
MUTUAL					INCREASED INTERFERENCE TO RADIO ASTRONOMY (HAVE CCIR STUDY)
NAS	NOW	FUTURE?	NOW	TIMETABLE	IMPLEMENT 2° ONLY AFTER ADEQUATE GROUND SEGMENT
NCTA					NEUTRAL/USES MAINLY LARGE ANTENNAS
PBS	NOW	YES	YES	YES	OPEN 55-70° ARC/ADVISORY GROUP TO PLAN FOR 2°
RCA	NOW	YES	NOW	G'FATHER	12 GHZ FM OR TV MAY TAKE WIDER SPACING
SBS		DEFER	YES	YES	XPONDER ABUNDANCE FAR OUTWEIGH 2° COSTS/LIMIT # OF SATS/COMP.
SSSI	YES	YES		MODIFY	OPEN 55-70° ARC/DEFER < 3° DECISION UNTIL SATURATED AT 3°
S-A			DEFER		OPPOSE ACTION MAKING SETS HARDER TO LICENSE/SET ASIDE SET ARC
SCHLUMBERGER					3.0 M TVRO OK FOR HOME QUALITY BUT NOT FOR CATV QUALITY
SPACE	PRESENT				VARIABLE SPACING IN INTERIM/2° LEAVES LITTLE FLEXIBILITY
SPCC		?		G'FATHER	REVERSE FREQUENCY BANDS/TIMETABLE FOR ANT. STD – '88 & '90
TMCT	NO	DELAY			SERIOUS CONCERNS FOR IMPACT ON RECEIVE-ONLY SETS
UPI		NO			IMPROVED ANT. STD NOT NEEDED FOR 2.5°/SET MIN. SAT DESIGN STDS.
USAT	NOW	EVOLVE TO REVISIT	WAIT	PR. DENS. '83/'90	STRONGLY OPPOSE NEW ENTRANT SETASIDE
WUTCO	POSTPONE	NO			FULLY USE 12/14 GHZ/OPEN 55-70° ARC
GROUP W	YES				2° SPACE HAS TOO LARGE AN IMPACT
WOLD					

		3°	2°	12/14
NOW	YES	16	7	8
	NO	2	7	3
FUTURE		1	11	4

Figure 3-3 Orbit/Spectrum Planning Scheme Proposed for Region 2
Source: Federal Communications Commission, *Common Carrier Docket N81-704*, Washington, D.C.: October 1; November 18, 1981.

sufficient resources (economic, technical, legal) could apply for and obtain a license to operate in the 6/4 GHz or 12/14 GHz frequency bands using a satellite in the GSO. This policy was established in 1973-74. In addition, a policy was established in which U.S. satellites were required to stay 5° away from Canadian satellites and the Canadian satellites themselves were spaced 5° apart (e.g., 5°.5°.5°). This had the effect of making unavailable 20° of prime GSO arc for U.S. use. The useful arc for the U.S. extends from approximately 136 to 70° West longtitude. Sixty-four degrees of orbit minus 20 for Canada left 44 for the U.S. carriers who entered the U.S. DOMSAT market. Those initial entrants were Western Union, RCA, and AT&T. Each requested and received three orbit locations. With the FCC adopting a policy of 4° spacing between the U.S. satellites, 44° of orbital arc proved to be more than enough.

For approximately two to three years after launch trandponder use languished, and several satellite companies were close to having severe economic losses on their satellite investments. However, in 1976 and 1979 the FCC took steps to "deregulate" the commercial earth stations. The first step (1976) was to remove the restriction on the size of earth stations which could be used. The original restriction was based on the belief that large earth stations (i.e., 30 feet and greater) had better radiation patterns (e.g., better than 32-25 log θ) and as a result could make better use of the GSO. It was demonstrated that many smaller receive-only terminals were quite capable of this objective. As might be expected, this resulted in a dramatic drop in the price of receive-only terminals, and thus came well within the budget of cable television systems. Within several years, several thousands of such stations were in operation.

In 1979 a further and more dramatic deregulatory step was taken. Whereas the previous step permitted smaller stations, they nonetheless had to be coordinated under FCC rules in order to ensure protection from harmful interference. This next step completely eliminated this somewhat time consuming and costly requirement. The policy reasoning was that it was in the self-interest of the user to purchase a terminal and to locate it so as to obtain a high quality picture.

This action resulted in an explosive growth in small receive-only earth terminals (TVROs) and a corresponding improvement in the economic viability of U.S. domestic satellites. Table 3-3 shows the type of use on RCA and Western Union satellites; the present estimate is that there are close to 50,000 TVROs and growing. Because of the numbers, the price for a TVRO has dropped dramatically also; today being in the range of 3-10 thousand, depending on quality and electronics.

The second wave of DOMSAT growth was based on the FCC's authorization of Satellite Business Systems to operate in the 12/14 GHz band in January 1977. This was the first *all-digital* satellite system which would provide *network* service on a *premise-to-premise basis*. These three characteristics were a significant departure from the previous type of system. Because of the higher frequency, terminals could be smaller, there were no terrestrial services in the band that required coordination. Therefore, a terminal could be located right at a plant or business. It was envisioned that networks would be sold to "Fortune 500" companies.

When it became clear, after WARC 79, that the band 11.7-12.2 GHz would be made available exclusively for development of Fixed Satellite Services, a number of additional firms decided to enter the field; other firms saw the appeal of TV distribution to the growing CATV and home terminal market. GTE, Southern Pacific, and Hughes Communications formed the second wave of U.S. DOMSAT applicants; these too could be accommodated under the 4° spacing policy.

However, as DOMSAT popularity grew and other services besides TV became viable, it became increasingly apparent that additional demand for orbit locations was in the offing:

In March 1981, the U.S. and Canada, at a meeting of senior telecommunications officials, established a special working group on orbit spacing.

Brazil, Colombia, and Mexico expressed interest in launching domestic satellites.

RCA, Western Union, asked for additional orbital locations and American Satellite, Hughes, GTE, Southern Pacific and U.S. Satellite Corp. indicated a desire for additional positions.

Clearly steps needed to be taken to increase capacity. In the Fall of 1981 the FCC issued its Docket 81-704 on orbital spacing, a rulemaking proposing 2° spacing at both 4/6 GHz and 12/14 GHz. The responses are summarized in Table 3-2. The overwhelming majority of responses supported closer spacing, about 2.5-3° but not 2°. The docket achieved the desired result, namely, a commitment on the part of industry to move toward closer spacing.

In the Spring of 1982, the Mexican Government approached both the U.S. and Canada, seeking a guaranteed commitment of access to two orbit locations in the North American arc between 100° W and 120° W. It desired to place hybrid satellites, i.e., operating at both 4/6 GHz and 12/14 GHz at these locations. This occurred simultaneously with the convening of a second meeting of high level telecommunication officials from the U.S. and Canada on a number of telecommunication policy issues, including stalemated discussions with Canada on orbit spacing.

In this context, the subject of North American orbit spacing received extraordinary attention at the meeting of U.S. and Canadian officials. A commitment was made to solve the "Can/US/Mex" orbit spacing problem by mid-June.

What resulted from this set of circumstances was a major breakthrough in orbit spacing policy, and an indication of the methodology which can be successfully used for ensuring guaranteed access to the GSO in the Fixed Satellite Service. The resulting orbit spacing scheme is found in Figure 3-4. It established the following new GSO orbit spacing policy:

3.5° spacing between Canadian satellites at 4/6 GHz;

2.5° spacing between Canadian satellites at 12/14 GHz;

2.5-3.0° spacing between U.S. satellites at 4/6 GHz;

2-2.5° spacing between U.S. satellites at 12/14 GHz;

2° spacing between Mexican and Canadian satellites at 4/6 GHz;

1° spacing between Mexican and Canadian satellites at 12/14 GHz;

3° spacing between Mexican satellites at both 4/6 GHz and 12/14 GHz (hybrid);

3.5° spacing between Canadian and U.S. satellites at 4/6 GHz; and

2.5° spacing between Canadian and U.S. satellites at 12/14 GHz.

This agreement, which probably will be the orbit spacing plan for the North American arc for the rest of this decade, does not preclude other satellites serving Central and South American countries. These countries are separated geographically both in longitude and latitude and, therefore, should have no problem realizing access to the GSO in the aforementioned bands; particularly, if princriples, such as those indicated above, are followed among contiguous contries.

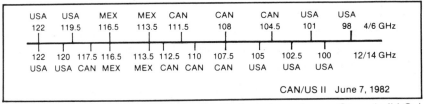

Figure 3-4   Orbit Spacing for the North American Arc — Canada/U.S./ Mexico

3.4 Concluding Observations

The principal problem facing the effective use of the GSO is evolving an international framework for its continued use which meets the needs of both the developed and developing countries. The needs of the developing countries include a guarantee of access and a stability of technology which will ensure the lowering of the cost of access. The needs of developed countries include the flexibility to introduce technological change and to adapt to changing market demands.

Although it may appear that there are conflicts in these needs, the evidence to date seems to indicate that there are ample resources in imagination, institutions, technology, and experience to find solutions. The solutions lie in the following directions:

(1.) Multi-Administration Systems — The immediate aspiration for access to the GSO by developing countries can be realized by existing and planned international and regional systems. These include:

INTELSAT — 106 members, the majority of which are developing nations, over 20 of which are now using INTELSAT for domestic service. These include such countries as Algeria, Nigeria, Brazil, Mexico, Phillipines, and soon China.

INTERSPUTNIK — this 20-plus nation, Soviet Union-based system includes several developing countries such as Cuba and Ethiopia, in addition to Eastern European countries.

PALAPA II — this Indonesian based, second generation domestic system is expected to provide service to a number of Asian counties.

ARABSAT — can provide service to developing countries in the Middle East.

INSAT — this Indian satellite can also provide service to Asian countries such a s Sri Lanka.

European Region Satellite — even this satellite serving basically developed countries could service several developing countries.

AFRO-SAT — this system, still in the conceptual stage, could serve many developing countries.

"AMIGOSAT" — such could be the name of a still-to-be designed Latin American Satellite System; but, in the interim, the Brazilian and Colombian domestic systems could certainly serve regional needs.

Taken together, it appears that there will be ample capacity of GSO Fixed Satellite Service to satisfy the near-term need of any developing country.

(2.) Analytical Techniques — During the last decade many analytical techniques have been developed which can be used to analyze ways of optimizing orbit use. These include:

The SOUP programs — originally developed by General Electric under contract to the Office of Telecommunication Policy/EOP. Successive modifications to this program have a means for analyzing interference interaction of provided dual space satellites.

The ORBIT-I program — developed in Japan provides an iterative sive method for determining the subsequent addition of satellites.

Canadian Systems program — probably the most comprehensive of analytic computer models can be used to improve orbit efficiency.

These and other analytical methods provide the bases for determining the impact of various technical improvements, and providing information on how to adjust orbit positions to meet various requirements.

(3.) Institutional Arrangements — The pros and cons of various orbit planning methods have been described. The ITU remains the single, most viable international institution for ensuring equitable management of the GSO. While its present coordinating methods may by confusing and complicated, it has nonetheless provided all countries with the framework for accommodating the spectacular growth of COMSATs.

Given the experience of North America where there is the highest density of FS/GSO satellites, it is envisioned that the most realistic, practical means of improving management and use of the GSO would be to adopt some form of periodic multinational planning mechanism. Such a mechanism could meet every five years to plan orbit use for the following five years. This would constitute a mini-treaty, and every country and organization specifying real needs would have to be accommodated taking advantage of successive generations of improved technology. There is no logical reason to believe that such a mechanism would not work. The evidence to date is to the contrary.

TABLE 3-1
MILESTONES IN THE DOMESTIC SATELLITE AT THE FCC

21 September 1965	ABC-TV requests authorization for a TV distribution satellite: FCC returns application "without prejudice"
2 March 1966	FCC establishes Notice of Inquire: asks for opinions on policy, impact on INTELSAT, spectrum and technical questions, sharing, and ecomomic impact on carriers
1 August 1966	19 Comments filed, plus proposal of Ford Foundation for "Broadcasters Non-profit System" (BNS)
20 October 1966	FCC establishes "supplemental inquiry": asks for system plans, is there should be any legal restrictions on market entry, and whether the BNS should be approved
16 December 1966	21 Supplemental Comments filed: carriers state that carriers should be authorized, non-carriers state that non-carriers should be authorized (no surprises there); four filings included system proposals (ABC, AT&T, COMSAT, and the Ford Foundation)
3 April 1967	19 Reply Comments filed: COMSAT proposes Pilot Demonstration Program (with COMSAT as manager)
18 September 1967	Supplemental Comments filed on COMSAT proposal
20 September 1967	COMSAT Comments on Supplemental Comments
14 August 1967	President Johnson's Taks Force established; FCC delays action until:
7 December 1968	Task Force submits Final Report recommending multi-purpose system—pilot program w/COMSAT as trustee of space segment and system mgr. But new administration (Nixon's) asks for delay until it can study the question. FCC delays.
23 January 1970	Administration labors and brings forth: The Flanigan Letter (to FCC Chairman Burch) suggests "open entry"
24 March 1970	Report & Order asks for applications

15 March 1972	Memorandum Opinion & Order sets forth staff opinion proposing limited—or structured—entry, and describes the eight applications that had been received.
1-2 March 1972	Comments, Reply Comments and Oral Argument completed on 3/15/72 and the staff opinion therein: carriers do not support the staff proposal, seek open entry Memorandum Opinion and Order
16 June 1972	FCC rules for open entry, subject to applicants being financially, legally, and technically qualified: Commission imposes conditions on certain applicants (no specialized services via satellite for AT&T for three years; COMSAT must choose between supplying space segment to AT&T or competing with AT&T in its "multi-purpose system"; FCC Order also approves separate GTE system for long-distance telephone service; requires rate integration for Alaska, Hawaii, and Puerto Rico; does not require lower rates for CPB or educational users, and does not require competitive bidding for procurement.
21 December 1972	Six Petitions for Reconsideration filed: with the exception that AT&T may provide specialized services to government via satellite, all conditions ar upheld COMSAT's new partnership with Lockheed and MCI (MCIL) is approved (this partnership has been changed into the present SBS venture). FCC states "This proceeding is terminated."

Source: Richard G. Gould, Telecommunications Systems, Washington, D.C.: September 1979.

TABLE 3-2

ISSUE	DEVELOPING	UNITED STATES
1. Limited Natural Resource	Requires Planning	Effective Use & Management
2. GSO Goes With RF	Requires Planning	Effective Use & Management
3. Access to 4/6 GHz	Developed Countries Should Vacate	Technology Gives Access
4. Sovereignty of Orbit	Part of Equatorial Nation (some) *Sui Generis* Region	Outerspace Treaty Applies
5. Efficiency of Use	Not Barrier to Technological Self Reliance	Use Better Technology for Orbit Use
6. Planning Methods	Accommodate Future Needs of Developing Countries	Don't Inhibit Technological Development
7. Orbit Debris	Remove Unused Satellite From Orbit	User Responsibility Not Launcher
8. Uses in Planning	Emphasis on Development and Education	For Whatever Country Wants
9. Alternate to GSO	Consider Value of Elliptical	Use What Viable
10. Large Space Platforms	Give Consideration for GSO	Evaluate Before Use
11. New Technology	Replace High Capacity on Optical Fiber	Planning Should be Flexible Enough to Accommodate
12. Terrestrial Services	Eliminate from Sharing	Cannot be Eliminated From Sharing

TABLE 3-3

SATELLITE TRANSPONDER OCCUPANCY

(as of August 1980)

A. Satellite transponder assignments on the Western Union satellites

Westar I		Westar II		Westar III	
Trans.	*Service*	*Trans.*	*Service*	*Trans.*	*Service*
#1	*Occasional video	#1	Message traffic	#1	Single-channel-per-carrier (high-power audio)
#2	*National Public Radio (Mutual Broadcasting)	#2	*Occasional video	#2	*Hughes Television Network
#3	Satellite Comm. Network	#3	American Sat. Corp.	#3	*Mexico
#4	Single-channel-per-carrier (high-power audio)	#4	American Sat. Corp.	#4	Single-channel-per-carrier (high-power audio)
#5	*Bonneville International	#5	American Sat. Corp.	#5	*Occasional video
#6	*Occasional video	#6	Message traffic	#6	*CBS
#7	Inoperative	#7	American Sat. Corp.	#7	*Robert Wold Co.
#8	*Public Broadcasting System	#8	Message traffic	#8	*Spanish International Network
#9	*Public Broadcasting System	#9	Message traffic	#9	*Satellite Programming Network
#10	Inoperative	#10	American Sat. Corp.	#10	*ABC
#11	*Public Broadcasting System	#11	Message traffic	#11	*Cable News Network
#12	*PBS (after 1-1-81)	#12	Inoperative	#12	*Occasional video

B. Satellite transponder assignments on the RCA Americom satellites

Satcom I		Satcom II		Comstar D-2**	
Trans.	*Service*	*Trans.*	*Service*	*Trans.*	*Service*
#1	KTVU distant signal	#1	Programs to Alaska	#1V	Message traffic
#2	PTL Network	#2	Message traffic	#1H	Message traffic
#3	WGN distant signal	#3	Alaska "bush" system	#2V	Message traffic
#4	Inoperative	#4	Message traffic	#2H	Message traffic
#5	The Movie Channel (Warner-Amex)	#5	Alascom	#3V	Message traffic
#6	WTBS distant signal	#6	Alascom	#3H	Message traffic
#7	*Entertainment & Sports Programming Network	#7	Alascom	#4V	Message traffic
#8	*Christian Broadcasting Network	#8	Alascom	#4H	Message traffic
#9	*US Air Network (C-SPAN, BET, sports)	#9	*NBC	#5V	Entertainment & Sports Programming Network
#10	Front Row—East (Showtime Entertainment)	#10	Alascom	#5H	Message traffic
#11	Nickelodeon	#11	Alascom	#6V	Cineamerica (Rainbow Comm.)
#12	Front Row—West (Showtime Entertainment)	#12	Inoperative	#6H	Message traffic
#13	Trinity Broadcasting Network	#13	Government	#7V	Not decided
#14	*Cable News Network	#14	Government	#7H	Message traffic
#15	Message traffic	#15	Alascom	#8V	Showtime Entertaiment
#16	*Community Service Network	#16	Network audio feeds	#8H	Message traffic
#17	WOR distant signal	#17	Inoperative	#9V	Home Box Office
#18	Galavision	#18	Alascom	#9H	Home Box Office
#19	Message traffic	#19	Alascom	#10V	Satellite Comm. Network
#20	Cinemax	#20	Alascom	#10H	Spanish Int. Network
#21	*Satellite Programming Network	#20	Alascom	#11V	Message traffic
#22	Home Box Office	#22	Inoperative	#11H	Warner-Amex
#23	Cinemax	#23	Alaska "bush" system	#12V	United Video
#24	Home Box Office	#24	Alascom	#12H	National Christian Network

*Denotes services in which broadcasters may have an interest **until Satcom III launch

Source: *Television/Radio Age,* August 11, 1980.

REFERENCES / PART ONE

1.   *UNISPACE '82*, preparatory report, International Aeronautical Federation, 1981.

2.   CCIR: XVth Plenary Assembly, Geneva, 1982, Study Group 4, *Report 453-3* (Technical factors influencing the efficiency of use of the geostationary orbit by radiocommunication satellites sharing the same frequency band), Geneva, 1982.

3.   CCIR: XVth Plenary Assembly, Geneva, 1982, Study Group 4, *Report 556-1* (Factors affecting station-keeping of geostationary satellites of the fixed satellite service), Geneva, 1982.

4.   CCIR: XVth Plenary Assembly, Geneva, 1982, Study Group 4, *Report 454* (Method of calculation to determine whether two geostationary satellite systems require coordination), Geneva, 1982.

5.   CCIR: XVth Plenary Assembly, Geneva, 1982, Study Group 4, *Report 455-3* (Frequency sharing between networks of the fixed satellite service), Geneva, 1982.

6.   General Electric Co., Space Systems Organization, Valley Forge Space Center, "Orbit/Spectrum Utilization Study," 4 vol., 1969-1970, prepared by D.A. Kane and M.C. Jeruchim for D.M. Jansky, Office of Telecommunications Management, Executive Office of the President.

7.   ITU, *Final Acts of the Extraordinary Administrative Radio Conference to Allocate Frequency Bands for Space Radiocommunication Purposes*, Geneva, 1963.

8.   ITU, *Final Acts of the World Administrative Radio Conference for Space Telecommunications*, Geneva, 1971.

9.   ITU, "Final Acts of the 1979 World Administrative Radio Conference," *Resolution 3*, Geneva, 1979.

10.   Report of the Second United Nations Conference on the Peaceful Uses of Outer Space, Section G, para. 277-288.

11.   CCIR: XVth Plenary Assembly, Geneva, 1982, *Recommendation 466-3* (Maximum permissible level of interference in a telephone channel of a geostationary satellite network in the

fixed satellite service employing frequency modulation with frequency-division multiplex, caused by other networks of this service), Geneva, 1982.

12. (a) Final Report of the Conference Preparatory Meeting (CPM) for RARC '83, Geneva, July 1982.

(b) ITU Administrative Circular A.C./241, "Conference Preparatory Meeting of the CCIR for RARC for Planning the BSS in Region 2."

(c) Report of CCIR IWP 10-11/2-15 (Rev. 1).

(d) "Inquiry into the Development of the Regulatory Policy in Regard to Direct Broadcast Satellites," (informal document).

13. (a) CCIR: XVth Plenary Assembly, Geneva, 1982, *Recommendation 353* (Allowable noise power in the hypothetical reference circuit for frequency-division multiplex telephony in the fixed satellite service), Geneva, 1982.

(b) CCIR: XVth Plenary Assembly, Geneva, 1982, *Recommendation 522* (Use of frequency bands above 10 GHz in the fixed satellite service), Geneva, 1982.

14. (a) CCIR: XVth Plenary Assembly, Geneva, 1982, *Recommendation 356* (Maximum allowable values of interference from line-of-sight radio-relay systems in a telephone channel of a system in the fixed satellite service employing FM when frequency bands shared), Geneva, 1982.

(b) CCIR: XVth Plenary Assembly, Geneva, 1982, *Recommendation 357* (Maximum allowable values of interference in a telephone channel of an analogue angle-modulated radio-relay system sharing same frequency bands as systems in the fixed satellite service), Geneva, 1982.

(c) CCIR: XVth Plenary Assembly, Geneva, 1982, *Recommendation 358* (Maximum permissible values of pfd at surface of the Earth produced by satellites in the fixed satellite service using the same frequency bands above 1 GHz as line-of-sight radio-relay systems), Geneva, 1982.

(d) CCIR: XVth Plenary Assembly, Geneva, 1982, *Recommendation 558* (Maximum allowable interference from terrestrial radio-relay links to systems in the fixed satellite service employing 8k bit + PCM encoded telephony and sharing the same frequency bands), Geneva, 1982.

(e) CCIR: XVth Plenary Assembly, Geneva, 1982, *Report 793* (Derivation of interference criteria for digital systems in FSS sharing with TS), Geneva, 1982.

(f) CCIR: XVth Plenary Assembly, Geneva, 1982, *Report 710*, (Interference allocations in digital systems operating at frequencies greater than 10 GHz in the FSS), Geneva, 1982.

15. CCIR: XVth Plenary Assembly, Geneva, 1982, *Report 393* (Intersection of radio-relay antenna beams with orbits used by space stations in the fixed satellite service), Geneva, 1982.

16. CCIR: XVth Plenary Assembly, Geneva, 1982, *Report 382-3* (Determination of coordination area), Geneva, 1982.

17. ITU, *Radio Regulations*, Art. 11; World Administrative Radio Conference (WARC '79), Geneva, 1979.

18. (a) CCIR: XVth Plenary Assembly, Geneva, 1982, *Report 388* (Methods for determining interference in terrestrial radio-relay systems and systems in the fixed satellite service), Geneva, 1982.

(b) CCIR: XVth Plenary Assembly, Geneva, 1982, *Report 449* (Measured interference with FM TV systems using frequencies shared within systems in the fixed satellite service or between these systems and terrestrial systems), Geneva, 1982.

(c) CCIR: XVth Plenary Assembly, Geneva, 1982, *Report 448* (Determination of interference potential between earth stations and terrestrial stations), Geneva, 1982.

19. (a) CCIR: XVth Plenary Assembly, Geneva, 1982, *Report 391* (Radiation diagrams of antenna for earth stations in the fixed satellite service for use in interference studies and for determination of design objectives), Geneva, 1982.

(b) CCIR: XVth Plenary Assembly, Geneva, 1982, *Recommendation 465* (Reference Earth station patterns for use in coordination and intereference assessment in the frequency range from 2 to about 10 GHz), Geneva, 1982.

20. CCIR: XVth Plenary Assembly, Geneva, 1982, *Report 387-3* (Probability of terrestrial line-of-sight radio-relay systems against interference due to emissions from space stations in the fixed satellite service in shared frequency bands between 1 and 25 GHz), Geneva, 1982.

21. CCIR: XVth Plenary Assembly, Geneva, 1982, *Report 792* (Calculation of the maximum power density averaged over 4 kHz of an angle-modulated carrier), Geneva, 1982.

22. CCIR: XVth Plenary Assembly, Geneva, 1982, *Recommendation 406* (Maximum equivalent istropically radiated power of line-of-sight radio-relay system transmitters operating in frequency bands shared with the fixed satellite service), Geneva, 1982.

23. CCIR: XVth Plenary Assembly, Geneva, 1982, *Report 790* (E.I.R.P. and power limits for terrestrial radio-relay transmitters sharing with digital satellite systems in bands between 11 to 14 GHz and around 30 GHz), Geneva, 1982.

24. CCIR: XVth Plenary Assembly, Geneva, 1982, *Recommendation 524* (System models for evaluation of interference co-site analysis model), Geneva, 1982.

25. CCIR: XVth Plenary Assembly, Geneva, 1982, *Recommendation 496* (Circuits for high quality monophonic, and stereophonic transmisstion), Geneva, 1982.

26. CCIR: XVth Plenary Assembly, Geneva, 1982, *Report 560* (Sharing criteria for the protection of space stations in the fixed satellite service receiving in the band 14.0 to 14.4 GHz), Geneva, 1982.

27. ITU, Plenipotentiary Conference, Malaga-Torremolinos, *International Telecommunication Convention* (1973), Art. 33.

28. CCIR: XIVth Plenary Assembly, Kyoto, Japan, Vol. IV, *Report 453-2* (Technical factors influencing the efficiency of use of the geostationary satellite orbit.by radiocommunication satellites sharing the same frequency bands. General Summary), Geneva, 1978.

29. Final Acts of the World Administrative Radio Conference on Broadcasting Satellites, Geneva, 1977.

30. FCC, "U.S. Proposals to WARC '79," Memorandum Opinion and Order, Washington, D.C.: 1979.

31. Summary Record of Part Two of the Tenth Meeting of Committee 6, "Regulatory Procedures," WARC 1979, Doc. No. 846, Washington, D.C.: 26 November 1979.

32. CCIR: XVth Plenary Assembly, Geneva, Switzerland, February 1982.

33. (a) FCC, "An Inquiry Relating to Preparation for ITU WARC on Use of GSO and Planning of the Space Services Utilizing It," Doc. 81-317, Washington, D.C.: 4 August 1981.

    (b) FCC, Doc. 81-317, Second Notice of Inquiry, Washington D.C.: 1 June 1982 (includes: 1) U.S. Policy Objectives; 2) Identification of U.S. Interest; 3) Objectives of Conference; and 4) List of Related ITU Recommendations).

34. ITU, Interim Working Party 4/1, *Provisional Technical Report for WARC '84*, Geneva, 1981.

35. ITU, *Radio Regulations*, Art. 27, para. 2501-2511; Art. 28, para. 2539-2579, Geneva, 1981.

36. FCC, "Orbit/Spectrum Planning Scheme for Region 2," Doc. 81-704, Washington, D.C.: 1981.

PART TWO

Chapter Four

FACTORS AFFECTING
ORBIT/SPECTRUM UTILIZATION:
THE HOMOGENEOUS CASE

4.0 Introduction

From a technical standpoint, orbit/spectrum utilization is concerned with the relationships between system design factors and the communications capacity of the geostationary orbit and (some part of) the radio frequency spectrum. Although our primary focus will be on the orbit itself, communication cannot take place without using spectrum. Hence, whenever we speak about orbit utilization, as we often do for brevity, it should be kept in mind that the joint orbit/spectrum resource is always implied. In the present context, capacity does not have a strict definition. Clearly, in a broad sense capacity is related to the amount of information that can be transmitted from a given arc of the orbit within a specified bandwidth. Indeed, we shall often adopt a normalized unit of this type as a measure of orbit/spectrum utilization (i.e., capacity), for example, the number of telephone channels per degree of arc and per MHz (which we shall denote $\tilde{n}$). This is not necessarily the only measure of interest. For instance, the number of satellites that may be placed in a given arc is often of importance. Equivalently we can deal with the inverse, namely the average intersatellite spacing, $\Delta\theta$. In the particular case where the transmissions are fixed, the latter variable contains all the relevant information. In this case, incidentally, orbit utilization becomes a very appropriate term. By and large we shall deal with $\tilde{n}$ or $\Delta\theta$ or some variation thereof as the context warrants; we shall also frequently deal with other, directly related quantities, such as the carrier-to-interference ratio, which in some situations has a one-to-one relationship with intersatellite spacing. These measures are sufficient for most purposes. However, the interested reader may find it useful to peruse Section 5.11 which contains a more detailed discussion of the possible choices for measures, as well as a related discussion on the concept of efficiency of use of the orbit/spectrum resource.

Our objective in this chapter and the next is to relate, quantitatively if possible, one or another of the measures of orbit utilization to the design factors of individual systems as well as to those factors which reflect the environment; i.e., the joint properties of systems created by their interaction. A list of the most significant of these factors follows:

1. Antenna properties
   antenna size
   sidelobe radiation levels
2. Interference level
3. Polarization discrimination
4. Satellite station-keeping
5. Carrier frequency plans (carrier interleaving)
6. Modulation characteristics
   analog vs. digital
   modulation index/bandwidth utilization
   interference immunity
7. Network inhomogeneity
8. Frequency re-use at the satellite
   spot beams
   pointing accuracy
9. Sharing constraints
   flux density limits
10. Usage of frequency bands
    bidirectional frequency bands
    frequency band (up/down) pairing
    intersatellite links
11. Propagation factors
12. Criteria of efficiency
13. Operational factors
14. Miscellaneous factors

We must also mention here a very important factor which is, *per se*, not technical in the sense of those above, namely the regulatory environment. However, this environment can have a profound influence on the technical aspects. Conceptually, at least, there are different possible regulatory structures depending perhaps on the service, the frequency band, and other considerations. The nature of this structure can have a major role in orbit utilization; e.g., by encouraging or denying the application of certain techniques, setting standards, etc. However, this issue is not within the intended scope of this chapter or the next, but has been taken up in Chapter 3.

The list of factors above does not lend itself to neat categorization. In actuality, all or most of these factors are simultaneously in force and interact in a complex way to create the existing environment. However, for purposes of exposition, it is for the most part possible to study the effect of one or a few at a time, all other factors fixed. Among the possible ways of organizing the exposition, a major division can be conceptually made between what we shall call the *homogeneous* and *non-homogeneous* cases. Basically, the definition of homogeneity in this context refers to the sameness of technical parameters from one network to another. In general, when considering a set of interacting networks; i.e., a set of networks mutually intefering with one another, we can expect that the characteristics of each network can be more or less arbitrarily different than those of every other network. In that environment it is problematical how to relate a measure of orbit utilization to the technical parameters, or at least how to interpret this relationship. Conceptually this relationship could be visualized as a multidimensional surface, with one axis defined for each parameter of every network. Aside from difficulties of presentation, it would be equally difficult to draw inferences from such a display. For example, it we wished to infer the significance of antenna size, it may not be possible to make a single unequivocal statement valid for the entire range of every antenna.

In view of the difficulties described, there has emerged an idealized model of orbit utilization, called the homogeneous model, in which it is assumed that all networks are identical. While a certain degree of homogeneity may exist in some arcs and frequency bands, the actual case is not so simple. Nevertheless, the homogeneous model turns out to be quite useful in developing an understanding of the trade-offs involved in orbit utilization. Perhaps the chief utility of this model is that it enables us to pose an analytically tractable problem, one in which we can draw unambiguous conclusions regarding the effect of individual technical parameters. This is because when we speak about the influence of a particular parameter, it has the same value for all networks. As stated, this model is very useful for developing intuition for less constrained environments. Thus, in this chapter we confine outselves to the homogeneous case, and in particular to the first six items in the preceding list of factors because a meaningful analysis and interpretation can be given within the homogeneous context. Items 7 through 14 in the list of factors are taken up in the next chapter where we depart from the assumptions of our homogeneous model and consider the nonhomogeneous case.

The constraint in orbit utilization is the interference which geostationary satellite networks inevitably induce in one another. It is this mutual interaction which prevents satellites from being arbitrarily closely spaced. There are two aspects to this problem. One concerns the physical level of interference at the input to a receiver and the other deals with the way a particular receiver's mechanism transforms this interference from an input to an output quantity. (This latter aspect is treated in some detail in Chapters 6 and 7.) Before we can develop the homogeneous model, it is necessary to establish the relevant equations which will be used to connect interference to orbit utilization. Therefore, in Section 1 of this chapter we do not yet address the homogeneous model, but set down some general relationships concerning interference. These relationships are subsequently specialized to the homogeneous case, so that the results-oriented reader may skip the first section.

The next section then defines precisely what we mean by the homogeneous model and explains several simplifying assumptions that are usually made. Then, a number of equations (special cases of those in the preceding section) that we need for the sequel are developed. These equations basically connect the amount of interference to satellite spacing.

Sections 3 through 7 then consider a number of factors that have in common the fact that their influence is manifested at the input to a receiver; i.e., their effect can be characterized by a carrier-to-interference ratio (CIR). The CIR is then straightforwardly related to the intersatellite spacing $\Delta\theta$. It is assumed in the analysis of any given factor that others are held constant. Consequently we shall often deal with relative effects since absolute values will depend on the specific values taken on by all the factors. The first and probably most significant factor we study is the earth station antenna, which has two complementary properties: the on-axis or main beam gain, which is directly determined by its size (diameter for a paraboloid reflector) in relation to wavelength and the off-axis or sidelobe gain. The relation between antenna size and $\Delta\theta$ is examined in section 4.3, while the influence on $\Delta\theta$ of the sidelobe gain is taken up in section 4.4.

Perhaps next in significance to the antenna is the actual amount of interference permitted to be induced in one geostationary satellite network from other such networks. Section 4.5 shows the relation between the interference allowance and intersatellite spacing. The next factor studied is the use of orthogonal polarization, which can be a powerful technique for improving orbit utilization. There are two principal ways in which orthogonal polarizations can be used. One is where alternate satellites use orthogonal polarizations, and the other is where each satellite uses both polarizations. Both of these approaches are discussed in section 4.6, following an introductory discussion of cross-polarized reference patterns and the way in which uplink and downlink polarization discrimination combine. In the following section (4.7), we study the effect of imperfect satellite station-keeping, which necessitates a larger $\Delta\theta$ than would have ideally been required in order to maintain a given CIR.

The remaining sections consider factors related to the modulation characteristics of mutually interfering carriers. In section 4.8 we develop an initial feeling for the significance of these characteristics by relating the modulation index of FDM/FM signals to intersatellite spacing. In the next section (4.9), we consider the beneficial effects of carrier interleaving; i.e., the influence of the relative carrier frequencies of wanted and interfering signal. This effect is a function of the modulation index for analog signals. In Section 4.10 we revisit polarization discrimination, but this time adding a new variable, namely interleaving. The combination of these techniques can bring about improvements superior to that available with either acting alone.

In Sections 3 through 10 we implicitly adopted intersatellite spacing as the relevant measure of orbit utilization. In the remaining sections we broaden our perspective and take as our measure one or another of a normalized unit of information per degree of arc and per unit of bandwidth. We also look at the trade between this orbit use measure and the information transmission capacity of individual networks. Section 4.11 provides a short introduction to the following two sections. Section 4.12 then studies the matters just mentioned for one important class of analog signals, FDM/FM telephony signals. The next, and last, section examines the same types of questions for digital signals, specifically coherent phase shift keyed signals, both with and without the use of error-correction coding.

4.1 Some Analytical Preliminaries

It has previously been pointed out that the fundamental limitation to orbit utilization stems from the unavoidable existence of mutual interference among geostationary satellite networks. Consequently, we begin in this section by establishing in fairly general terms some of the relationships that will be used to link interference to orbit utilization.

4.1.1 Carrier-to-Interference Ratios

Figure 4-1 shows the interference paths that may arise between two networks, and also illustrates nomenclature to be defined below. The two networks form part of an arbitrary assemblage of M networks. The carrier power received at the ground station associated with the jth signal, due to satellite transmission of the ith signal, is given by the standard range equation, namely

$$C_{grji} = P_{sti} G_{stij} G_{grji} L_{stij} \tag{4-1}$$

where the "space loss" factor, L_{stij}, is defined by

$$L_{stij} = \left(\frac{\lambda_{sti}}{4 \pi d_{stij}} \right)^2 \tag{4-2}$$

While the multiple subscript notation may at first seem forbidding, it is actually quite straightforward and intuitive. This notation adheres to the following general convention: the first letter indicates the location, s for satellite and g for ground, of the parameter in question; the second letter indicates transmit (t) or receive (r) function; the third letter identifies the signal associated with the first two items; and the fourth subscript indexes the terminal at the other end of the link (whose function is necessarily complementary to the one identified by the second subscript). Thus, the terms in (4-1) and (4-2) are interpreted as follows:

C_{grji} = carrier power received at the ground station of the jth signal due to the ith signal;

P_{sti} = satellite transmitted power of the ith signal;

G_{stij} = gain of the satellite antenna transmitting the ith signal, in the direction of the earth station associated with the jth signal;

G_{grji} = gain of the ground antenna intended for receiving the jth signal, in the direction of the satellite transmitting the ith signal;

λ_{sti} = wavelength of the ith signal, transmitted from a satellite;

d_{stij} = slant range from the satellite transmitting the ith signal to the ground station intended for receiving the jth signal;

L_{stij} = the space (or spreading) loss between the satellite transmitting the ith signal and the ground station intended for receiving the jth signal.

A small notational simplification is available by denoting

$$E_{stij} = P_{sti} G_{stij} \tag{4-3}$$

which is simply the e.i.r.p. of the ith signal in the indicated direction.

The wanted jth carrier power is obtained simply by letting $i = j$ in the previous equations. The unwanted carrier power (interference), denoted I, interfering with signal j is obtained from (4-1) when $i \neq j$. Thus, the ground (i.e., downlink) jth carrier-to-ith interferer ratio is given by

$$\left(\frac{C}{I}\right)_{gji} = \frac{E_{stjj} G_{grjj} L_{stjj}}{E_{stij} G_{grji} L_{stij}} Y_{stij} \tag{4-4}$$

where we have added the factor Y_{stij}, to account for other isolation mechanisms between the ith and jth signals; for example, polarization discrimination or propagation conditions.

An equation similar to (4-4) is easily inferred for the uplink jth carrier-to-ith interferer ratio, namely

$$\left(\frac{C}{I}\right)_{sji} = \frac{E_{gtjj}G_{srjj}L_{gtjj}}{E_{gtij}G_{srji}L_{gtij}}Y_{gtij} \tag{4-5}$$

Generally, we identify by the same subscript an uplink and a downlink signal wherever the latter is a "continuation" of the former, whether it is merely repeatered or remodulated.

In some situations, to be discussed in the next section, the carrier-to-interference ratio (CIR) is sufficient to evaluate whether or not an interference environment is satisfactory, in which case (4-4) and (4-5) completely define the computations required. If, in these same situations, the modulation characteristics and frequency plans of all mutually interfering signals are identical, then the total carrier-to-interference ratio is relevant. (This quantity is also useful when, as an approximation we assume that the interference is equivalent to thermal noise of the same power). Assuming that the various interferers arriving at a given terminal are uncorrelated (a generally valid assumption), the total interference power at that terminal is the sum of the individual interference powers. Thus, the total CIR for the jth signal on the downlink is given by

$$\left(\frac{C}{I}\right)_{gj} = \left\{ \sum_{i\varepsilon Lj} (C/I)_{gji}^{-1} \right\}^{-1} \tag{4-6a}$$

and that for the uplink by

$$\left(\frac{C}{I}\right)_{sj} = \left\{ \sum_{i\varepsilon Lj} (C/I)_{sji}^{-1} \right\}^{-1} \tag{4-6b}$$

where L_j is an integer set identifying the particular signals interfering with signal j. We assume here that this set is identical for uplink and downlink, though this is perhaps not always the case. The size of L_j may exceed M, the number of networks considered, if, for example, one or more satellites re-use frequency by means of narrow-beam antennas, or if a staggered (interleaved) frequency plan is used.

The total CIR for the jth signal is

$$\left(\frac{C}{I}\right)_j = \left\{ \left(\frac{C}{I}\right)_{gj}^{-1} + \left(\frac{C}{I}\right)_{sj}^{-1} \right\}^{-1} \tag{4-7}*$$

This equation plays a central role in the homogeneous model of orbit utilization and relates to the latter the influence of the parameters embedded in it (e.g., antenna characteristics), which will be explicitly explored subsequently.

4.1.2 Protection Ratios

The effect of interference on a desired signal depends upon the amount, or level, of that interference at the input to the receiver of the wanted signal and upon the characteristics of both, the wanted and interfering signal. These characteristics include the nature of the modulating signals, the relative carrier frequencies, the modulation methods, and, by extension, the receiver structure. It was mentioned earlier that in some instances the CIR contained the essential information regarding the interference environment. This statement in light of the previous discussion, is to be interpreted as saying that *given* all the particulars of a wanted and interfering signal pair, the CIR is sufficient to assess the effect of interference.

To be more specific, consider a pair of signals *(i,j)* with defined characteristics. By the protection ratio, ρji, we mean that value of CIR deemed to be acceptable when signal *j* is being interfered with by signal *i*. When the roles of wanted and interfering signal are reversed, we denote the protection ratio by ρji. Of course, the value of ρ implies a criterion of acceptability, which we do not go into for the moment, but which can generally be interpreted as "sufficiently small." Generaly speaking, protection ratios are used when analytical methods (considered in the next section) are found wanting, in which case they are experimentally. For purposes of this discussion we can assume ρji to be given for all signal pairs *(i,j)* under consideration. The import of this assumption, as will be seen shortly, is simply that from the computation viewpoint the carrier-to-interference ratios comprise the essential quantities to be computed in order to assess orbit utilization.

---

*This commonly used equation is actually an approximation which neglects the cross-product between the two terms in (4-7). For typical operating conditions this approximation is amply justified. A more general version is shown in equation (4-19).

As a simple example, suppose all signals are identical, so that $\rho_{ij} = \rho_0, \forall\ i,j$. We compute the aggregate CIR, $(C/I)_j$, for the jth wanted signal, and simply compare $(C/I)_j$ to ρ_0. If $(C/I)_j \geq \rho_0$ we consider the interference level to be acceptable. Now $(C/I)_j$ is an implicit function of satellite spacing and all of the other parameters defined in (4-1) and (4-2); in this way we can compute a measure of orbit utilization.

The preceding example involves an implied assumption, namely that the effect of any number of identical (but uncorrelated) interferers with a certain total power is the same as that of a single interferer of like characteristics having the same power. While this is eminently reasonable in many cases, and quite usable as an approximation, it is not necessarily always exactly true. Nevertheless, we shall assume it to be so for the remainder of this discussion.

We wish now to generalize the above example to the case where ρ_{if} is not necessarily the same for all (i,j). Consider signals i, j, k, and suppose j is the wanted one, with protection ratios ρ_{ji} and ρ_{jk}. Suppose $\rho_{jk} < \rho_{ji}$ and let $10\ log\ (\rho_{ji}/\rho_{jk}) = (dB)$. This implies that signal j is less sensitive to interference from signal k than to signal i by an amount $S\ dB$. Equivalently, we can postulate that $p_k = p_{ij}$ but increase the CIR, $(C/I)_{jk}$, by $S\ dB$. In other words, for the purpose of computation we can temporarily assume the protection ratio is the same as that with respect to a different interferer, but increase (or decrease) the carrier-to-interference ratio accordingly. We saw in the first example that if all signals are identical, we need merely obtain the composite CIR. The idea, then, is to construct an equivalent set of "identical" signals from the given set by appropriately modifying the carrier-to-interference ratios.

For wanted signal j, then, we define the sensitivity factor

$$s_{jk} \triangleq p_{jj}/p_{jk}, \ \forall k \tag{4-8}$$

whence we obtain an equivalent CIR

$$\left(\frac{C}{I}\right)_{jk,e} = s_{jk}\left(\frac{C}{I}\right)_{jk}$$

and hence a composite equivalent CIR equal to

$$\left(\frac{C}{I} \right)_{j,e} = \left\{ \sum_{\substack{\downarrow k}} s_{jk}^{-1} \left(\frac{C}{I} \right)_{jk}^{-1} \right\}^{-1} \tag{4-9}$$

In this equation, in effect, we have considered all signals to be identical to signal j, but with modified CIR to account for the difference in sensitivity between the actual interferers and an interferer which is identical. The computed number $(C/I)_{j,e}$ is then to be compared to ρ_{jj} to determine whether or not the interference level is acceptable. Equation (4-9) applies to both uplink and downlink, as well as their sum, and forms the general basis for orbit utilization calculations when the interference effect can be described by protection ratios. Generally, protection ratios tend to be used when the wanted signal is television. In Chapter 6 we shall return to this topic and indicate some typical measured values of protection ratio for TV as well as empirical relationships for the protection ratio and, from that, for the sensitivity factor.

4.1.3 Baseband Analog Performance Measure

For analog signals the measure of performance that is almost universally employed is an appropriately defined (power) signal-to-noise ratio (SNR) at the output of the system receiver; i.e., at baseband. It is generally convenient to separate the effects due to the link parameters (i.e., slant range, e.i.r.p., antenna gains, etc.) from those due to the modulation parameters since these two sets of parameters are physically independent. The first set of parameters we have already considered and lead, of course, to the carrier-to-interference ratios. These same parameters for the wanted satellite, in combination with the noise temperatures, also lead to the carrier-to-thermal noise ratios. We may thus ascribe the influence of the modulation (and by inference, demodulation) method to a quantity, termed the receiver transfer characteristic (RTC), which, by definition, serves to convert predetection to baseband performance measures. Thus, for an arbitrary type of unwanted signal, X, the RTC is denoted R_x, and the baseband SNR, S/X, is related to the carrier-to-X ratio, (C/X), by

$$(S/X) = R_x(C/X)$$

Generally speaking, we must consider two types of unwanted signal, thermal noise and interference, both of which contribute to the baseband SNR. Since a network is normally engineered to provide a specified performance, there is a trade between the thermal noise and interference contributions, which is central to the economics of individual networks as well as to orbit utilization, the latter being embedded in the amount of interference that is permitted.

Hence, for thermal noise acting alone we have

$$(S/N)_t = R_t\,(C/N) \tag{4-10}$$

where C/N is the composite carrier-to-noise ratio, composed of downlink and uplink contributions in the fashion

$$(C/N)^{-1} = \{\,(C/N)_d^{-1} + (C/N)_u^{-1}\}^{-1} \tag{4-11}$$

In Appendix A we shall obtain R_t for several types of signals.

The RTC for interference is a function of all the particulars of both the wanted and interfering signals. Thus, the SNR of the jth wanted signal due to the presence of the ith interferer can be expressed as

$$\left(\frac{S}{N}\right)_{ji} = R_{ji}\left(\frac{C}{I}\right)_{ji} \tag{4-12}$$

where

$$\left(\frac{C}{I}\right)_{ji} = \left\{\left(\frac{C}{I}\right)_{gji}^{-1} + \left(\frac{C}{I}\right)_{sji}^{-1}\right\}^{-1} \tag{4-13}$$

is the total CIR due to the ith interferer. It is implied, of course, that R_{ji} is so defined to produce equivalence between $(S/N)_{ji}$ and $(S/N)_t$ when the two are numerically equal. Equation (4-12) assumes that the interfering signal is simply translated at its own satellite. If it is "processed" in the satellite so that the downlink has different modulation characteristics than the uplink, then we have

$$\left(\frac{S}{N}\right)_{ji} = \left\{\left(\frac{S}{N}\right)_{gji}^{-1} + \left(\frac{S}{N}\right)_{sji}^{-1}\right\}^{-1} \tag{4-14}$$

where

$$\left(\frac{S}{N}\right)_{gji} = R_{gji} \left(\frac{C}{I}\right)_{gji} \tag{4-15a}$$

$$\left(\frac{S}{N}\right)_{sji} = R_{sji} \left(\frac{C}{I}\right)_{sji} \tag{4-15b}$$

If the wanted signal is also processed in its satellite, that is demodulated and remodulated, additional considerations are required that would take us somewhat far afield from our intended course. We will take up these considerations in Section A.4 of Appendix A. For present purposes, as well as for the remainder of this chapter, it will be sufficient to confine our attention strictly to translating repeaters both for wanted and interfering satellites.

The total, or combined SNR, can then be written as

$$(S/N)_j^{-1} = R_t^{-1} (C/N)^{-1} + \Sigma R_{ji}^{-1} (C/I)_{ji}^{-1} + R_t^{-1} (C/N)_d^{-1} \{(C/N)_u^{-1} + (C/I)_{sj}^{-1}\}$$

$$+ \Sigma R_{ji}^{-1} (C/I)_{gji}^{-1} [(C/N)_u^{-1} + (C/I)_{sj}^{-1}] \tag{4-16}$$

Equation (4-16) is based on the assumption that the ratio of carrier-to-noise plus interference transmitted by the satellite is the same as that at its input; i.e., the transponder is linear, and also on the assumption that there is no interaction in the detector among the various signals present. These assumptions will be satisfactory for our present purposes. Actually, we can simplify (4-16) considerably with negligible error in most instances because we can normally assume that the CNR and CIR are both reasonably large. Hence, the third and fourth terms, being the product of reciprocals of large quantities, can typically be ignored in relation to the first two terms. Thus, (4-16) reduces to

$$(S/N)_j^{-1} = R_t^{-1} (C/N)^{-1} + \sum_i R_{ji}^{-1} (C/I)_{ji}^{-1} \tag{4-17}$$

which is the form in which this equation is normally encountered. The situation in which (4-17) may not be adequate is during periods of heavy precipitation when the neglected terms could actually dominate. However, it can be presumed that when there is high enough likelihood of heavy rain, sufficient margin will have been built into the system that even under such conditions Equation (4-17) is still adequate. Thus, we shall consider this equation to govern the trade-offs of interest. The orbit utilization problem in the case at hand amounts to constraining the left-hand side of (4-17) to a fixed value, or a set of such values, while trading off the parameters contained in the right-hand side, the measure of orbit utilization (as yet undefined) being some function of these parameters. The sections to follow will continue the current line of thought in a context where further assumptions will permit a quantitative analysis to take place.

4.1.4 Digital Systems

The situation for digital systems differs markedly from that in the previous section because the performance measure of interest, universally the bit (or symbol) error rate, cannot be expressed as the sum of components due to individual noise or interference sources. Indeed, the BER is as well a sensitive function of all the particulars of a system's design. Thus, the most that can be said as a general statement is

$$P_e = f\ (C/N;\ I) \tag{4-18}$$

where $f(\bullet)$ is a functional that depends on the specific case examined and I stands for all the interferences. Since I contains all of the information regarding the inter-satellite spacings, the orbit utilization problem involves the trade between C/N and I for given P_e constraint. [Actually, even (4-18) is a somewhat simplified statement since in a nonlinear repeater P_e is a function of the uplink and downlink contributions to C/N; however, (4-18) will suffice for present purposes]. We shall later examine specific forms of $f(\bullet)$. Here we shall formulate a frequently used approximation which is often adequate in this context.

This approximation involves the assumption that interference has the same effect as thermal noise of the same power. The usefulness of the approximation derives both from its simplicity and from the fact brought to light in Chapter 7 that it usually gives an upper bound on error probability, thus leading to conservative results.* The implication of the foregoing is that we are interested in the carrier-to-noise

---

*Under certain conditions the approximation may not give an upper bound, as discussed in Chapter 7.

plus interference ratio, denoted C/NPI. This can be readily derived as

$$(C/NPI)^{-1}_j = (C/N)^{-1} + (C/I)^{-1}_j + (C/N)^{-1}_d \left\{ (C/N)^{-1}_u + (C/I)^{-1}_{sj} \right\}$$

$$+ (C/I)^{-1}_{gj} \left\{ (C/N)^{-1}_u + (C/I)^{-1}_{sj} \right\} \qquad (4\text{-}19)$$

Considerations similar to those discussed in the previous section apply with respect to the relative importance of the terms in (4-19). That is, in most cases we can safely approximate this equation by

$$(C/NPI)^{-1}_j = (C/N)^{-1} + (C/I)^{-1}_j \qquad (4\text{-}20)$$

We can somewhat generalize (4-20) by introducing a noise-equivalence factor (NEF), δ, to the carrier-to-interference ratio, the purpose of which is to account for the difference in the actual effect of the interference and that implied by the thermal noise approximation. Thus, (4-20) takes the slightly altered form

$$(C/NPI)^{-1}_j = (C/N)^{-1} + \delta^{-1} (C/I)^{-1}_j \qquad (4\text{-}21)$$

Evidently, $\delta < 1$ implies the interference has a lesser effect than thermal noise of the same power. The NEF might be found empirically or it might be varied as a parameter to determine how sensitive the results are to the thermal noise assumption. The BER would then be evaluated as

$$P_e = f (C/NPI) \qquad (4\text{-}22)$$

where $f(\bullet)$ would be interpreted as the function that relates the BER to the carrier-to-noise ratio for the modulation method in question.

4.2 Homogeneous Orbit Utilization

The equations given previously provide a basis for investigating the influence of the various parameters listed earlier on orbit utilization. This, however, cannot be done for the general case, by which we mean an arbitrary collection of satellite networks, each of which can be independently specified with respect to all operating parameters (e.g., modulation method, type of signal, type of service, e.i.r.p., etc.). The reason for this, basically, is that the general case is so unconstrained as to beg even a meaningful formulation. Furthermore, under such conditions, the choice of the proper measure of orbit utilization is problematical. It is clear, therefore, that a tractable formulation requires certain constraints to be imposed. A set of constraints which allows us to pose both a meaningful and tractable problem is embodied in what is known as the homogeneous orbit utilization model. The essential assumptions in this model are:

all satellite networks are identical to one another;

satellites are spaced at equal increments in geostationary orbit;

satellite antennas are sufficiently broad-beam to encompass all earth stations within their beamwidths;

propagation conditions are clear weather.

The uniformity of characteristics over all networks thus permits us to relate orbit utilization unambiguously to any parameter of interest. In this context it is also straightforward to define intuitively pleasing measures of orbit utilization; for example the amount of information transmitted per unit of bandwidth and orbital arc, subject to certain constraints;* or the intersatellite spacing. Some of the above conditions can be relaxed if desired, but this definition of the model is sufficiently useful for our purposes.

The homogeneous model is obviously an approximation to reality when considering the entire geostationary orbit. It may become a better approximation when applied to moderately restricted arcs of the orbit because it then becomes more plausible to find such uniformity. Clustering techniques, discussed later, when applied also tend to create more homogeneity within an arc. Thus, a perhaps more satisfying visualization of the homogeneous model is that the entire orbit can be considered as the sum of homogeneous segments. In the analyses to follow we will in effect consider only one such segment.

The emphasis on the homogeneous model in this chapter does not imply that the general case is unimportant nor broachable in some fashion. Indeed, this case conforms with reality. In the usual situation, however, one is faced with a set of networks with predetermined characteristics, and the typical problem is to ensure that all net-

---

*Such constraints might be: a minimum per satellite capacity, a maximum ratio of interference to thermal noise, or a power flux density constraint at the earth.

works can coexist within the mandated regulatory constraints or to determine the (minimal) modifications that will make it so. For these objectives it is possible by computer (see Section 5.6) to "solve" the general case; i.e., handle an arbitrary but given set of networks. But to make general inferences one needs the structure imposed by the homogeneous model.

4.2.1 Analytical Formulation

In addition to the conditions above defining the homogeneous model, a number of subsidiary assumptions are normally made that simplify greatly the calculations and notation, with but a minor or negligible effect on accuracy.

Assumption 1: the path loss of all downlinks is identical, i.e.,

$L_{stij} = L_{st}, \forall_{i,j}$

Assumption 2: the path loss of all uplinks is identical, i.e.,

$L_{gtij} = L_{gt}, \forall_{i,j}$

Assumption 3: the angular separation $\Delta\theta_t$ between two satellites in geostationary orbit as seen from the earth's surface, i.e., the topocentric angle, is sufficiently well approximated by the actual angular separation $\Delta\theta$, that is, the geocentric angle, which is measured from the center of the earth. This approximation is always conservative in the sense that $\Delta\theta_t > \Delta\theta$.

Assumption 4: the sidelobe gain of earth station antennas can be expressed as

$$G(\theta) = A/\theta^\beta \ (numeric) \tag{4-23}$$

where θ is the angle off boresight. Together with Assumption 3, Equation (4-23) implies that the gain towards an interfering satellite separated from the wanted one by $\Delta\theta$ can be written as $A/(\Delta\theta)^\beta$. This is a significant convenience in that it brings out the satellite spacing explicitly rather than implicitly, as an inverse function of antenna gain, which would be the case for an arbitrary sidelobe pattern. However, Equation (4-23) has sufficient flexibility, through suitable choices of A and β, to approximate closely enough most any actual sidelobe pattern. An antenna pattern such as (4-23) is called a "reference" pattern because it does not necessarily represent any single antenna but is a reasonably good abstraction for a class of antennas. The standard reference pattern, established by the CCIR for earth station antennas, uses $A = 10^{3.2}$ and $\beta = 2.5$ for ratios of diameter-to-wavelength exceeding 100. (See reference [1].)

The magnitude of the approximations in Assumptions (1-3) in any given case can be estimated by the reader with the help of Appendix B, which gives relevant geometrical data.

Consider now the frequency plan shown in Figure 4-2. Each satellite emits the same number of (identical) carriers separated by some guard band, b_g, and the RF signal bandwidth is W. The frequency plan from satellite to satellite is symmetrically alternated, as shown in the figure, and the frequency plans of adjacent satellites are relatively shifted so that the carrier offset between a typical carrier and the left and right interfering carriers is f_{d_1} and f_{d_2} respectively. The frequency plan on the uplink is assumed identical but translated to some other frequency band. Assume, furthermore, that alternate satellites are oppositely polarized, resulting in $Y_p = 10 \; log \; (1/y_p)$ decibels of isolation between any satellite and its odd-ordered removed neighbors.

Thus, a typical carrier is interfered with by two overlapping carriers from every odd-ordered removed satellite, both on the downlink and the uplink. (Carriers at the edge of the band are somewhat better off, having only one interfering carrier.) Also, each carrier is subjected to co-channel interference from every even-ordered removed satellite and its uplink.

Since all carriers are identical, we need not explicitly identify the wanted carrier. We can thus somewhat simplify earlier notation by dropping the index j identifying the jth wanted carrier. We further denote by the ith interferer an interfering carrier from a satellite $|i| \Delta\theta$ degrees from the wanted one, where $\Delta\theta$ is the common intersatellite spacing. (When i is odd and the frequency plan is such that there are two offset interferers, we will still index these two interferers jointly with i; this will not lead to any confusion.) We will take $i = \pm 1, \pm 2, \ldots, \pm N$, so that the wanted satellite will be assumed to be in the center of a cluster of $2N + 1$ satellites; note that the center satellite is the constraining one since it is subject to the greatest interference. As will be seen, the results are not sensitive to N beyond a small number.

With these preliminaries aside we can now specialize the equations to the homogeneous case.

4.2.2 Carrier-to-Interference Ratio

For the homogeneous model, and assuming for the moment co-channel co-polarized conditions, we can immediately write for the carrier-to-interference ratio on the downlink,

$$\left(\frac{C}{I}\right)_{gi} = \frac{G_r}{G(i\Delta\theta)} \tag{4-24}$$

which is a special case of (4-4), and where G_r is the earth station receiving antenna on-axis gain. Typically, we express G_r as

$$G_r = \eta \, (\pi D/\lambda_d)^2 \tag{4-25}$$

where η is the antenna efficiency, D is the antenna diameter, and λ_d is the downlink wavelength. The expectedly simple result (4-24) simply says that under the homogeneous conditions, the CIR is just the ratio (or difference, in dB) of the gain in the direction of the wanted signal (normally the main beam gain) to the sidelobe gain of the receiving antenna in the direction of the interfering satellite. This conclusion is also easier to visualize by referring to Figure 4-1.

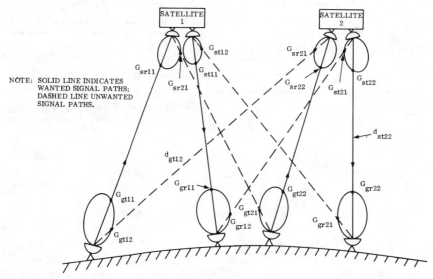

Figure 4-1 Illustrating the Interference Geometry

If we now invoke the reference pattern (4-23), Equation (4-24) becomes

$$\left(\frac{C}{I}\right)_{gi} = \frac{G_r \, (\Delta\theta)^{\beta} |i|^{\beta}}{A} \qquad (4\text{-}26a)$$

with the absolute value sign indicating that the interfering satellite could be either east or west of the wanted one.

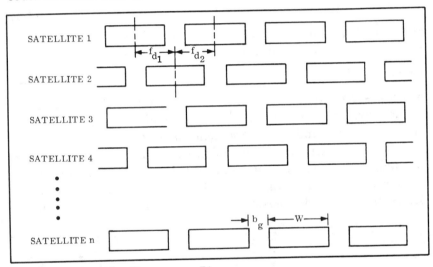

Figure 4-2 Satellite Frequency Plan

Now, we extend (4-26a) to take account of the possible use of carrier interleaving (Fig. 4-2) and orthogonal polarization. Basically, these factors are encompassed by the symbol Y in the general formulation (4-4). Hence, the extension is given by

$$\left(\frac{C}{I}\right)_{gi} = \frac{G_r(\Delta\theta)^{\beta} |i|^{\beta}}{A \, h_i p_i} \qquad (4\text{-}26b)$$

where p_i is the polarization discrimination, or isolation:

$$p_i = \begin{cases} 1, \; i \text{ even} \\ y_i, \; i \text{ odd} \end{cases} \qquad (4\text{-}27a)$$

h_i represents the fraction of the interfering power intercepted by the wanted carrier's receiver,

$$h_i = \begin{cases} h_1, \text{ for the left offset carrier, and } i \text{ odd} \\ h_2, \text{ for the right offset carrier, and } i \text{ odd} \\ 1, i \text{ even} \end{cases} \qquad (4\text{-}27b)$$

For the uplink it follows straightforwardly that*

$$\left(\frac{C}{I}\right)_{si} = \frac{G_t\,(\Delta\theta)^\beta|i|^\beta}{A\;h_i p_i} \qquad (4\text{-}28)$$

where G_t, the earth station transmit antenna on-axis gain, is

$$G_t = \eta\,(\pi D/\lambda_u)^2$$

where λ_u is the uplink wavelength. It follows, then, that

$$(C/I)_{gi}\,/\,(C/I)_{si} = (\lambda_u/\lambda_d)^2$$

hence the total (uplink + downlink) carrier-to-ith interference ratio can be put into the compact form

$$(C/I)_i^{-1} = (C/I)_{gi}^{-1}\,[1 + (\lambda_u/\lambda_d)^2]$$

Finally, we can write for the total CIR,

$$\left(\frac{C}{I}\right)^{-1} = [1 + (\lambda_u/\lambda_d)^2]\,\frac{2A}{G_r\,(\Delta\theta)^\beta}\,\sum_{i=1}^{N} H_i p_i i^{-\beta} \qquad (4\text{-}29)$$

where $H_i = h_1 + h_2$, for i odd, and advantage has been taken of the fact that (4-26) and (4-28) are symmetrical in i. Also implicit in (4-29) is the assumption that the uplink and downlink interference are not coherent, i.e., can be added on a power basis. This assumption is normally justified. We will deal frequently with the co-channel co-polarized case as a convenient point of reference; i.e., $H_i = 1$, and $p_i = 1$, $\forall i$. For this case, which we denote with subscript 0, we have

---

*As is implied in (4-28) we take it as a given of the homogeneous model that the polarization isolation on the uplink is the same as on the downlink. This is not a critical assumption, but it simplifies the notation. Polarization isolation is discussed at greater length in Section 4.6.

$$\left(\frac{C}{I}\right)_0^{-1} = [1 + (\lambda_u/\lambda_d)^2] \; \frac{A \, Z(\beta, N)}{G_r \, (\Delta\theta)^\beta} \tag{4-30}$$

where

$$Z(\beta, N) = 2 \sum_{i=1}^{N} (i)^{-\beta}$$

is twice the truncated Riemann zeta function, and is plotted in Figure 4-3; of course, only integer calues of N have meaning, though the points were smoothly connected for visual ease. As can be seen, after 3 or 4 interferers there is virtually no effect from increasing N, and in fact it is clear that the flanking interferers ($N = 1$) dominate. In numerical work we shall standardly assume a value of $N = 5$, i.e., 10 interferers.

We can therefore rewrite (4-29) as

$$\left(\frac{C}{I}\right)^{-1} = \frac{2 \, (C/I)_0^{-1}}{Z(\beta, N)} \sum_{i=1}^{N} H_i p_i i^{-\beta} \tag{4-31}$$

which isolates in an easily visualized way the effect of polarization isolation and carrier offset.* It should be noted that in this section, and henceforth unless otherwise mentioned, we neglect the product terms when formulating the performance measure.

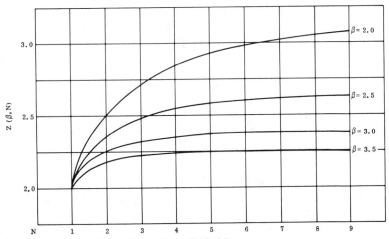

Figure 4-3 Plot of the Function $Z(\beta, N)$

---

*In Section 4.6 we shall compare this method of using polarization isolation to another.

4.2.3 Analog Baseband Signal-to-Noise Ratio

The general form for SNR given in (4-17) will be recast here for the homogeneous model. From that equation we see that the SNR has two components,

$$(S/N)^{-1} = (S/N)^{-1}_t = (S/N)^{-1}_I$$

the first due to thermal noise and the second due to interference. The thermal noise contribution is obtainable from standard theory and will be taken into account later. Here we consider only the interference contribution.

Using the homogeneous assumptions, in a development quite parallel to that yielding (4-26), we obtain for the SNR due to the ith downlink interferer,

$$\left(\frac{S}{N}\right)_{gi} = \frac{G_r \, (\Delta\theta)^\beta |i|^\beta}{A \, f_i p_i} \, R_0 \tag{4-32}$$

where R_o is the RTC for a co-channel carrier, and f_i is the interleaving advantage,

$$f_i = \begin{cases} f_1, \text{ for the left offset carrier, and } i \text{ odd} \\ f_2, \text{ for the right offset carrier, and } i \text{ odd} \\ 1, \text{ for } i \text{ even.} \end{cases}$$

Equivalently, the RTC is R_0/f_1 and R_0/f_2 for the left and right interleaved carriers, respectively. The factors f_1 and f_2 also include the effect of filtering. Thus, following the course of the preceding section, we obtain for the total interference SNR,

$$\left(\frac{S}{N}\right)_I^{-1} = [1 + (\lambda_u/\lambda_d)^2] \, \frac{2 \, A R_0^{-1}}{G_r \, (\Delta\theta)^\beta} \sum_{i=1}^{N} F_i p_i i^{-\beta} \tag{4-33}$$

where $F_i = f_1 + f_2$ for i odd.

As before, we shall often be interested in the co-channel co-polarized conditions as a reference case, for which

$$\left(\frac{S}{N}\right)_{I,0}^{-1} = [1 + (\lambda_u/\lambda_d)^2] \, \frac{A R_0^{-1} \, Z(\beta, N)}{G_r \, (\Delta\theta)^\beta} \tag{4-34}$$

so that

$$\left(\frac{S}{N}\right)_1^{-1} = \frac{2 \ (S/N)_{1,0}^{-1}}{Z \ (\beta, \ N)} \ \sum_{i=1}^{N} \ F_i p_i i^{-\beta} \tag{4-35}$$

which effectively brings out the effect of interleaving and polarization. Note that this effect depends upon the product $F_i p_i$, which means that functionally (4-35) behaves in the exact same way whether either of these measures is taken alone or in concert. This also implies that the application of either measure may negate the use of the other; for example, if the polarization discrimination were ideal — i.e., $p_i = o$, i odd — the use of frequency interleaving would provide no advantage. However, in practice both may prove useful.

4.2.4 Carrier-to-Noise Plus Interference Ratio

The C/NPI can be set down immediately from (4-21) and (4-31) as

$$\left(\frac{C}{NPI}\right)^{-1} = \left(\frac{C}{N}\right)^{-1} + \frac{2 \ (C/I)_0^{-1}}{Z \ (\beta, \ N) \cdot \delta} \ \sum_{i=1}^{N} \ H_i p_i i^{-\beta} \tag{4-36}$$

The equations of this section, (4-23) to (4-36) provide us with an initial basis for studying the influence of several technical parameters on orbit utilization. Some of these equations will be further expanded in subsequent sections, but for the moment we can begin to get an idea of some of the trade-offs involved, as developed in the next several sections.

4.3 Antenna Size and Satellite Spacing

For simplicity we focus our attention on the co-channel co-polarized case. For a fixed amount of interference, whether it is characterized by a predetection or post-detection quantity, it is clear from the foregoing equations [say, (4-30)] that there exists a trade between antenna size and satellite spacing embodied by the relation

$$\frac{Z \ (\beta, \ N)}{G_r \ (\Delta\theta)^{\beta}} = constant \tag{4-37}$$

or, using (4-25)

$$\Delta\theta = \left\{ \frac{Z \ (\beta, \ N)}{\eta \ (\pi D/\lambda)^2 \times constant} \right\}^{1/\beta} \tag{4-38}$$

and we shall measure antenna "size" as the ratio of diameter to wavelength, D/λ. (For notational simplicity we drop here the subscript on λ.) The term $Z(\beta,N)$ is not directly a function of either satellite spacing not antenna size. However, it is a function of the parameter β which affects the trade between the two parameters in question. As we have seen, however, [e.g., Fig. 4-3] the function $Z(\beta,N)$ is weakly dependent on β, so for convenience we shall ignore it here and in the next few sections. This approximation will render at most a few percent error when comparing results for two different values of β. Hence the relationship of interest will be

$$\Delta\theta = [(D/\lambda)^2 \times constant]^{1/\beta} \tag{4-39}$$

with a constant possibly different from that in (4-38). In fact, for plotting purposes we choose the constant in (4-39) so that $\Delta\theta = 5°$ for $D/\lambda = 100$ and $\beta = 2.5$. The results obtained therefrom are displayed in Figure 4-4, which shows $\Delta\theta$ vs. D/λ for several values of β.

The interpretation of Figure 4-4 is straightforward. We see from (4-39) that spacing decreases as $D^{-2/\beta}$, which means that for all $\beta > 2$, spacing deceases more slowly than $1/D$. Hence, while we do get a

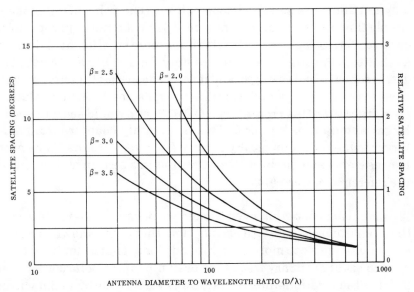

Figure 4-4 Satellite Spacing as a Function of Antenna Size

decrease in spacing as the diameter increases, as shown in the figure, this trade is inefficient in the sense that the improvement is not as rapid as that of the growth in antenna size. Also, as can be seen from the figure, there are diminishing returns once we get to a certain size. It should be pointed out that if we dealt with a fixed value of β there would be no need to choose a specific value for the constant in (4-39) in the sense that the relative spacing for different value of D/λ would be invariant. However, it is clear from (4-39) that the value of the constant will affect the relative values of $\Delta\theta$ for given D/λ as we go from one β to another (though the effect is moderate). Therefore, strictly speaking the position of the several curves on Figure 4-4 relative to one another is dependent on the actual spacing, but if we stay on a given curve (fixed β) the relative values do not vary, hence the right-hand scale on Figure 4-4 is applicable.

The parameter β actually reflects a different aspect of antenna behavior, namely the sidelobe pattern, which will be treated separately in the next section. In this context, however, we should note two relevant points. One is that β and D/λ are not necessarily independent. For example, an antenna operated at two different frequencies may not show the same sidelobe pattern. The point being made is simply that different ranges of value of β may be more appropriate to use in different regions of D/λ. Perhaps more significant is the fact, as plotted, that Figure 4-4 applies only for $\Delta\theta>1°$. For $\Delta\theta<1$, $\Delta\theta^{\beta}$ decreases with increasing β, and it can be seen that the curves would cross over at $\Delta\theta = 1$, which would occur for some D/λ. For values of D/λ larger than this crossover; i.e., $\Delta\theta>1$, one would seem to need a larger antenna for higher β. This is certainly counter-intuitive as larger values of β supposedly signify a "better antenna." The problem, as elaborated in the next section, lies in a certain artificiality of the reference pattern (4-23).

We should also note that even though we began, for simplicity, with the co-channel co-polarized assumption, the results shown in Figure 4-4 are extendable to more general conditions. That is, for fixed β and a given set of arbitrary frequency assignments or polarization conditions, the relationships of interest is still given by (4-39) with a different constant. Therefore, the relative spacing ordinate still applies under general conditions to each curve separately.

Finally, it should be kept in mind that Figure 4-4 was based on the idealized gain relationship (4-27) which cannot be indefinitely extended for arbitrarily large D/λ. It is well-known that for a given D and surface tolerance there is actually a maximum available gain as a function of frequency [see Reference 2 for further discussion].

4.4 Earth Station Antenna Sidelobes and Satellite Spacing

For a given amount of interference it is clear from the equations in Section 4.2 that, everything else being fixed, there is a trade between orbit utilization; i.e., satellite spacing, and the level of the earth station antenna sidelobes. This trade is embodied in the term $A/(\Delta\theta)^\beta$ = constant. The reference pattern, it will be recalled, is of the form

$$G(\theta) = A/\theta^\beta$$

where A will be referred to as the intercept and β as the slope. It is perhaps more customary to use the decibel equivalent,

$$10 \log G(\theta) = 10 \log A - 10 \beta \log \theta, \, (dB) \tag{4-40}$$

which for the CCIR reference pattern takes on the values $10 \log A = 32$ and $10\beta = 25$. It is clear from (4-37) and the sketch in [Figure 4-5(a)] that for fixed A the straight lines (4-40) meet at $\theta = 1$ and reverse positions on either side. In using a pattern such as (4-40) for general studies we are obviously motivated by showing the trade-off possibilities between orbit utilization and improved sidelobe patterns. Our intention is clear that the larger β is the better the antenna performance. Evidently, for $\theta < 1$ the function behaves inversely to our intentions.

The reason for this behavior clearly stems from the choice of function, which is a straightforward extension of the long used and familiar CCIR pattern. That pattern in the CCIR literature is attached to the proviso that is should be used for only $\theta \geq 1$ degree. For large antennas the sidelobe region may well extend to lesser angles. For the purpose at hand, it is significant to note that a problem arises only if the inter-satellite spacing is less than one degree: if not, no anomaly occurs. Thus, one could avoid the difficulty mentioned if, for $\Delta\theta < 1$, one shifted the pattern (4-40) so that for any β the straight lines intersect at $\Delta\theta$; i.e., at the position of the first interfering satellite. Such a procedure is illustrated in Figure 4-5(b) and could be easily mechanized.

The purpose of the preceding discussion has been to point out a possible anomaly and indicate one way of removing it. In fact, for our purposes it will be satisfactory to confine ourselves to the case

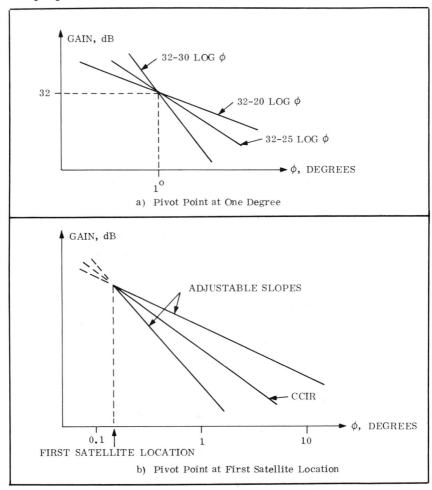

Figure 4-5 Illustrating the Sidelobe Pattern for Fixed Intercept and Variable Slopes

$\Delta\theta\geq 1$. We do not in the near future expect to see lesser statellite separations for co-coverage satellites, and in any event the numerical data given will provide typical relative results. No contradictions arise if we maintain β constant but change A. This amount to a shift in the pattern as a whole.

We can now observe the influence of (A,β) on orbit utilization. Figure 4-6 shows how $\Delta\theta$ varies as a function of $10 \log A$ for several values of β. The constants were chosen so that $\Delta\theta = 5$ for $10 \log A = 32$ and $\beta = 2.5$. The relative values shown on the right-hand scale will be a function of the reference when comparing two different curves, but for fixed β the right-hand ordinate is invariant. It can be seen that decreasing the sidelobe level or increasing the sidelobe decay exponent can both be effective means of improving orbit utilization. For example, the spacing can be halved for an approximately 7dB reduction in A, and going from $\beta = 2.5$ to $\beta = 3.0$ brings about roughly 30% improvement at $A = 32$ dB.

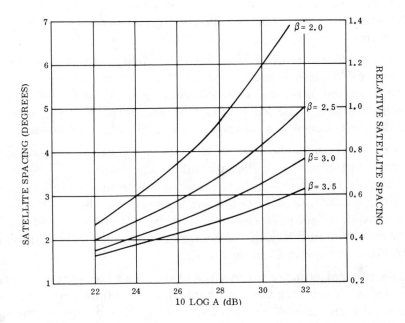

Figure 4-6 Satellite Spacing as a Function of Earth Station Antenna Sidelobes

We mention in passing another antenna pattern which is sometimes used in orbit utilization studies, namely

$$G(\theta) = \frac{G_0}{1 + (\theta/\Omega)^\beta}$$

where G_0 is on-axis gain and Ω is half the 3 dB beamwidth. Actually this pattern is also usable in the mainlobe. But in the sidelobes, it reduces to a form equivalent to (4-40). For in the sidelobes $(\theta/\Omega)^\beta >> 1$ and

$$G(\theta) \approx G_0/(\theta/\Omega)^\beta$$

or

$$10 \log G(\theta) = 10 \log A' - 10 \beta \log \theta$$

(4-40b)

where $A' = G_0 \Omega^\beta$. Equation (4-40b) is sometimes slightly more convenient to use than the earlier pattern (4-23), and in fact will be employed in Section 5.5 and Appendix G.

More elaborate reference antenna patterns are possible, perhaps comforming more closely to reality but lacking the flexibility and convenience of (4-23). One such pattern has been evolved for the broadcasting satellite service [3(e)] using (4-40a) for a portion of the pattern.

A separate question concerns the achievable values of A and β in practice. The design and physical factors influencing these values are discussed in [4] and more elaborately in [5]. In the latter it is estimated that $29\text{-}25 \log \theta$ should be attainable in the near term, while in the longer run we should expect $26\text{-}33 \log \theta$ to be feasible.

Reducing the sidelobes, *per se*, is not the only means of improving earth station antenna discrimination. The possibility of placing "nulls" in the direction of interference, which in principle is more efficacious than sidelobe reduction, has been considered. This, however, requires "active" means which are more properly considered in the context of null-steering or interference cancellers; this subject is briefly addressed in Chapter 8.

Reference patterns are one attempt to arrive at a general representation of sidelobe gain behavior. That behavior is a function not only of many design factors, but also of uncontrollable variations in implementation. An ultimate description of sidelobe behavior must, therefore, be statistical in nature. In Appendix C we discuss further this statistical aspect and its implication on satellite spacing.

4.5 Interference Noise Allowance and Satellite Spacing

All other things being fixed, the equations of Section 4.2 show that $\Delta\theta^\beta$ is proportional to *(C/I)* or to *(S/N)$_I$*. Therefore a trade exists between the amount of interference noise permitted and satellite spacing. This trade is shown in Figure 4-7, for which, as before, the constant of proportionality is chosen so that a reference *(C/I)* or *(S/N)$_I$* is achieved for $\Delta\theta = 5$ degrees and $\beta = 2.5$. As with the preceding figures the relative values are invariant with actual spacing so long as β is held constant. Otherwise, the spacing for one value of β relative to another will vary somewhat with the reference condition.

As one might expect, appreciable orbit utilization improvement may be had by allowing increasing amounts of interference. This trade, as pointed out above, is given by the *(1/β)* root of *(C/I)* or *(S/N)$_I$*; numerically this yields about 32% decrease in $\Delta\theta$ for each doubling of allowable interference, for $\beta = 2.5$. Evidently, Figure 4-7 does not give a complete picture of the interference noise trade-off because increasing the interference must imply a cost in some sense. Either the performance must suffer, or the power must be increased to keep it the same by lowering the thermal noise contribution, or the information rate must decrease. Thus, a more global approach to the influence of the interference noise must perform this examination under a set of meaningful constraints. This problem is evidently multifaceted and will be developed in stages in the course of this chapter.

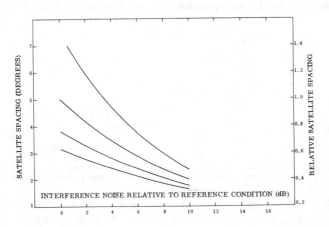

Figure 4-7 Satellite Spacing as a Function of Interference Noise Allowance

4.6 Polarization Isolation and Orbit Utilization

In this section we shall study the beneficial effect that polarization isolation* can have on orbit utilization. Two principal situations are usually considered, one where alternate satellites use opposite polarizations**, and the other where all satellites may use both polarizations. Co-channel operation will be assumed in both cases. A third situation, which involves the joint use of dual polarization and carrier interleaving, will be deferred to Section 4.10, after we have considered interleaving in Section 4.9. Before we proceed, however, it will be necessary to clarify the meaning of polarization isolation as we use it here.

4.6.1. Definition of Polarization Isolation

Our initial nomenclature in Section 4.2.1. for polarization isolation, y_i, implies that $y_i = y(i\Delta\theta)$ will generally be a function of satellite separation. This is because the cross-polarized response of an antenna, as well as its principal response, is a function of the off-axis angle. An illustrative representation of principal and cross-polarized reference patterns is shown in the sketch below; of course, the principal reference pattern is just what we have until now merely referred to as the reference antenna pattern. The pictorial description below applies in broad outline to either transmit or receive antennas. The polarization isolation y_i is a function of the ratio (or difference in dB) of the principal and cross-polarized responses for both transmit and receive antennas, and would be found as follows.

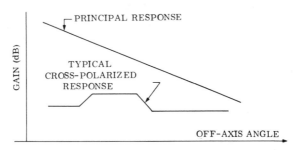

*Equivalent terms often encountered are polarization discrimination or cross-polarization isolation (or discrimination).

**The terms opposite and orthogonal (polarization) shall be used synonymously. When two co-frequency (or, generally, overlapping spectrum) carriers on orthogonal polarizations are transmitted from the same satellite, this is frequently referred to as dual-polarized transmission.

It will be instructive to develop the general case, from which we can reduce to the homogeneous situation. Consider two satellites spaced θ degrees apart, and their cooperating earth stations. Also suppose that the unwanted network uses for its principal polarization one that is nominally orthogonal to that of the wanted one. Now define the following quantities:

P_w = wanted satellite transmit power;

P_i = ith interfering satellite transmit power;

$G_{swp}(\delta_w)$ = gain of the wanted transmit antenna in its principal polarization, toward its own earth station (at an angle δ_w from boresight);

$G_{sip}(\delta_i)$ = gain of the unwanted transmit satellite antenna in its principal polarization, toward the wanted earth station (at an angle δ_i from boresight);

$G_{sic}(\delta_i)$ = gain of the unwanted transmit satellite antenna in its orthogonal polarization, toward the wanted earth station;

$G_{gwp}(\theta)$ = gain of the wanted receiving earth station in the principal polarization toward the unwanted satellite;

$G_{gwc}(\theta)$ = gain of the wanted receiving station in the orthogonal polarization toward the unwanted satellite;

$G_r = G_{gwp}(0)$, the on-axis gain of the wanted receiving antenna in the principal polarization, toward the wanted satellite.

The interfering satellite will produce interference in two distinct ways. One is due to the interaction of the principal polarization of the interference with the cross-polarized response of the wanted earth station antenna, and the other arises from the principal polarization response of that antenna to the cross-polarized interference component. The power (in clear weather) in the wanted receiver due to these two interference contributors is given by, respectively, [ignoring the space loss]

$$I_1 = P_i\, G_{sip}(\delta_i)\, G_{gwc}(\theta) \tag{4-41a}$$

$$I_2 = P_i\, G_{sic}(\delta_i)\, G_{gwp}(\theta) \tag{4-41b}$$

The total interfering power is a function of the phasing of these two components, which is difficult to predict. We can account for this abstractly by a phasing factor, φ_i, such that the interference power is given by

$$I_c = \varphi_i \, (I_1 + I_2) \tag{4-41c}$$

The value $\varphi_i = 1$ implies incoherence, (power addition); if $I_1 = I_2$, for example, $\varphi_i = 2$ would imply voltage addition.

The interfering power, *if* the interfering satellite's principal polarization were the same as that of the wanted one, would be

$$I_p = P_i \, G_{sip} \, (\delta) \, G_{gwp} \, (\theta) \tag{4-42}$$

The preceding equation assumes that the gain function $[G_{sip}]$ is independent of which specific polarization is the principal one; e.g., left-hand or right-hand circular. When dealing with reference patterns, we essentially define this to be true. In practice this will not be the case, but asymmetries of this kind should not be very significant.

The development in Section 4.2 essentially defined the polarization isolation as the ratio of co-polarized to cross-polarized carrier-to-interference ratios. Since the carrier power, $C = P_w \, G_{swp} \, (\delta_w) \, G_r$, is common to both, we have

$$y_{i,d} = I_c \, / I_p \tag{4-43}$$

$$= \varphi_i \, \left| \frac{G_{gwc} \, (\theta)}{G_{gwp} \, (\theta)} + \frac{G_{sic} \, (\delta_i)}{G_{sip} \, (\delta_i)} \right| \quad \textit{(numeric)}$$

Thus, the isolation is basically the sum of the individual antennas' isolations; this sum will be dominated by the poorer of the two antennas in this respect. Typically these isolations are given in decibels, in which case they must first be converted to numerics before adding.

The subscript d in (4-43) indicates downlink, which is the only path considered thus far. An indentical development for the uplink, using obviously corrresponding parameters, would lead to the analogous quantity $y_{i,u}$. The total CIR would therefore be

$$(C/I)^{-1} \underset{=}{\Delta} (C/I)^{-1}_c = (C/I)^{-1}_{c,d} + (C/I)^{-1}_{c,u}$$
$$= y_{i,d} (C/I)^{-1}_{o,d} + y_{i,u} (C/I)^{-1}_{o,u} \tag{4-44}$$

where the subscripts c and d indicate cross-polarized and co-polarized, respectively. Implicit in this formulation is the assumption that uplink and downlink interference powers are additive.

Turning now to the homogeneous model, it can be seen that the basic definition (4-43) still applies, but the numerical value of individual terms would reflect the nature of the model. For example, since in this model the satellite antenna is assumed to cover all of the earth stations of interest within its main beam, it could reasonably be assumed that the term $G_{sic} (\delta_i)/G_{sic} (\delta_i)$ is approximately constant, independent of i. In any case, we can proceed without having to make specific assumption of this type.

Assuming possibly different isolations on the downlink and uplink, the homogeneous model equation, (4-29), is generalized to [recalling the definition (4-27a)]:

$$\left(\frac{C}{I}\right)^{-1} = \frac{2A}{G_r (\Delta\theta)^\beta} \sum_{i=1}^{N} H_i p_i i^{-\beta} \tag{4-45}$$

where
$$p_i = p_{i,d} + p_{i,u} (\lambda_u/\lambda_d)^2$$

We shall see that (4-45) is functionally identical to (4-29), and for a given value of p_i differs from the latter only by a constant. Thus, insofar as the trade between p_i and $\Delta\theta$ is concerned, we are justified in speaking of *the* polarization isolation without further reference to relative uplink or downlink values. Of course, is the latter two are equal, (4-45) reduces identically to (4-29).

From the definition of polarization isolation, (4-43), and from the sketch of principal and cross-polarized responses, we can see that y_i will generally be an increasing function (decreasing isolation) of off-axis angle of an antenna, although the function may not be strictly increasing. For an earth station antenna, this means that progressively less polarization isolation is to be expected as the satellite separation increases. For satellite antennas the isolation will not be strictly a function of the satellite separation but will depend as well on the separation of the respective coverage areas.

The actual cross-polarized pattern of an antenna depends upon a multiplicity of design factors. Reference [6] provides a discussion of these factors and shows some actual measurements. From the regulatory point of view it is desirable to have a reference pattern as a standard which would have to be met for any actual pattern. Such a reference pattern for the cross-polarized response was developed (for both transmit and receive antennas) for the broadcasting satellite service around 12GHz [3(e)], and these patterns can be regarded as more or less typical in general. However, no generally accepted cross-polarized reference pattern does in fact exist, unlike the case for the principal response.

In view of the latter fact, and rather than making somewhat arbitrary assumptions about the behavior of y_i as a function of $i\Delta\theta$, we shall make the simplifying assumption in the computations to follow that y_i is independent of i. While this is a gross simplication, it will not lead us numerically far astray because of the dominance of the closest spaced interfering satellites.

The preceding discussion has assumed that the interference emanated from other satellites. In some situations it may be desirable to use both polarizations from the same satellite (dual-polarization transmission), in which case the network will generate cross-polarization self-interference. The preceding development still applies in form to this situation, with the proper parameter values. For example, the angles θ and δ_i are now both nominally zero, or at most, in the case of the latter, equal to half the 3 dB beamwidth. Thus the single path isolation is nominally the sum of the on-axis isolations for each antenna, with the phasing factor equal to unity. In this case, however, there is greater likelihood that this factor will be other than unity, than when the adjacent satellites are cross-polarized.

For the dual-polarized case we can thus still define a downlink isolation, computed from (4-43) with the appropriate values. We denote this isolation y_d. (This notation derives from the fact that the subscript i in (4-43) is asociated in the homogeneous model with the interfering satellite $i\Delta\theta$ degrees from the wanted one; in the dual-polarized case, therefore, one would have $i = 0$, and so we drop the i subscript.) For the uplink we define the corresponding isolation y_u. Note that y_u could take on somewhat different values depending upon whether or not the dual-polarized uplinks emanate from the same earth station. In the homogeneous model, all carriers have equal power. Therefore, the carrier-to-interference ratios in the co-polarized case would be unity, when the actual (cross-polarized) CIRs are

$$(C/I)_{c,d} = y_d \ and \ (C/I)_{c,u} = y_u \qquad (4\text{-}46)$$

The assumption that these two are additive may not, in this case, always be justified. We can therefore define a path phase factor, φ_p (analogous to φ_i) such that the actual total isolation, y_t, is given by

$$y_t \triangleq (C/I)_c = \varphi_p \ (y_d + y_u) \qquad (4\text{-}47)$$

and where, as before, $\theta_i = 1$ implies power addition.

4.6.2 Polarization Isolation and Satellite Spacing

(a) Alternate Satellites Oppositely Polarized

We first consider the case where alternate satellites use opposite polarizations. Assuming co-channel operation, we find from Section 4.2 or Equation (4-45) above, that, all other things being fixed, the trade between satellite spacing and polarization isolation is embodied in the relation

$$\frac{1}{(\Delta\theta)^\beta} \sum_{i=1}^{N} p_i i^{-\beta} = constant \qquad (4\text{-}48)$$

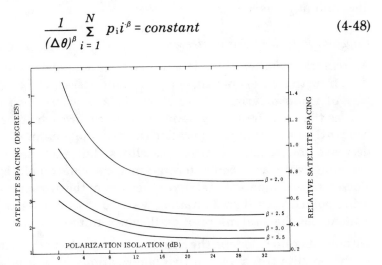

Figure 4-8 Satellite Spacing as a Function of Polarization Isolation

where $p_i = 1$ for i even and $p_i = y_i$ for i odd. As mentioned earlier, rather than assume a specific dependence of y_i on i, we assume that y_i is independent of i, i.e., $y_i = y_p$, $\forall\ i$. In effect, this is equivalent to assuming $y_i = y_p$ for the nearest cross-polarized satellite since the further removed cross-polarized satellites will have a relatively small influence in the homogeneous model. From (4-48) we plot in Figure 4-8 the resulting $\Delta\theta$ as a function of $Y_p = 10 \log (1/y_p)$. As before, we choose the constant so that $\Delta\theta = 5$ for $y_p = 1$ and $\beta = 2.5$. Also as before, the relative spacing scale applies irrespective of the reference condition if β is held constant, but comparing results for different β will to some extent depend on the absolute (left-hand) scale.

The ultimate spacing improvement possible, which occurs for $y_p = 0$, is immediately deducible as a factor of two since for this condition the odd-numbered satellites are "transparent," hence the even-numbered satellites can be placed at the positions occupied by the odd-numbered satellites with no polarization isolation. (The previous statement applies, strictly speaking, for an infinite string of satellites; in practice the actual ratio will depart slightly from 2:1 because of edge effects depending upon the number of satellites considered.) It can be seen that the maximum possible improvement is reached fairly rapidly and most of it is already obtained at a discrimination as low as 10 dB.

(b) Dual-Polarized Transmission

As remarked earlier, there is another situation, operationally distinct from the one examined, which in some cases may be a superior way of using polarization isolation. This is where each satellite re-uses the same frequency band twice, once for each polarization. Evidently, for co-channel operation this technique cannot be used to decrease the spacing between satellites, but each satellite can transmit twice as much information as a singly polarized one. Therefore, to compare the relative efficiency of the two approaches to using polarization discrimination, we need to examine not just the spacing but the total information flow per unit arc.

We now, therefore, extend the homogeneous model to the case where every satellite transmits on two polarizations. It will be recalled from subsection 4.6.1. that we denoted by y_t the total polarization isolation between co-channel cross-polarized carriers of the same network, and that for such carriers, $(C/I)^{-1} = y_t$. We assume, as before, that the polarization isolation from one satellite to another is y_p. To simplify notation, without materially detracting from generality, we shall assume that intersatellite isolation is the same for uplink and downlink, both equal to y_p. Hence a given carrier is interfered with by three types of entries: (1) the same satellite co-channel cross-polarized carrier, (2) the nominally co-polarized carriers on every other satellite; and, (3) the cross-polarized carriers on every other satellite. With our previous assumptions, therefore, the CIR for a typical carrier is given by

$$(C/I)^{-1} = y_t + [1 + (\lambda_u/\lambda_d)^2 \frac{A (1 + y_p) Z(\beta, N)}{G_r (\Delta\theta)^\beta} \qquad (4\text{-}49)$$

In order to compare the two polarization-use methods we must do so for equal carrier-to-interference ratio. For the usual reference condition ($\Delta\theta = 5$, $\beta = 2.5$, co-polarized co-channel carriers), the required CIR can be written as

$$(C/I)_{req}^{-1} = [1 + (\lambda_u/\lambda_d)^2 \frac{A Z(2.5, N)}{G_r (5)^{2.5}}$$

or,

$$[1 + (\lambda_u/\lambda_d)^2 \frac{A}{G_r} (C/I)_{req}^{-1} 5^{2.5}/ZX(2.5, N)$$

Substituting into the RHS of (4-49), and making use of the fact the left-hand side of (4-49) must also be $(C/I)_{req}^{-1}$, we obtain

$$(C/I)_{req}^{-1} \left\{ 1 - \frac{Z(\beta, N) (1 + y_p) 5^{2.5}}{Z(2.5, n) (\Delta\theta)^\beta} \right\} = y_t \qquad (4\text{-}50)$$

We see from (4-50) and for $y_t = y_p = 0$, $\beta = 2.5$, that $\Delta\theta = 5$. Since the information per unit bandwidth is twice the reference case, the limiting condition of ideal polarization isolation leads to the same information per unit arc for both methods of polarization use. But for finite polarization isolation the relation between y_p, y_t, and $\Delta\theta$ is C/I dependent, unlike the case previously studied. We illustrate this relationship in Figure 4-9 for $(C/I)_{req} = 20\ dB$ and $\beta = 2.5$. In order to make an equitable comparison with Figure 4-8 we plot the "equivalent" spacing $\Delta\theta_e = \Delta\theta/2$ to account for the fact that twice the information per unit bandwidth is transmitted. The abscissa, as before, is $Y_p = 10\ log\ (1/y_p)\ dB$, and the parameter is $Y_t = 10\ log\ (1/y_t)\ dB$. For a somewhat different perspective of the trade-offs involved, Figure 4-10, shows the relationship between $\Delta\theta_e$ and y_p for different β and y_t fixed at 26 dB, all for the same value of CIR as before.

Figures 4-9 and 4-10 show that orbit utilization is fairly insensitive to the intersatellite isolation Y_p, especially for high β. On the other hand, it is a fairly strong function of Y_t. Note that, of necessity, $y_t \leq (C/I)_{req}^{-1}$. Hence, for high CIR requirements the intra-satellite isolation may impose stringent design constraints.

No definite conclusions can be drawn with respect to the relative desirability of the two polarization-use methods. Orbit utilization improvements are potentially of the same order. The dual polarization mode on the same satellite generally requires a higher polarization isolation, for a given improvement, than does the single polarization per satellite mode. However, as implied by the sketch in Section 4.6.1, this does not represent different degrees of difficulty since it is known [6] that polarization purity is easier to achieve in the main lobe of an antenna than in the sidelobes. The dual polarization mode may be preferred by the system operator since twice the amount of information can then be sent from a given satellite. On the other hand, polarization isolation may be required between satellites in order to permit them to coexist within a moderately small arc.

It has been implied here that polarization isolation can be used to improve orbit utilization about equally by reducing the spacing

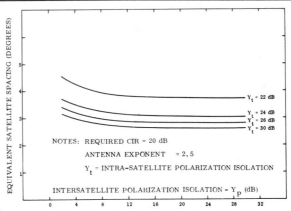

Figure 4-9 Equivalent Satellite Spacing for Dual Polarization Use

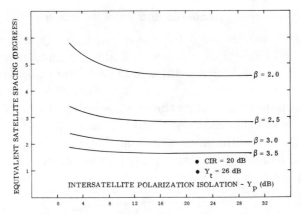

Figure 4-10 Sensitivity of Dual Polarization Use to Antenna Sidelobes

between satellites or increasing the satellite capacity, but that both cannot be done. This is true if all carriers are co-channel as has been assumed. However, it is possible to obtain both a capacity increase and a spacing reduction if carrier interleaving is used simultaneously in the proper manner. This will be discussed in Section 4.10.

Lastly, we should remark that polarization isolation will be degraded by the propagation medium, particularly by the presence of rain. While we have so far in this section explicitly considered only the isolation due to the antenna systems, the results (curves) can be applied more generally by interpreting the symbols Y_p or Y_t as the polarization isolation available in the presence of a depolarizing medium. Generally this isolation will be given by the appropriate sum* of the isolation of the antenna systems and that of the medium, and will be dominated by the latter during periods of moderately heavy rain. Of course, the homogeneous model is defined in clear weather, and in order to extend the results to the situation discussed, we would have to assume the same amount of depolarization on all paths; while this is unlikely in actuality, this still allows us to obtain some appreciation for the influence of the medium. The nature of the medium-induced degradations has been extensively discussed in the literature [see, e.g., Ref. 9, Part 3]. Methods for compensating the depolarization have been devised [and discussed in Chapter 8], but these are generally applicable or practical only in dual-polarized systems.

---

*In the worst case this will be given by the voltage sum.

4.7 Station-Keeping Accuracy and Satellite Spacing

In the previous sections it was assumed implicitly that unique positions could be assigned to spacecraft along the geostationary orbit. As is well-known [7,8] satellites drift from their nominal position, both in the equatorial plane and perpendicular to it. Here, we shall consider motion only along the orbit, which is the most significant from the orbit utilization standpoint. This longitudinal error causes a loss in the orbit utilization potential. Paradoxically, an intentional north-south inclination can be used to advantage, as will be discussed in Section 5.13.5

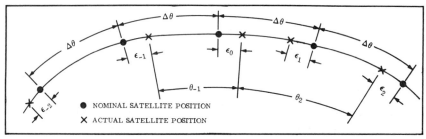

Figure 4-11 Typical Satellite Deployment with Station-Keeping Error

The situation at hand is illustrated in Figure 4-11. The homogeneous model still applies except that the ith satellite is assumed to have station-keeping error ε_i degrees (ε_i is the error for the wanted satellite), and the various ε_i are assumed to take on values randomly and independent of each other, within a specified station-keeping tolerance band. We assume this band is identical for all satellites, so that

$$|\varepsilon_i| \le \varepsilon_{max}, \; \forall \; i \tag{4-51}$$

The spacing between the wanted satellites and the ith satellite can be written as

$$\theta_i = |i| \, \Delta\theta + (\varepsilon_i - \varepsilon_0) \, sgn(i) \tag{4-52}$$

where $sgn \, (\cdot)$ is the signum function

$$sgn \, (x) = \begin{cases} 1, \, x > 0 \\ -1, \, x < 0 \end{cases}$$

Using (4-52) in the development of Section 4.2, one arrives at

$$\left(\frac{C}{I}\right) = [1 + (\lambda_u/\lambda_d)^2] \frac{A}{G_r (\Delta\theta)^\beta} \sum_{i=N}^{N} H_i p_i |i|^{-\beta} \tag{4-53}$$

$$\left\{ 1 + \frac{[\varepsilon_i - \varepsilon_0] sgn(i)}{|i| \Delta\theta} \right\}^{-\beta}$$

for the carrier-to-interface ratio expression extending (4-29). Since a similar extension applies to (4-33) or (4-36) we shall concentrate on (4-53).

In order to simplify our assessment of the effect of station-keeping error we shall assume the co-channel co-polarized case, for which (4-53) reduces to

$$\left(\frac{C}{I}\right)^{-1} = \left(\frac{C}{I}\right)_0^{-1} \sum_{i=N}^{N} \frac{|i|^{-\beta}}{Z(\beta, N)} \left\{ 1 + \frac{[\varepsilon_i - \varepsilon_0] sgn(i)}{|i| \Delta\theta} \right\}^{-\beta} \tag{4-54}$$

where $(C/I)_0$, defined in (4-30), is the CIR that would result in the event of ideal station-keeping. In Appendix D we shall reconsider (4-54) from a statistical viewpoint, treating the ε_i as random variables. For present purposes it will be sufficient to evaluate the worst-case effect of station-keeping error. This worst case occurs when the wanted satellite is at either extreme of its tolerance band, say the positive extreme, and the other satellites are at their respective band limits such as to place them closer to the wanted satellite. If we set

$$\varepsilon_{max} = \varepsilon_r \Delta\theta$$

where ε_r is a specified fraction of $\Delta\theta$, then we have for this worst case

$$\left(\frac{C}{I}\right)^{-1} = \left(\frac{C}{I}\right)_0^{-1} \sum_{i=-N}^{N} \frac{|i|^{-\beta}}{Z(\beta, N)} \left\{ 1 - \frac{2\varepsilon_r}{i} \right\}^{-\beta} = \left(\frac{C}{I}\right)_0^{-1} S(\varepsilon_r) \tag{4-55}$$

where $S(\varepsilon_r)$ stands for the summation indicated, and represents the degradation in CIR for fixed $(C/I)_o$. The decrease in CIR in dB is *10 log* $S(\varepsilon_r)$ and is plotted in Figure 4-12 as a function of ε_r for several values of β. If we wish to maintain the CIR at a fixed value $(C/I)_o$ under the worst-case placement, then the original nominal spacing $\Delta\theta$ must be increased to, say, $\Delta\theta'$, which is the solution to

$$\left(\frac{\Delta\theta'}{\Delta\theta}\right)^\beta = S(\varepsilon_r') \tag{4-56}$$

where $\varepsilon'_r = \varepsilon_{max} / \Delta\theta'$. Notice that at the new, larger spacing, a given station-keeping accuracy in absolute terms becomes a smaller relative maximum error. Equation (4-56) is transcendental but can be solved rapidly by iteration. One can otain from this the factor $[S(\varepsilon'_r)]^{1/\beta}$ by which the original nominal spacing must be increased, and the result is shown also on Figure 4-12, using the right-hand ordinate. The abscissa, of course, corresponds to the "old" spacing; if this factor were to be plotted against the new relative error, ε'_r, the latter would be obtained simply by dividing ε_r by the right-hand ordinate.

As can be seen, rather serious impairment of the orbit utilization potential can be incurred from poor station-keeping accuracy. The decrease in CIR, in particular, can be dramatic, especially as the antenna decay exponent increases. On the other hand, in order to keep the CIR constant a spacing greater than originally envisioned is required, on the order of one-and-a-half times that spacing as the relative error reaches the higher values; this increase in spacing appears milder than one .might expect from the left-hand scale. Basically this is because of the β^{-1} root relation between power and spacing; and because, as we increase the spacing, the relative error decreases and the corresponding improvement accelerates with β. The latter property counteracts the greater sensitivity to β of a given

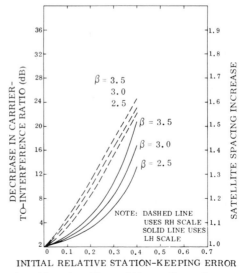

Figure 4-12 Effect of Station-Keeping Error (Worst-Case Assumption)

relative error and explains why the spacing increase is essentially independent of β.

Actually, the possible loss resulting from station-keeping inaccuracy has been in large measure limited by the existing *Radio Regulations* [3(a)] which stipulate that for the fixed satellite service, the largest class of users of the geostationary orbit, the inaccuracy shall not exceed ± 0.1 degrees. This means, for example, that for a desired nominal spacing of 1 degree, maintaining the interference budget would necessitate an actual spacing of about 1.1 degrees. For a two degree desired spacing, 2.1 degrees would actually be needed.

In actuality, satellites will drift independently within their tolerance bands, subject to natural as well as restoring forces. Thus, at any instant satellites can be considered independently and randomly placed with respect to their nominal positions. The effect of imperfect station-keeping will then take on a statistical character which is discussed at some length in Appendix D.

4.8 FDM/FM Telephony Modulation Index and Satellite Spacing

Up to this point the trade-offs examined have involved physical parameters of the transmitting and receiving systems, without making any specific assumptions as to the nature of the modulation itself. The modulation characteristics have a major influence on orbit utilization because they determine at the same time the amount of information that a signal of prescribed bandwidth can carry and the relative immunity to interference. The latter aspect directly controls the intersatellite spacing and it is only this aspect that we shall deal with here briefly. The relationship between satellite spacing and modulation characteristics would in general depend upon the modulation method itself. To keep the discussion brief we shall confine our attention to FDM/FM telephony signals because the trade-off to be examined can be formulated in terms of a simple smoothly continuous equation, hence easily visualizable (unlike the case for some other types of signal). However, our findings will be generally applicable in a qualitative way because the immunity to interference increases with the ratio of occupied bandwidth to information rate, which is a variable property of any modulation scheme, except AM and related systems.

For FDM/FM signals this property is inherent in the rms multi-channel modulation index M_o, which is common to all signals in the homogenous model. The modulation index exercises its influence through R_o in Equation (4-32), to which it is related by the formula

$$R_o \propto 1 + 9.5\,M_o^3 \qquad\qquad (4\text{-}57)$$

a relation which will be derived in Chapter 6.

For simplicity we consider now the co-channel case, given by (4-34). For a given interference SNR, and other things being fixed, we find from the latter equation and (4-57) that satellite spacing and modulation index are connected through

$$\Delta\theta^{-\beta} = const \; x \; (1 + 9.5\,M_o^3) \qquad\qquad (4\text{-}58)$$

For the reference condition, i.e., choice of the constant, we let $\Delta\theta = 5, \beta = 2.5$, correspond to $M_o = 1$. Thus we can plot the variation of $\Delta\theta$ as a function of M_o and β, as has been done in Figure 2-13. The abscissa could also be labeled in terms of the ratio of occupied to information bandwidth, which can be simply found from the formula $2\,(\lambda M_o + 1)$ where λ the peak-to-rms ratio of the baseband. The curves indicate the substantial benefits possible from increasing M_0, in terms of spacing. For example, for large M_0 we can see from (4-58) that $\Delta\theta \propto M_0^{-3/\beta}$, so that for $\beta = 2.5$ the spacing can be halved with M_0 increasing to about 1.75 times its former value. But at the same time, as was pointed out earlier, there is an accompanying loss of information transmission per unit bandwidth. The relation between the two under some constraints will be studied shortly.

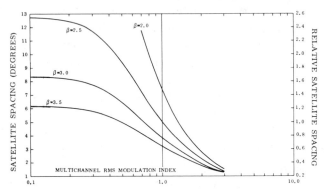

Figure 4-13 Satellite Spacing as a Function of Modulation Index for FDM/FM Transmission: Co-Channel Case

4.9 Carrier Interleaving and Satellite Spacing

The frequency difference between two mutually interfering carriers is one of the very few conditions that may ameliorate the interference environment without affecting major design parameters and that can be set more or less arbitrarily, within limits, at essentially no cost or inconvenience. In fact, for FM signals there is something to be gained by operating at other than co-channel conditions, the gain depending on the modulation index and the type of signal. In a practical system the receiver invariably incorporates some degree of front-end filtering matched to the wanted signal's bandwidth. Hence an interfering signal offset in frequency from the wanted one will inevitably experience some filtering, depending on its bandwidth relative to that wanted signal. The degree and type of filtering will affect the numerical value of the improvement. However, the fact that there is an improvement for FM signals is not dependent on the presence of filtering; it is inherent in the nature of the modulation method. In other words, there is an improvement, depending on the modulation index, even if the wanted signal's receiver front-end has no filtering at all. This property is not shared by digital signals, whose performance will be invariant to the carrier offset of an interfering signal so long as it is wholly intercepted by the receiver. This is no longer quite the case if the intererence is filtered. In fact, in the latter situation it is possible for the performance to degrade. Broadly speaking, however, we can expect the performance of a digital signal to be only mildly affected by the specific interleaving condition of an interfering signal.

To illustrate the potential beneficial effect of interleaving we consider here the case of FDM/FM telephony signals. The interleaving improvement will be taken to be that in the absence of filtering whose effects will be explicitly addressed in Chapter 6. Referring to the frequency plan of Figure 4-2, the most common and sensible interleaving arrangement is a symmetrical one, whereby the two offset frequencies are equal, i.e., $f_{d_1} = f_{d_2} = f_d$. Assuming co-polarized carriers, Equation (4-33) becomes

$$\left(\frac{S}{N}\right)_I^{-1} = [1 + (\lambda_u/\lambda_d)^2] \frac{2AR_0^{-1}}{G_r(\Delta\theta)^\beta} \sum_{i=1}^{N} F_i i^{-\beta} \qquad (4\text{-}59)$$

where $F_i = 2F_o$ for i odd and F_o is the single-interferer interleaving improvement.

Assuming the frequency separation $f_d = 0.6 \, x$ signal RF bandwidth, which includes a 20% guardband, it will be shown in Chapter 6 that the improvement is approximately given by

$$10 \log F_0^{-1} = 18.6 + 10 \log exp \left(\frac{0.38}{M_0} - \frac{0.24}{M_2^0}\right) , M_0 \gtrsim 0.5 \quad (4\text{-}60)$$

The improvement vanishes at $M_o \approx 0.1$ and for smaller modulation indices it becomes negative. The behavior for $0.1 \le M_o \le 0.5$ is shown in Section 6.3.6.

All other things being fixed, the trade-off between satellite spacing and carrier interleaving is therefore given by

$$\Delta \theta^\beta = const \times R_0^{-1} \times \sum_{i=1}^{N} F_i i^{-\beta} \quad (4\text{-}61)$$

Choosing the constant so that $\Delta \theta = 5$ for $\beta = 2.5$, $M_o = 1.0$, and no interleaving, Figure 4-14 shows the $\Delta \theta$ required as a function of the modulation index. As can be seen by comparing with Figure 4-13, carrier interleaving provides an improvement in spacing of about 2:1 for all but very small indices. This is entirely expected from the results previously obtained for polarization isolation since the corresponding expressions are functionally identical, and the inter-

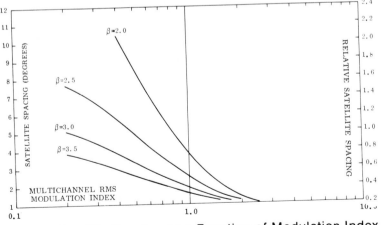

Figure 4-14 Satellite Spacing as a Function of Modulation Index for FDM/FM Transmissions: Interleaved Characters

leaving improvement is large enough even at $M_0 = 0.2$ (≈ 11 dB for F_o) to achieve nearly the asymptotically possible ($F_o \rightarrow 0.0$) improvement.

4.10 Combined Polarization/Carrier Interleaving and Satellite Spacing

Here we look into the potential benefits to be gained by combining polarization isolation and carrier interleaving. Consider Figure 4-15(a) which shows a possible and not uncommon frequency/polarization plan on a satellite. In it, a sequence of channels or transponders (*1, 3, 5, . . ., N-1*) have the same polarization and are separated in frequency as shown. Then, a sequence of oppositely polarized channels (*2, 4, 6, . . ., N*) is interleaved in frequency. We refer to this arrangement as Plan A. Another type of plan possible is just the polarization complement of the preceding one, which we shall refer to as Plan B and is illustrated in Figure 4-15(b).

Consider now a homogeneous environment and suppose that all networks use the same frequency/polarization plan (A or B). For a typical carrier there will be interference generated by the co-polarized co-frequency carriers on other satellites and by the oppositely polarized, interleaved carriers both from other satellites and

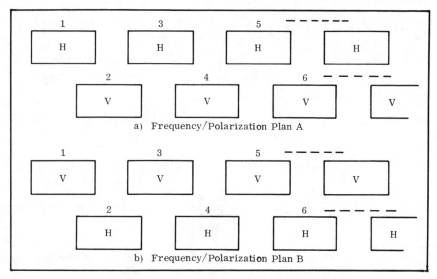

Figure 4-15 Illustration of Possible Frequency/Polarization Plans

from its own satellite. Thus, using the same notation as in Sections 4.2 and 4.6, we find the following types of contributions to the baseband signal-to-interference ratio*

$(S/N)_1 = (y_t F)^{-1} R_0$; opposite-polarization; interleaved carriers; same satellite

$(S/N)_1 = \dfrac{G_r R_0 (\Delta\theta)^\beta |i|^\beta}{q\, A F y_i}$; opposite-polarization; interleaved carriers; ith satellite

$(S/N)_I = \dfrac{G_r R_0 (\Delta\theta)^\beta |i|^\beta}{q\, A}$; co-polarized; co-channel; ith satellite.

In these formulas we have set

$$q = 1 + (\lambda_u / \lambda_d)^2$$

for simplicity, which implies that y_i is the same for uplink and downlink. This restriction could easily be lifted, as shown following (4-45), but we ignore this refinement here. The symbol F accounts for the carrier interleaving advantage against both left and right overlapping channels; as noted previously, this will attain its optimum value when the overlap from these channels is symmetrical.

Thus, the baseband signal-to-interference ratio for Plan A or B is given by

$$(S/N)_1^{-1} = R_0^{-1} y_t F + \frac{q\, A}{G_r R_0 (\Delta\theta)^\beta} \left\{ \sum_{\forall i} |i|^{-\beta} + F \sum_{\forall i} y_i |i|^{-\beta} \right\} \qquad (4\text{-}62)$$

Now suppose that frequency Plans A and B are alternated along the orbit. The following contributions to the baseband signal-to-interference ratio can be identified:

$(S/N)_1 = (y_t F)^{-1} R_0$; opposite polarization; interleaved carriers; same satellite

$(S/N)_1 = \dfrac{G_r R_0 (\Delta\theta)^\beta |i|^\beta}{q\, A}$; co-channel; co-polarized; ith satellite, i even

$(S/N)_1 = \dfrac{G_r R_0 (\Delta\theta)^\beta |i|^\beta}{q\, A F y_i}$; opposite polarization; interleaved carriers; ith satellite, i even

---

*Note that we are obliged to deal here with demodulated quantities because the effect of carrier interleaving is not manifested until the signal is at baseband; it should also be clear that our present discussion implies analog modulation.

$$(S/N)_I = \frac{G_r\,R_0\,(\Delta\theta)^\beta|i|^\beta}{q\,Ay_i}\,; \text{opposite polarization; co-channel;} i\text{th satellite, } i \text{ even}$$

$$(S/N)_I = \frac{G_r\,R_0\,(\Delta\theta)^\beta|i|^\beta}{q\,A\,F}\,; \text{co-polarized; interleaved carriers; } i\text{th satellite, } i \text{ odd.}$$

The total $(S/N)_I$ is therefore given by

$$(S/N)_I^{-1} = R_0^{-1}\,y_t\,F + \frac{q\,A}{G_r\,R_0\,(\Delta\theta)^\beta}\left\{ \Sigma\,(1 + Fy_i)|i|^{-\beta} \right. \tag{4-63}$$

$$= \sum_{i\ odd}\,(F + y_i)|i|^{-\beta}$$

We now compare these frequency/polarization plans to the case discussed in Section 4.6, where dual polarization is used without carrier interleaving. This case is characterized by (4-50) where we had assumed the polarization isolation between two satellites, y_p, was independent of satellite spacing. For simplicity we continue this assumption here, i.e., $y_i = y_p, \forall_i$. With this assumption, (4-62) reduces to

$$(C/I)_e^{-1}\,\triangleq\,(S/N)_I R_0 = y_t F + \frac{q\,AZ(\beta, N)(1 + y_p F)}{G_r\,(\Delta\theta)^\beta} \tag{4-64}$$

Note that the RHS can be regarded as an equivalent interference-to-carrier ratio, $(C/I)_e^{-1}$, with respect to the co-channel case. Hence, to compare this situation to (4-50) we need to set (4-64) equal to the same $(C/I)_{req}^{-1}$ as in the former equation. Invoking (4-49) and repeating the steps leading to (4-50) yields the rearrangement of (4-64):

$$(C/I)_{req}^{-1}\left\{ 1 - \frac{Z(\beta, N)(1 + y_p F)\,5^{2.5}}{Z(2.5, N)(\Delta\theta)^\beta} \right\} = y_t F \tag{4-65}$$

By comparing with (4-50) we see that asymptotically $(y_p = y_t = 0$, or $F = 0)$ no advantage will accrue from Plan A or B with respect to the co-channel case. However, these plans are advantageous in practice because for desired isolations $y_p F$ and $y_t F$ it is clear that any amount of interleaving improvement $(F < 1)$ will lighten the requirements on polarization discrimination. The advantage due to using frequency/polarization Plan A or B can therefore be deduced from Figure 4-9 if instead of $Y_p\,(= 10\,\log y_p^{-1})$ the abscissa is read as $Y_p + 10\,\log F^{-1}$, and the parameter $Y_t\,(= 10\,\log y_p^{-1})$ is interpreted as $Y_t + 10\,\log F^{-1}$.

To examine the improvement available from the alternating frequency/polarization plan (Plan AB) characterized by (4-63) it is more straightforward to compare it directly to Plan A or B rather than to the co-channel case. Thus, we set (4-62) equal to (4-63), still assuming $y_i = y_p$, and the result is

$$\left(\frac{\Delta\theta_{AB}}{\Delta\theta_{AA}}\right)^{\beta} = \frac{(1+Fy_p)\sum_{i\ even}|i|^{-\beta} + (F+y_p)\sum_{i\ odd}|i|^{-\beta}}{(1+Fy_p)\,Z(\beta,N)} \tag{4-66}$$

where $\Delta\theta_{AB}$ and $\Delta\theta_{AA}$ are the spacings associated with Plan AB and Plan A, respectively. Equation (4-66) has been evaluated for $\beta = 2.5$ and several values of F (in dB) as a function of Y_p. The result is shown in Figure 4-16, which indicates that significant spacing reduction is potentially available depending upon the interleaving improvement that may be available. It might be noticed that the result is independent of the intra-network isolation, and this is because the self-generated interference within a network is the same irrespective of the plan on other satellites. Of course, the realization of a frequency/polarization plan such as Plan AB implies a high degree of coordination or cooperation among diverse networks, which might be difficult in practice, and the improvement obtainable thereby also depends upon a high degree of homogeneity among the various carriers, which might be equally difficult to expect.

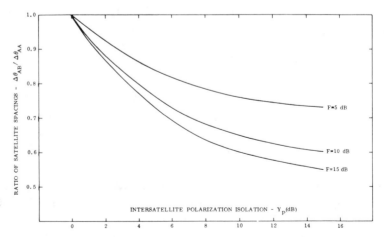

Figure 4-16 Comparison of Satellite Spacing for Two Methods of Carrier Interleaving/Polarization Reuse

4.11 Homogeneous Orbit Utilization and Modulation Characteristics

To this point we have examined basically only a single dimension of the orbit utilization problem, namely the dependence of satellite spacing on various parameters. The satellite spacing obviously establishes the number of satellites that can be placed within a given arc, and therefore determines whether or not all of the potential demand for satellite locations can be accommodated. The issue, however, is larger than this. Of more immediate concern to individual network operators is the optimization of the rate of information transmission (the "capacity") at reasonable cost, the latter being closely bound up with the implementation parameters which, in turn, impact the spacing. An orbit utilization measure (the "orbit capacity") which combines both of the aspects discussed, and which immediately suggests itself, is the information rate per unit bandwidth and unit arc. While this is not the most general such measure, as will be seen in Section 5.11, it is intuitively pleasing and is quite satisfactory for our purposes.

Maximizing the orbit capacity does not necessarily either minimize satellite spacing nor maximize network capacity. It is clear, nevertheless, that it is a goal to be generally encouraged even if subject to certain constraints. The interrelationship among orbit capacity, satellite spacing, network capacity, and implementation constraints, is intimately dependent on the modulation characteristics, by which we mean not only the modulation method itself, but the nature of the modulating signal and any processing that it may encounter prior to modulation and after demodulation. Examples of this signal processing are pre-emphasis/de-emphasis for analog signals, (forward) error-correction coding (FEC) for digital signals, and energy dispersal. Evidently, myriad possible modulation characteristics exist. In the next two sections we shall study two of these possibilities in some detail, and the development therein should provide useful guidance on the methodology to be used in arbitrary cases. Section 4.12 will take up FDM/FM telephony signals, while in Section 4.13 we shall address digital signals with emphasis on coherent phase-shift-keying (CPSK). As in earlier sections, in order to expose the complex interrelationships in question, we need the structured environment of the homogeneous model, which we shall assume to be still in effect. Furthermore, to simplify the notation, without essential loss, we shall assume the co-channel co-polarized conditions.

4.12 Homogeneous Orbit Utilization for FDM/FM Telephony Signals

Before proceeding to the problem at hand, we set down a few additional relationships and discuss certain considerations necessary for the sequel. We defined earlier a thermal noise SNR as

$$(S/N)_t = (C/N) R_t \tag{4-67}$$

The RTC for thermal noise, R_t, will be shown in Appendix A to have the form

$$R_t = 380 M_0^2 (\propto M_0 + 1),\ n \geq 240 \tag{4-68a}$$

$$R_t = 15 n^{0.6} M_0^2 (\propto M_0 + 1),\ 12 \leq n < 240 \tag{4-68b}$$

where

n = number of channels per carrier;

$\propto^2$ = peak-to-average power ratio in the baseband (generally a function of n); and, as before,

M_o = multichannel rms modulation index.

Equations (4-68) also include, as is customary, a 4dB pre-emphasis advantage as well as a 2.5 dB psophometric weighting factor, and were, further, based upon Carson's rule for FM bandwidth, namely

$$B = 8400 n (\alpha M_o + 1) \tag{4-69}$$

In this context several (obviously interrelated) quality measures are used, as convenience or the specific application dictate. Perhaps most common for our immediate problem is the specification of baseband noise (in $pWOp$), N, which is related to the SNR through

$$N = 10^9/(S/N) \tag{4-70}$$

which applies irrespective of the type of noise, assuming of course that the RTC has been so defined in each case to make N subjectively equivalent. Thus, the baseband noise thermal contribution is

$$N_t = 10^9/(S/N)_t \tag{4-71a}$$

For the interference we may therefore immediately write for the baseband noise contribution

$$N_I = 10^9/(S/N)_I \tag{4-71b}$$

where

$$(S/N)_I = (C/I) R_o \tag{4-72}$$

and, as will be shown subsequently, the common RTC for interference is given by

$$R_o = 76 \ (1 + 9.5 \ M_o^3) \ , \ n \geq 240 \tag{4-73a}$$

$$R_o = 3n^{0.6} \ (1 + 9.5 \ M_o^3), \ 12 \leq n < 240 \tag{4-73b}$$

As with (4-68), equations (4-73) also include 4dB pre-emphasis advantage and 2.5 dB psophometric weighting.

A constraint which must be considered basic in any trade-off study is the provision of a standard level of signal quality. Thus, in any telephone channel, the total baseband noise

$$N_T = N_t + N_I + N_a \tag{4-74}$$

must adhere to a minimum requirement. The standard of performance most widely used is that established by the CCIR [9], namely that N_T shall not exceed the following values:

1. 10,000 pW psophometrically-weighted mean power in any hour;

2. 10,000 pW psophometrically-weighted one-minute mean power for more that 20% of any month;

3. 50,000 pW psophometrically-weighted one-minute mean power for more than 0.3% of any month;

4. 1,000,000 pW unweighted (with an integrating time of 5 ms) for more than 0.03% of any month.

Since a given of the homogeneous model is that clear weather conditions apply, the signal quality will not vary with time. Hence the 10,000 $pWOp$ specification is controlling.*

In (4-74) we have introduced a new component, N_a, of the total baseband noise, which is intended to account for any additional sources of noise beyond the two main components that we have

---

*The abbreviation $pWOp$ stands for *picoWatts* at a point of O relative level, *psophometrically* weighted.

identified. There are in fact several such noise sources. One large contributor, up to $10,000\,pWOp$ (long-term) has been set aside by the CCIR [10] for interference from radio-relay (terrestrial) transmitters. There are several individually identifiable equipment and design related noise sources, for example imperfections in the modulator, non-ideal passband transfer functions, intermodulation caused by satellite nonlinearities, and so on. Some of these do have a role in the orbit utilization trade-offs in that their magnitudes are not independent of other parameters that enter into these trade-offs. For example, the bandwidth utilization affects the filtering (hence, the associated noise) as well as the orbit capacity; the intermodulation noise is dependent on the power amplifier's operating point, which determines the required eirp. A detailed accounting of these interrelationships would bring us to a level of specificity in system design that is not intended here; nor is it really necessary. We can consider the design to be a given and hence N_a to be fixed at some value. Thus, the "total" noise available N_T' for the trade between thermal and interference noise is simply

$$N_T' = N_T - N_a = N_t + N_I \qquad (4\text{-}75)$$

which allows us to proceed in straightforward fashion.

In general, N_T or N_T' are functions of the position of the channel within the baseband. The worst-located channel is constraining since the quality standard must be maintained for any channel (some, of course, may be better). This has already been taken care of in (4-68) and (4-73) which have been derived for the worst channel.

We need now to address one more auxiliary consideration, namely that of threshold. The performance equations of this section are valid only above the threshold, T, which is a minimum carrier-to-noise ratio. In an interference plus noise environment, it is not clear what the correct value of T is, but we shall assume it to be invariant to the ratio of noise to interference. This assumption is convenient and not critical since threshold has a somewhat fuzzy definition, and our use of it is merely to identify the range of valid solutions. These solutions satisfy

$$C/NPI = [(C/N)^{-1} + (C/I)^{-1}]^{-1} \geq T \qquad (4\text{-}76)$$

where C/NPI is measured in the predetection bandwidth, and for T we adopt the definition [11]

$$T = (250\,\alpha M_o)^{1/3} \tag{4-77}$$

which applies for a conventional discriminator. For threshold extension demodulators, T is lower than that given by (4-77), depending exactly upon the design. The net conclusion drawn later, however, is that for most cases of interest the threshold will not be the limiting condition.

4.12.1 Orbit Capacity as a Function of Network Capacity

The natural unit of information (rate) in this context is a telephone channel. Thus, the orbit capacity $\ddot{n}$ can for now be defined as

$$\ddot{n} = \frac{\dot{n}}{(\Delta\theta)} \tag{4-78}$$

where $n \triangleq n(B_r)$ is the number of telephone channels in a bandwidth equal to B_r, i.e., n is by definition the (normalized) network capacity with respect to a reference bandwidth, B_r. We wish to investigate $\ddot{n}$ as a function of n for fixed link performance N_T' and given implementation complexity which we take here to mean C/N.

From Equation (4-30),

$$(C/I)^{-1} = [1 + (\lambda_u/\lambda_d)^2]\,\frac{AZ(\beta, N)}{G_r\,(\Delta\theta)^\beta}$$

and combining this with previous expressions, we obtain

$$\frac{1}{\Delta\theta} = \left\{ \left(\frac{N_T'}{10^9} - \frac{1}{(C/N)R_t} \right)\,\frac{R_o G_r}{AZ(\beta, N)[1 + (\lambda_u/\lambda_d)^2]} \right\}^{1/\beta} \tag{4-79}$$

which is implicitly a function of M_o through (4-68) and (4-73).

For the current discussion it is convenient to take the reference bandwidth, B_r, as 1 MHz. The network (satellite) capacity is then very simply obtained from (4-69) as

$$n = 119/(\alpha M_o + 1)\ channels/MHz \tag{4-80}$$

Hence,

$$\ddot{n} = \frac{119}{(\alpha M_0 + 1)\,\Delta\theta}\ channels/degree/MHz \tag{4-81}$$

with $\Delta\theta$ from (4-79).

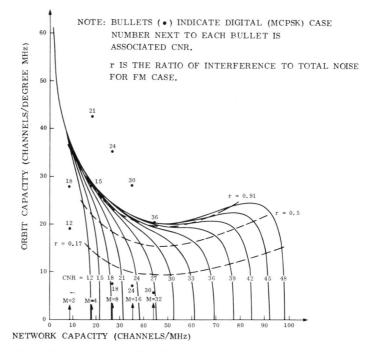

Figure 4-17 Orbit Capacity for FDM/FM Telephony Signals

Equation (4-81) has been plotted in Figure 4-17* for $D/\lambda = 300$ and $\eta = 0.5$ [see (4-25)], $N_T' = 7500$, $(\lambda_u/\lambda_d)^2 = 0$, $\beta = 2.5$, and $n \geq 240$ channels. The value of $\bar{n}$ is affected by these particular assumptions but the general character of the curves is not. One can see that generally orbit capacity increases with decreasing network capacity (or increasing modulation index) indicating that this loss of individual capacity is more than compensated by the ability to place more satellites in orbit. However, there is obviously a point of diminishing economic sense so that the modulation index cannot be increased indefinitely, even if threshold were not limiting.† We see a somewhat curious behavior in the curves of Figure 4-17 as the CNR becomes large, namely that the orbit capacity becomes double-valued. We also see as the CNR attains large values that the curves tend to flatten out over a fairly broad range of $\bar{n}$ so that, in this range, there is some flexibility in choosing the satellite capacity without compromising orbit capacity. Of course, there is then a trade-off between

---

*Also indicated on Figure 4-17 are "bullets" which represent the results for digital telephony, which will be derived in Section 4.13.2. These results can be ignored for the moment.

satellite capacity and the number of satellites. We may further observe from the figure that as the CNR decreases so does the possible range for $\dot{n}$. This is explained by the fact that for given CNR there is an $\dot{n}$ for which the thermal noise approaches the total noise allowance N_T', at which point there can be no interference noise which, in turn, requires "infinite" spacing. This explanation accounts for the precipitous drop to zero for each CNR at some $\dot{n}$.

4.12.2 Orbit Capacity as a Function of Interference Noise Allowance [12]

Figure 4-17 shows loci of constant $N_I/N_T' = r$. The parameter r is very important in two respects. On the one hand, it is the only non-implementation-specific quantity available to the regulator for insuring reasonably efficient orbit utilization; this can be seen from Fig. 4-17 by noticing that a given value of r implies a lower bound on orbit capacity, the bound increasing with r. On the other hand, r wholly embodies the (economic) penalty that must be paid by network designers in order for any sharing to take place, the penalty obviously increasing with r. Thus, perhaps even more interesting than the variation of orbit capacity with $\dot{n}$ is its functional dependence on r. The objective is to see whether an optimum with respect to r exists or if close to optimum orbit capacity is achievable for values of r small enough to be acceptable to system designers. It should be noted that the CCIR has recommended [13] a value for N_I of 2000 $pWOp$ for non-frequency re-use satellites and 1500 $pWOp$ for those that employ frequency re-use; this amounts to specifying N_I/N_T which, in regulatory terms, is more sensible since N_T' is rather implementation specific. However, for given N_T' the relation between the two ratios follows trivially.

We shall generalize somewhat here over the previous section in that we shall assume the occupied bandwidth B is not necessarily as large as the allocated bandwidth W. Thus, denoting $(C/N)_o$ the CNR referred to W, we have

---

†One should also note that since the assumption $n \geq 240$ channels was made in obtaining this figure, any actual bandwidth should be consistent with this assumption. For example, $\dot{n} = 10\ channels\ /MHz$ implies a bandwidth of at least 24 MHz. Conversely, starting with a fixed bandwidth allocation, one could obtain a figure similar to 4-17 without any restriction on n, but for any given modulation index one would first have to make a test on n and use the appropriate equation for R_t and R_o depending on whether $n \geq 240$ or $n < 240$.

$$C/N = (C/N)_o\,(W/B)$$

We shall also find it convenient to write

$$C/I = K\,(\Delta\theta)^\beta \tag{4-82}$$

where $K = K\,(G_r,\,A,\,N,\,\beta,\,\lambda_u/\lambda_d)$ is obviously identified by matching terms in (4-30), and varies from 10 to 100 for the expected range of the parameters.

Defining $r = N_I/N_T'$, whence $(1\text{-}r) = N_t/N_T'$ we find that the constraint on N_T' can be written as the pair of equations

$$1\text{-}r = \frac{10^9/N_T'}{(C/N)_0(W/B)R_t} \tag{4-83a}$$

$$r = \frac{10^9/N_T'}{K(\Delta\theta)^\beta\,R_0} \tag{4-83b}$$

Our objective is to study $\ddot{n}$, still given by (4-78), as a function of r subject to the constraints (4-83). This is done through a straightforward computational algorithm, as follows:

(1) Fix $(C/N)_o$, W, W/B, N_T', K, β, α;

(2) Select a value for M_o;

(3) Compute n through (4-69);

(4) Compute $R_t\,(M_o,\,n)$, $R_o\,(M_o,\,n)$ from (4-68) and (4-73);

(5) Compute r through (4-83a): check that $(1\text{-}r) \leq 1$;

(6) For the r computed in step 5, find $\Delta\theta$ from (4-83b);

(7) From steps 3 and 6, compute $\ddot{n}$ from (4-78);*

(8) Compute (C/I) from step 6 and (4-82);

(9) Compute C/NPI from (4-76) and check threshold;

(10) Cycle through M_o to span an appropriate range of r;

(11) Repeat for a different set of parameters in step 1.

Using the above procedure we have computed a number of cases (for $N_T' = 7500$), the results of which are plotted in Figures 4-18 to 4-22.**

---

*For the purpose of this exercise, we have chosen $B_r = B$, the occupied bandwidth, so that the network capacity is not normalized here with respect to spectrum utilization.

**Strictly speaking the algorithm should contain an iterative loop for computing n because the parameter α in (4-69) is a function of n for $n<240$. However, we ignore this refinement here, particularly since for the cases illustrated it turns out that $n\geq240$, i.e., α is constant.

Figure 4-18 shows orbit capacity as a function of r, with $(C/N)_o$ in dB as a parameter and the CCIR antenna sidelobe decay exponent, $\beta = 2.5$. For this figure, as the others following, the parameter K was taken equal to 50; this corresponds approximately to $(D/\lambda) = 150$ for the earth station antenna. An allocated band, W, of 40 MHz was assumed, and the occupied bandwidth B was taken to be the same. Also shown on the figure, for each $(C/N)_o$, are the values of r corresponding to a satellite spacing of 1 degree. There are reasons, discussed subsequently, to consider solutions valid, or realistic, only for satellite spacings ≥ 1 degree; the corresponding range of permissible values of r lie below the triangular 1 degree markers.

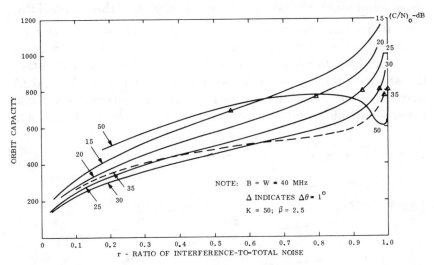

Figure 4-18 Orbit Capacity for FDM/FM Systems

The nature of the curves of Figure 4-18 amply demonstrates the analytical untractability of the plotted relations. As can be seen, there is no value of r which yields maximum capacity. As $(C/N)_o$ increases, the behavior becomes more idiosyncratic, producing at $(C/N)_o = 50\,dB$, for example, a local maximum and a local minimum. It is interesting to note, however, that within the constraint $\Delta\theta \geq 1^o$, the maximum available capacity does not vary much with $(C/N)_o$.

The orbit capacity generally increases with r, except as noted. There is a trade to be made, here, between orbit capacity, $(C/N)_o$, and the individual satellite capacity, which is shown in Figure 4-19. The trend here, versus r, is counter to that for the orbit capacity. Hence a judicious choice must be made to balance the two types of requirements.

We now inquire whether it may be advantageous to reduce the occupied bandwidth to less than the allocated bandwidth, the idea being that the resulting higher C/N will permit closer satellite spacings. However, as indicated in Figure 4-20, it turns out that the loss in individual satellite capacity cannot be offset by the possible decrease in satellite spacing. Hence, if the power is available to use the allocated bandwidth, the best strategy is to use all of that bandwidth.

Now, if there is not enough power to use the allocated bandwidth, the so-called power-limited case, some bandwidth reduction must be accepted, both to permit some level of interference and to satisfy threshold. Figure 4-21 illustrates one case, in which $(C/N)_o = 10\,dB$. Recalling that this is the carrier-to-noise ratio referred to the allocated bandwidth, and assuming threshold $T = 10\,dB$, Figure 4-21

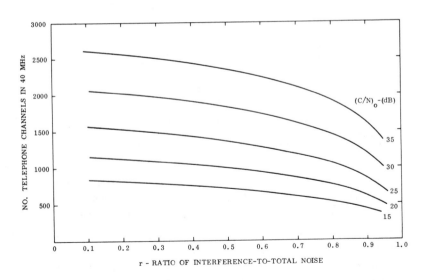

Figure 4-19 Number of Telephone Channels per Transponder as a Function of Interference Level

shows the region of operation that meets this condition is to the left of the dashed curve. There is thus a definite *r* which maximizes orbit capacity; however, the optimum is fairly broad and is fortunately biased towards values of *r* which also tend to maximize satellite capacity. Notice, in this instance, that the condition $\Delta\theta \geq 1^{\circ}$ generally supersedes the threshold constraint.

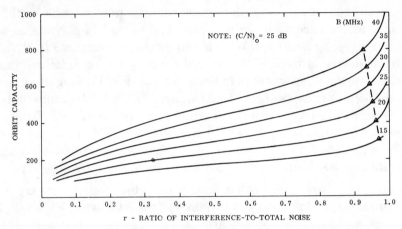

Figure 4-20 Orbit Capacity as a Function of Occupied Bandwidth

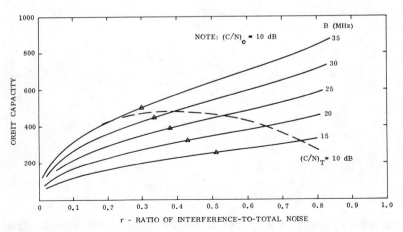

Figure 4-21 Orbit Capacity for Power-Limited FDM/FM Systems

The condition $\Delta\theta\geq1^\circ$ is intended only as an approximate indicator of the region where the solutions to the problems discussed are meaningful. There is, in any case, a lower limit in that vicinity. Three factors combine to impose this limit. First, the reference antenna pattern is defined only for $\Delta\theta\geq1^\circ$. Although, as discussed in Section 4.4, the pattern could be meaningfully extended into somewhat smaller angles, depending upon D/λ, there is not much point in doing so because these smaller angles correspond to a range of r that would not be acceptable in practice. The second factor relates to station-keeping accuracy. With a mandated accuracy of $\pm 0.1^\circ$, the orbit capacity is negligibly reduced for $\Delta\theta>1^\circ$; for smaller separations, however, as was seen in Section 4.7 station-keeping tolerance might begin to have a significant impact and alter the nature of the orbit capacity curves, actually starting to bring them down. Finally, there is the question of threshold. Since $(C/I) = K (\Delta\theta)^\beta$, we see that $C/I = K$ for $\Delta\theta = 1$. For $D/\lambda = 100$, $K \approx 13$, hence $C/I \approx 10 \, dB$, which is about the lower limit of threshold for conventional FM receivers.

Finally, we also observe the influence on orbit capacity of varying the antenna sidelobe decay exponent, β. This is shown in Figure 4-22 for one particular set of conditions. As expected, improved orbit capacity accompanies increasing β. Note that all the curves merge at $\Delta\theta = 1$. This could have been predicted from equation (4-82) for constant K, which we have assumed in the calculations. This represents a slight artificiality in the sense that if all the physical param-

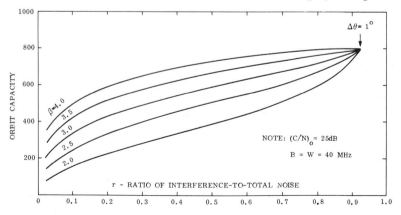

Figure 4-22 Effect of Antenna Sidelobe Decay Exponent on Orbit Capacity

eters were fixed, then K could not be constant because $Z(\beta, N)$, which it contains, must vary with β. As previously noted, however, this dependence is weak. If we had incorporated this refinement the nature of Figure 4-22 would remain substantially the same but the curves would not identically coincide at $\Delta\theta = 1$. The reversal of the curves for $\Delta\theta < 1^\circ$ is an artifact of the reference antenna pattern which we alluded to in section 4.4, and underscores the reasons cited for considering the solutions valid only for $\Delta\theta \geq 1^\circ$.

4.13 Homogeneous Orbit Utilization for Digital Signals with Emphasis on Coherent Phase-Shift Keying (CPSK)

Paralleling the development of the previous section we shall study the relationship between orbit capacity and network capacity and between orbit capacity and the (relative) interference noise allowance. The analytical approach, however, is necessarily quite different. Digital signals are inherently difficult to deal with because the performance measure of interest — invariably for our purposes the probability of bit error (or BER) — is a sensitive nonlinear function of the entire waveform present at the input to the detector. One immediate consequence of this is that, unlike analog signals, the output quality cannot be decomposed into interference and thermal noise contributions. In fact, this is true for all sources of degradation, i.e., the final performance is a complex function of their combined presence. This means that we cannot attribute a unique effect due to the presence of interference, independent of the system implementation. This implies, further, that orbit capacity calculations will depend on the assumptions made regarding the implementation. In order to obtain as clear a picture as possible of the basic constraints imposed by interference we would therefore like to remove the implementation dependence from our considerations. This we can do simply by assuming the transmission link to be essentially distortion-free.* However, we shall make this assumption somewhat

---

*In effect we have already made that assumption for the FDM/FM systems. To make the comparison with these systems equitable, it would seem that the same assumption should be made for digital systems. It should be realized, however, that this assumption does not represent an equal degree of realism for both cases. In part of our work here, in Section 4.13.3, we will not find it necessary to always assume an ideal system. In Chapters 6 and 7 we shall consider explicitly some of the implementation-induced effects.

self-correcting by associating a bandwidth occupancy consistent with it, which turns out to be larger than is customarily used. Furthermore, as will be seen shortly, the orbit capacity results can be easily adjusted for any desired additional degradation measured in terms of increased required power for the same BER.

The relationship between the BER and the noise and interference levels depends on the modulation technique as well as the detection mechanism and, further, on the type of coding that may be used. When it becomes necessary to make specific assumptions here, it will be sufficient for illustrative purposes to confine ourselves to multilevel (M-ary) coherent phase-shift-keying (CPSK or MCPSK), which constitutes perhaps the most important class of digital signaling schemes. Maximum-likelihood detection will be assumed, and coding will not be used except when specifically indicated. The effect of coding will be considered in section 4.13.4.

The remainder of this section is organized as follows. In 4.13.1, we compute orbit capacity versus network capacity for MCPSK systems using the approximate approach of Section 4.1.4. This simplification will make clearer the methodology and is adequate in many cases. In 4.13.2, we will consider the situation where multichannel telephony is transmitted by digital means and attempt thereby to make a one-to-one comparison of the orbit capacity attainable by FM and digital transmission. In section 4.13.3, we examine the behavior of orbit capacity as a function of the interference allowance and in section 4.13.4 we obtain estimates of the optimum capacity as a function of power, with and without coding.

4.13.1 Orbit Capacity vs Network Capacity (CPSK Signals)

In section 4.1.4 the approximate expression

$$P_e = f\,(C/NPI)$$

was given for the error probability, where

$$(C/NPI)^{-1} = (C/N)^{-1} + \delta^{-1}\,(C/I)^{-1}$$

and δ was defined as a noise-equivalent factor, $\delta = 1$ corresponding to complete equivalence between noise and interference. We wish now to convert $f\,(\cdot)$ to an explicit form.

For MCPSK signals we start out with the approximate formula for the symbol (or word or character) error probability, P_M, [14]

$$P_M \cong \varepsilon_M \, erfc \, [\sqrt{E_r} \, sin\frac{\pi}{M}] \qquad (4\text{-}84)$$

where $\varepsilon_M = 0.5$ for $M = 2$ and $\varepsilon_M = 1.0$ for $M > 2$; $erfc \, (\cdot)$ is the complementary error function, and E_r the signal symbol energy to noise density ratio is given by

$$E_r = PT/N_o \qquad (4\text{-}85)$$

where

P = *signal average power,*

T = *symbol duration,*

N_o = *noise density at the receiver.*

Equation (4-84) is exact for $M = 2$ and a tight approximation for $M > 2$ when the error probability is in the range of common interest, i.e., $< 10^{-3}$. It applies to maximum-likelihood detection in a white Gaussian noise (WGN) channel, i.e., an ideal one.

To establish a common basis for comparing the performance for various M, we shall express such performance in terms of the probability of bit error, which, under the same conditions that validate the approximation (4-84), can be approximated with an equal degree of accuracy by

$$P_e \triangleq P_2 = P_M/log_2 \, M = P_M/m \qquad (4\text{-}86)$$

where $m = log_2 \, M$ is the number of bits per symbol and will always be taken as an integer.

We now need to relate the occupied bandwidth to the symbol rate $R_s = (1/T)$. We shall use the conservative specification

$$B = 2R_S = 2R_b/m \qquad (4\text{-}87)$$

where R_b is the bit rate. In practical systems one frequently sees smaller B/R_s ratios, typically on the order of or less than $3/2$ but as mentioned earlier, (4-87) is in a sense required for consistency with the WGN assumption. Hence we have

$$E_r = \frac{PT}{N_o} = \frac{P}{N_o B}BT = 2(C/N)$$

Extending the interpretation of *(C/N)* to *(C/NPI)*, one obtains

$$P_e \approx \frac{1}{m}\, \varepsilon_M\, erfc\, [\sqrt{2(C/NPI)}\, sin\, \frac{\pi}{M}] \qquad (4\text{-}88)$$

For fixed *(C/N)* and *M*, $\Delta\theta$ is an implicit function of P_e; say, $\Delta\theta = g(P_e)$.
Also, the bit rate per MHz is obtainable from (4-87) as

$$\dot{R}_b = 5\, m \times 10^5\, bps/MHz \qquad (4\text{-}89)$$

The orbit capacity is therefore

$$\ddot{R}_b = \frac{5_m \times 10^5}{\Delta\theta}\, bps/degree/MHz \qquad (4\text{-}90)$$

For $P_e = 10^{-6}$, we have obtained (4-90) for $\beta = 2.5$ and for $M = 2,4,8,16, 32$
(m = 1,2,3,4,5), and the results are shown in Figures 4-23 and 4-24

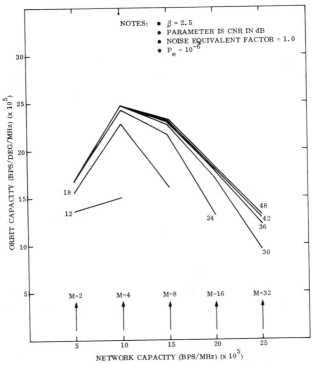

Figure 4-23 Orbit Capacity for MCPSK Signals (noise equivalence
factor = 1.0)

with the carrier-to-noise ratio (in dB) as a parameter. Of course, since *M* is discrete the results apply only to the values indicated, but the points have been joined into continuous curves for visual clarity. In order to obtain some appreciation of the sensitivity of the results to the assumptions characterizing the effect of interference, two different values of the noise-equivalent factor δ were assumed. Figure 4-23 corresponds to $\delta = 1$, which equates interference to thermal noise. For Figure 4-24, $\delta = 2$ was assumed, which implies that interference is half as damaging as an equal (power) level of thermal noise. This reduced sensitivity to interference yields orbit capacity which is about 25% higher than for the case $\delta = 1$.

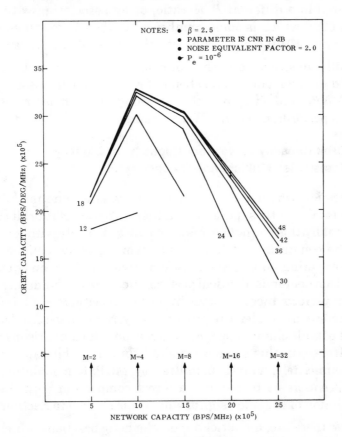

Figure 4-24 Orbit Capacity for MCPSK Signals (noise equivalence factor = 2.0)

The actual effect of interference relative to that of thermal noise depends on the specifics of the case, as remarked earlier, and will be further studied in Chapter 7. However, irrespective of the assumption on δ, it is interesting to note that orbit capacity shows a definite maximum as a function of network capacity, unlike the FM case of the preceding section. Furthermore, the maximum (or near-maximum) occurs for practical values of M and CNR thus suggesting the possibility of actually attaining near-optimum orbit capacity in practice.

Figures 4-23 and 4-24 were obtained under somewhat idealized conditions detailed earlier. The figures can be straightforwardly modified for more realistic conditions simply by shifting the abscissa to correspond to a different B/R_s ratio, or by associating with each curve a CNR which is some number of decibels higher (generally depending on M) to account for practical design-induced degradation.

We finally mention that the foregoing analysis can be easily extended to any modulation scheme if we continue the assumption $P_e = f(C/NPI)$ and if we know the function $f(\cdot)$ which describes that scheme's performance in the WGN channel.

4.13.2 Orbit Capacity vs Network Capacity for Digitally Transmitted Multichannel Telephony

One question which naturally arises is whether one modulation method is inherently superior to another with respect to orbit capacity. Normally this cannot be answered unequivocally without specifying the constraints at hand. Furthermore, complications arise when comparing analog and digital methods because an equitable comparison demands identical performance, and the fundamentally different performance measures in the two cases may not be simple to reconcile in a completely equivalent way. Nevertheless, at least in the case of multichannel telephony, a reasonable comparison can be made. Hence, in this section we shall find the orbit capacity for multichannel telephony transmitted by MCPSK modulation and this will enable us to perform a direct comparison with the FM transmission of the same signal which was studied in Section 4.12.

Consider, therefore, a multichannel telephone baseband which has been sampled and quantized to K levels, and transmitted over a digital link with $P_e = p$. If we also assume logarithmic companding, the total baseband noise in a telephone channel (after D/A conversion), psophometrically weighted, is then approximately given by

$$N_T' = \frac{10^9}{18.7} \left\{ 4p + \frac{1}{K^2} \right\} \;, n \geq 240 \; channels \qquad (4\text{-}91a)$$

$$N_T' = \frac{10^9}{0.75 \, n^{0.6}} \left\{ 4p + \frac{1}{K^2} \right\} \;, 12 \leq n < 240 \; channels \qquad (4\text{-}91b)$$

the derivation of which would take us somewhat far afield in the current context, but can be found, for example, in [15]. Equation (4-91a) is plotted in Figure 4-25 as a function of p with $k = log_2 K$ as a parameter. Assuming $N_T' = 7500$, as was assumed in section 4.12, we see that this requirement is met if $k \geq 7$ and $p \leq 3$ x 10^{-5}. We shall therefore assume $k = 7$ and, to provide a little margin, $p = 10^{-5}$. Notice that this is reasonably compatible with CCIR recommendation [16] that the error probability for PCM telephony be 10^{-6}. This would add very little in terms of improved noise performance, as can be seen from Figure 4-25, but would add an extra margin of safety.

Equation (4-91) is not specifically related to the modulation method, but for the value of p required the latter determines the locus of pairs $(C/N, \Delta\theta)$ which satisfy the requirement. An approximate equation interrelating these variables was given in (4-88). Here, in order to make our comparison as accurate as possible, we shall use more

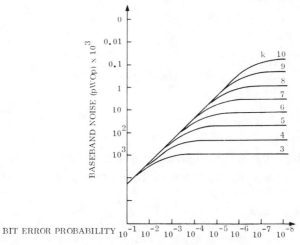

Figure 4-25 Baseband Performance for Digital Telephony

exact results whose derivation will be given in Chapter 7. These results appear (in part) in Figure 4-26 which shows the required satellite spacing for $\beta = 2.5$ as a function of C/N for fixed P_e and for $M = 2,4$.* Curves for $M > 4$ have a similar appearance. Thus, symbolically we can write $\Delta\theta = u\ (C/N)$, for each M and P_e.

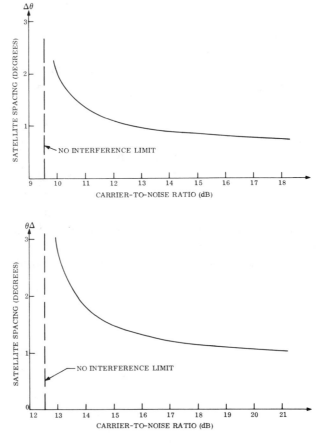

Figure 4-26 Satellite Spacing as a Function of Carrier-to-Noise Ratio in a Homogeneous Environment and an Ideal Channel (for D/λ) = 100).

---

*The computations were actually carried out for $D/\lambda = 100$, but from Section 4.3 they can be easily converted to apply to any other value simply by multiplying by $[(D/\lambda) \ / \ 100]^{2/\beta}$.

Now, assuming 4kHz bandwidth per voice channel, and sampling at twice this frequency, our previous choice of $k = 7$ implies that a telephone channel is equivalent to 56,000 bps. Hence, calling upon (4-89) the network capacity is given by

$$\dot{n} = \dot{R}_b/56 \times 10^3 = 8.93m \; channels/MHz,$$

whence the orbit capacity is

$$\ddot{n} = \frac{8.93m}{u(C/N)} \; channels/degrees/MHz,$$

which has been evaluated for $M = 2,4,8,16,32$. The results have been superimposed as "bullets" on Figure 4-17 in order to facilitate comparison with the FM case.

It can be seen that for given network capacity $(\dot{n})$ and sufficiently high CNR, CPSK offers potentially greater orbit capacity than FM, for all M except binary PSK. This observation should be tempered by the fact that idealized channels and receivers have been assumed in obtaining our results, but this assumption represents a greater departure from actuality for digital than for FM systems. So, a fair comparison would require a CNR adjustment; say, increasing that of the digital system by 2dB or so. Still, for the values of M indicated, potentially superior orbit capacity is available by using PSK. But FM can still surpass PSK if we go to sufficiently small $\dot{n}$, which may not be a practical assumption. In the comparison, however, co-channel operation has been assumed, while it is known that considerable improvement can be had with carrier interleaving for FM systems, but little if any such improvement for PSK. On the other hand, improved performance can be had for both digital and analog telephony by using a number of processing techniques, most of which are either in use or advanced stage of development [see, e.g., 17,18,19]. Examples of these techniques are error-correction coding, syllabic companding, and speech interpolation. Therefore, it is difficult to unequivocally assert the superiority of one form of modulation over another, at least from the orbit utilization standpoint, without carefully specifying all of the conditions attending the comparison. In practice, the choice will depend on the given situation; e.g., C/N constraint, desired network capacity, implementation, and operational considerations.

4.13.3 Orbit Capacity as a Function of Interference Noise Allowance

An interesting perspective of the orbit capacity problem (using digital signals) emerges when we consider its dependence on the permitted amount of interference [12]. We consider this question in the present section, but expand our treatment somewhat to apply to any digital modulation/coding scheme. Furthermore, the formulation will allow for practical degradations and we will inquire into the existence of optimum orbit capacity (as a function of interference level) for both power-limited and bandwidth-limited cases.

As before, we assume that
  a) the performance requirement; i.e., the bit error probability, is to be held fixed at P_e,
and we further assume that
  b) there is an allocated bandwidth, W, not to be exceeded, but which need not be completely occupied.

For a given modulation/coding technique, performance curves will have the general appearance shown in Figure 4-27. (Particular examples of these types of curves will be given in Chapter 7) The abscissa in Figure 4-27, C/N, is the carrier-to-thermal noise ratio,

$$\frac{C}{N} = \frac{P_0}{N_0 B} = \frac{(P/N)_0}{B}$$

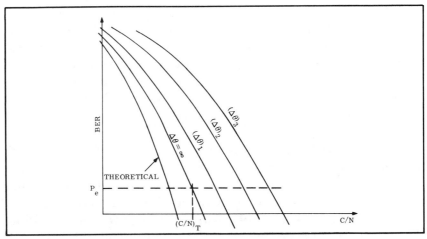

Figure 4-27 Characteristic Behavior of Digital System

where

P_0 = *wanted signal (received power level)*,
N_0 = *effective receiver noise power spectral density*,
B = *bandwidth of the transmission ("occupied" bandwidth)*.

When there is no interference, there is a minimum, or "threshold," C/N required, which we denote $(C/N)_T$, and we note that

$$(C/N)_T = F(P_e)$$

where the function $F(\cdot)$ depends not only upon the specifics of the modulation/coding method, but also upon implementation-induced degradations such as filtering, imperfect synchronization, and other impairments.

For the sequel, it will be useful to distinguish between power-limited and bandwidth-limited modes of operation. These are defined in the following way. First, for given $(P/N)_0$, we define

$$B_u = \frac{(P/N)_0}{(C/N)_T} \tag{4-92}$$

and

$$(C/N)_l = \frac{(P/N)_0}{W} \tag{4-93}$$

If $B_u \leq W$, then we say the network is power-limited, but we still require $(C/N) \geq (C/N)_T$. If $B_u \geq W$, then the network is bandwidth-limited, and of necessity $(C/N) \geq (C/N)_l$. Hence we can write for the occupied bandwidth,

$$B = \frac{(P/N)_0}{C/N}, \ C/N \geq (C/N)_{min}, \tag{4-94}$$

where $(C/N)_{min} = max\ [(C/N)_T, (C/N)_l]$.

In the case of digital signals, as mentioned earlier, we cannot assign part of the BER to any particular cause. We are thus obliged to define "allowable interference" in terms of predetection power levels. Thus, a convenient definition for the permissible interference is

$$r = (C/N)_T/(C/I) \tag{4-95}$$

in which we basically refer the interference level to that of the thermal noise which, by itself, would cause the network to reach its threshold BER. If the interference is assumed to behave like thermal noise (insofar as the BER is concerned), then $0 \le r \le 1$. For arbitrary interference, the upper limit is not necessarily unity. For the remainder of our discussion, however, we shall assume the noise-like interference case. A somewhat more general formulation is given in Appendix E.

The orbit capacity per unit arc is simply

$$\dot{C} = C / \Delta\theta \qquad (4\text{-}96)$$

where C is the individual network capacity (data rate would be a somewhat more correct term). The latter can be expressed in the form

$$\dot{C} = \gamma B \ (bits/sec) \qquad (4\text{-}97)$$

where γ is a constant depending on the modulation/coding method, and upon certain design choices. Generally, γ and $(C/N)_T$ are not independent but for purposes of the moment we need not be concerned about that particular relationship. Note that the orbit capacity, as defined in (4-96), is not normalized with respect to bandwidth because such a normalization could be misleading when the occupied bandwidth is variable.

It will be demonstrated in Appendix E that the capacity can be put in the form

$$\dot{C} = \frac{\gamma K^{1/\beta} (P/N)_0}{(C/N)_T^{1+1/\beta}} (1 - r) \, r^{1/\beta} \qquad (4\text{-}98)$$

where K is the same constant appearing in (4-82). Equation (4-98) is valid under the constraint

$$r \ge 1 - (C/N)_T/(C/N)_{min} \qquad (4\text{-}99)$$

which is obtained by combining (4-94) and (4-95). The optimum orbit capacity with respect to r is obtained by setting $d\dot{C}/dr = 0$. Performing this operation, we find the solution to be

$$\hat{r} = 1/(\beta + 1) \tag{4-100}$$

which exists if (4-99) is satisfied; i.e., if

$$(C/N)_{min} \leq (C/N)_T/(1-\hat{r}) \tag{4-101}$$

It is of interest to study the behavior of $\dot{C}$ as a whole, in order to determine its sensitivity to departures of r from the optimum value. It can be seen that, apart from constants, the essential behavior of $\dot{C}$ is embodied in the function

$$\zeta (r, \beta) = (1-r)r^{1/\beta} \tag{4-102}$$

which is plotted in **Figure 4-28** for $\beta = 2.5$. It can be observed that relatively little capacity loss is experienced for a substantial range of r about $\hat{r}$. It may also be noted that if a network is well into the bandwidth-limited region, that is $(C/N)_{min}$ does not satisfy the inequality (4-101), then $r > \hat{r}$, of necessity. The optimum strategy, then, is to use r satisfying the *RHS* of (4-99) with equality since $\dot{C}$ decreases monotonically to the right of $\hat{r}$. This means that there is nothing to be gained by reducing the bandwidth below W in an attempt to trade thermal noise for interference.

On the other hand, for power-limited systems $(B_u \leq W)$, some reduction in bandwidth is mandatory in order to permit some degree of interference. At the optimum value of r, $\hat{r}$, which always exists for power-limited systems, the bandwidth can be easily shown to be $B = B_u \cdot \beta/(\beta+1)$, that is, reduced by the factor $\beta/(\beta+1)$ from its maximum value. For $\beta = 2.5$, the interference-free available bandwidth is reduced by about 29%. However, reducing the bandwidth further will not produce any orbit capacity improvement.

The results just discussed, on the assumption that interference is noise-like, apply qualitatively more generally, but the quantitative conclusions, which would depend on the specific nature of the interference, should not be significantly different.

Earlier we introduced the notion of power-limited and bandwidth-limited regions. Based on the immediately preceding development, it will be convenient to slightly redefine these regions. From Equation

(4-98) the orbit capacity appears to increase linearly with $(P/N)_o$ for given r. However, this is not true for all values of r because of the constraint (4-99). Basically $\dot{C}$ increases linearly with $(P/N)_0$ in the power-limited region where a linear increase in power implies a linear increase in occupied bandwidth. This region is still defined by the condition $(P/N_o)/B \leq (C/N)_T$. It is useful, however, to introduce the idea of a transition region defined by $(C/N)_T \leq (C/N)_{min} \leq (C/N)_T / (1-\hat{r})$. In this region the bandwidth cannot be increased but the orbit capacity can still increase linearly with $(P/N)_0$ by choosing $r = \hat{r}$. We can thus define the bandwidth-limited region as that for which $(C/N)_{min} \geq (C/N)_T/(1-\hat{r})$, and in this region the minimum possible r increases with $(P/N)_o$. As mentioned earlier, this implies that maximum capacity is obtained by satisfying (4-99) with equality.

4.13.4 Orbit Capacity of Coded MCPSK Systems as a Function of Power Level [12]

The previous development assumed implicitly that we were dealing with an arbitrary but given modulation/coding scheme. It is of interest to study the influence on orbit capacity as a function of the scheme chosen. For this purpose we continue to assume that the effect of intereference is equivalent to that of thermal noise. Thus, Equation (4-98) shall be the focus of our considerations. It will first be useful, however, to re-phrase this equation in terms of the parameters of the code that may be selected. Quite generally, we may characterize a code by two parameters: R_c, the code rate; and G_c, the coding gain at the BER of interest, which is the ratio of the required E_b/N_o without coding to that with coding. Thus, we may express the network data rate

$$C = \eta m R_c B \qquad (4\text{-}103)$$

i.e.,
$$\gamma = \eta m R_c$$

where $B_o = R_c B$ is the "information" bandwidth, or the bandwidth that would be occupied if there were no coding; m is the number of levels transmitted (bits/symbol); and η is a constant depending upon the modulation method and certain related design features. In fact, there is a trade between η and $(C/N)_T$ but this particular optimization is not the one of concern here. We can further write

$$\left(\frac{C}{N}\right)_T = \left(\frac{C}{N}\right)_{T0} \quad \left(\frac{R_c}{G_c}\right) \tag{4-104}$$

where $(C/N)_{T,O}$ is the CNR required for an uncoded system. Thus, we may re-write (4-91) in the form

$$\dot{C} = \frac{\eta m K^{1/\beta} (P/N)_0}{(C/N)_{T,0}^{1+1/\beta}} \frac{G_c^{1+1/\beta}}{R_c^{1/\beta}} (1 - r) r^{1/\beta} \tag{4-105}$$

still subject to the range of solution (4-99), which also depends on (R_c, G_c). Hence, we may study the behavior of $\dot{C}$ as a function of modulation method, characterized by $[m, (C/N)_{T,O}]$, and as a function of the code parameters (R_c, G_c). A note is in order concerning the way in which R_c enters (4-105). On the surface, $\dot{C}$ increases as R_c decreases, which is counter-intuitive. In fact, this is the case in the power-limited regime. There, as R_c decreases, the bandwidth increases for a given data (information) rate. Hence, for a given C/I (fixed $\Delta\theta$) the equivalent noise spectral density N_0' due to the interference decreases. Therefore, for given available E_b, the ratio $E_b/(N_0 + N_0')$ can be maintained at a fixed value by decreasing $\Delta\theta$ at the same time that R_c decreases. This can take place until the bandwidth limitation sets in. At that stage, decreasing R_c further implies decreasing the information rate. There will then be excess power which can be taken advantage of by decreasing the satellite spacing. However, information rate decreases faster than does the spacing, resulting in a net capacity loss. These opposite trends are implicit in (4-105) because, as mentioned above, the region of r for which it holds is also a function of R_c. In fact, as $R_c \rightarrow 0$, one can see from (4-99) that $r \rightarrow 1$ and the term $1-r \rightarrow 0$ as the first power of R_c; the net result is that $\dot{C} \rightarrow 0$.

To obtain some appreciation of the behavior of (4-105), we provide some numerical results for *M-ary* coherent *PSK*, for M = 2,4,8,16, which correspond, respectively to m = 1,2,3,4. It is instructive to display $\dot{C}$ as a function of the available carrier-to-noise ratio, $(C/N)_l$ = $P/N_o W$, in the allocated bandwidth W. For this purpose we take $r = r_o$, namely the value which maximizes $\dot{C}$; it will be recalled that for power-limited systems, r_o = $1/(1 + \beta)$, and for bandwidth limited systems r_o is the value which satisfies (4-99) with equality. Setting $\zeta_o = \zeta(r_o, \beta)$, we get

$$\dot{C}_{max} = \eta K^{1/\beta} W m \quad \frac{(P/N_0 W)}{(C/N)_{T0}^{1+1/\beta}} \quad \frac{G_c^{1+1/\beta}}{R_c^{1/\beta}} \zeta_0 \qquad (4\text{-}106)$$

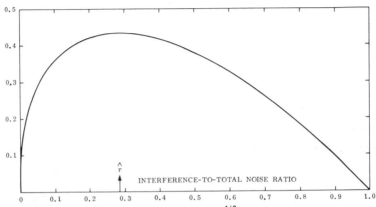

Figure 4-28 The Function $\zeta (r,\beta) = (1-r) r^{1/\beta}$ for $\beta = 2.5$

We can reasonably take η to be the same for all m. Hence, for plotting purposes we take the factor $\eta K^{1/\beta} W$ to be unity for convenience. We also set $\beta = 2.5$ in the calculations. Figures 4-29 to 4-31 show Equation (4-106) as a function of $(P/N_0 W)$ for no coding, a code with rate 1/2 and 5 dB coding gain; and for a code with rate 2/3 and 5 dB coding gain. Values of $(C/N)_{T0}$ were chosen to correspond to 10^{-10} bit error rate in an undistorted system; this implies that the relative values of $\dot{C}$ still apply in a distorted system if the degradation is the same for all M. In practical systems this is not likely to be the case; furthermore, the degradation is design-specific, so it would be difficult to present actual results without making rather detailed design assumptions. The curves all display initially a relatively rapid rise because lower values of $(P/N_0 W)$ correspond to the power-limited region where the capacity increases linearly with power. (Notice, however, that the abscissa is linear in the logarithm of power.) But, as might be expected, the curves exhibit an asymptotic behavior as $(P/N_0 W)$ becomes large. In fact, as $(P/N_0 W) \to \infty$ we get

$$\dot{C}_{max} = \frac{m G_c^{1/\beta} R_c^{1/\beta}}{(C/N)_{T,0}^{1 \times 1/\beta}}$$

This shows that the ultimate capacity is only weakly dependent on the code parameters for practical codes. The beneficial effect of coding arises from the fact that a relatively large fraction of the ultimate capacity is reached for reasonably small values of *(P/N₀W)*. Such values also imply relatively low values of r, which in general is desirable. Coding can also be used to maintain the throughput of a given network, while increasing orbit capacity. For example, with *(P/N₀W) = 15 dB*, $\dot{C}_{max} = 0.35$ for an uncoded system with *M = 4*; for the same CNR, *M = 8* and $R_c = 2/3$ imply the same network information rate but $\dot{C}_{max}$ has increased to 0.51.

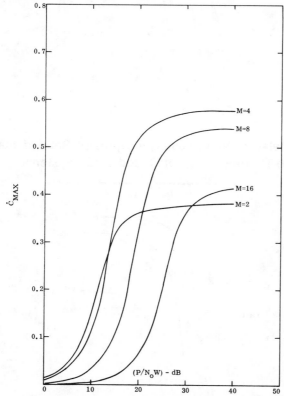

Figure 4-29 Relative Orbit Capacity at Optimum Ratio of Interference to Thermal Noise for Uncoded MPSK Systems

A question of interest in this context is how closely a given modulation/coding scheme approaches the theoretically optimimum limit dictated by information-theoretic considerations. The answer can be deduced from Figure 4-32, which shows the optimum normalized capacity $\dot{C}/[K^{1/\beta}W]$, the same quantity plotted in the three preceding

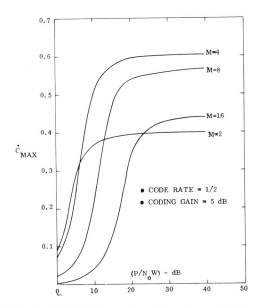

Figure 4-30 Relative Orbit Capacity at Optimum Ratio of Interference to Thermal Noise for Coded MPSK Systems

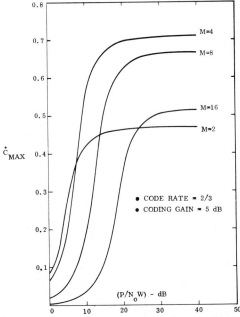

Figure 4-31 Relative Orbit Capacity at Optimum Ratio of Interference to Thermal Noise for Coded MPSK Systems

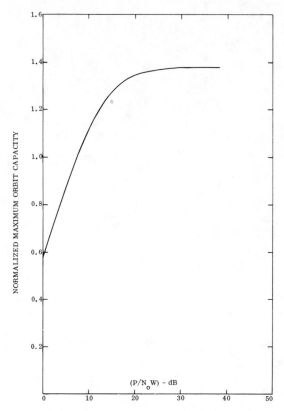

Figure 4-32 Theoretically Optimum Orbit Capacity

figures. The derivation of Figure 4-32, which is given in Appendix E, is based on the two central assumptions underlying Figures 4-29 to 4-31, namely that the interference is noise-like and that, for any given $(P/N_o W)$, r is set at its optimum value.

It can be seen that the optimum capacity is on the order of twice the capacity of the best coded MCPSK case considered earlier, in the normal operating range for $P/N_o W$. Considering the implied complexity of the code achieving the theoretical optimum, the practical codes used earlier (which were chosen more or less arbitrarily) actually deliver a reasonably large fraction of the theoretical limit. Of course, the comparison is not quite fair because the optimum code can achieve arbitrarily low error probability. Nevertheless, these results show that it is possible to obtain substantial capacity improvement with practical codes.

Chapter Five

FACTORS AFFECTING
ORBIT/SPECTRUM UTILIZATION:
THE NON-HOMOGENEOUS CASE

5.0 Introduction

In this chapter we extend the discussion of technical factors influencing orbit utilization, but broaden our perspective from the homogeneous model to a more general environment which we refer to as the non-homogeneous model. Simply put, a non-homogeneous situation is one that does not strictly meet the conditions that we have imposed on the homogeneous model (Section 4.2). In itself this does not necessarily mean that the satellite networks are not identical; for example, the occurrence of rain, even in an otherwise homogeneous environment, does not conform to the model definition. Evidently, this is an arbitrary division, but it serves the purpose of segregating an effect which would unnecessarily complicate the analysis of the homogeneous model. The most common interpretation of non-homogeneity, however, is that networks are dissimilar in one or more respects, and this is certainly subsumed under our definition. Thus, under the non-homogeneous heading we discuss in this chapter a number of topics to which we now give a brief introduction.

The homogeneous model is a useful idealization because it provides a structured framework which allows us to place some conceptual order on the complex interrelationships involved. In certain arcs of the orbit and in some frequency bands, the homogeneous model may even be a reasonable approximation to reality. But in general non-homogeneity of satellite networks will be the norm rather than the exception. We can carry over certain broad inferences from the homogeneous model to the general case. For example, we can infer that, all other things being fixed, increasing the size of all earth station antennas will be beneficial, even if they are not of the same size.

Similarly, increasing the sidelobe decay exponent, β, from its initial value for every antenna will improve orbit capacity. However, aside from qualitative statements of this type, it is all but impossible to establish a meaningful quantitative formulation for orbit utilization in the general non-homogeneous situation. By the latter, we mean that, within the broad range of values to be found in practice, every satellite network can take on an arbitrary set of parameter values, independently of every other network.

It should not be inferred from the preceding that an arbitrary but *given* non-homogeneous ensemble of networks is not analyzable in some sense. In fact, it is relatively straightforward to examine such an ensemble with respect to interference compatibility; i.e., to determine whether or not the interference into every carrier exceeds the mandated limits. This requires the computation of interference for all pair-wise combinations of carriers and, for each carrier, summing the contributions from every other carrier interfering with it. Thus, in this sense, the general case is reducible to a sequential analysis of pairs of interferers. Consequently, we begin, in the next section, with a discussion of the satellite spacing required between an arbitrary pair of interfering satellites. This discussion is instructive because it highlights the issue of "efficiency" of orbit utilization by non-homogeneous systems. We will point out that too great a disparity in characteristics between networks can cause difficulties that can be phrased as inefficiency. In Section 5.2, we shall give some specific examples of the extent of non-homogeneity that can exist between typical current implementations, and in both of these first two sections we shall discuss means of characterizing pair-wise non-homogeneity. In Section 5.3 we address two methods which have been proposed for limiting non-homogeneity, thereby ameliorating the difficulties associated with it. Another way to mitigate the effects of non-homogeneity is to make the interference allowance between networks flexible, subject only to a total maximum from all interfering networks; this is discussed in Section 5.4.

We pointed out above the difficulty in abstractly formulating the general non-homogeneous problem (as opposed to analyzing a given, arbitrary set of networks). In a somewhat more constrained setting, however, it may be possible to phrase an analytically reasonable problem. We shall, in fact, consider such a controlled non-homogeneity in Section 5.5. One benefit of this analysis is that it will give an inkling of the complexity of the unrestricted problem, and another side benefit is that it will uncover one technique, known as clustering, for relieving to some extent the undesirable effects of non-homogeneity.

A general but given non-homogeneous environment, as has been said, is analyzable through successive examination of pair-wise combinations of carriers. While this is conceptually straightforward, practicality requires computer mechanization for all but the most simple cases. In Section 5.6, we shall address the nature of such a computer program, and we shall also discuss the augmentation of such a program so as to yield optimum arrangements of satellites according to certain criteria.

The remainder of this chapter considers a number of topics that have orbit utilization in common, but otherwise are not logically connected in the way that the first six sections are. Thus, the order of presentation is to some extent arbitrary, and therefore these sections need not be read sequentially. Section 5.7 discusses one of the potentially most significant means for expanding the capacity of the geostationary orbit; namely, frequency re-use at the satellite through the use of narrow (or "spot") beam satellite antennas. The ensuing section, 5.8, examines the impact on satellite design of power flux density limits at the earth's surface. Although such limits arose historically to provide a sharing "interface" with terrestrial radio-relay systems, more recently it has been suggested that their use be extended as a possible means for limiting non-homogeneity as well. In the latter context, this section could be read in conjunction with Section 5.3.

One technique for improving orbit utilization which has received some attention is the use of the same frequency band for both uplinks and downlinks (also sometimes called bidirectional or reverse-frequency mode of operation), but of course the same frequencies are not used for both directions in a given network. This mode of operation is allowed on a limited basis by the *Radio Regulations*. In Section 5.9 we examine the potential benefits of bidirectional allocations from the orbit utilization standpoint.

As was pointed out earlier, one of the assumptions of the homogeneous model is that clear weather conditions prevail. In fact, irrespective of homogeneity, clear weather conditions do exist most of the time. However, for small but operationally significant fractions of the time, substantial attenuation may be caused by rain, the more so the higher the frequency. Since these higher frequencies ($\geq$ 10 GHz) are only now beginning to come into use, very little attention has generally been paid to the influence of rain on orbit utilization, and this is an area that requires study. As a first step in appreciating this influence, the reader may consult Section 5.10, which contains a simplified analysis of the potential impact of rain on orbit utilization.

In order to speak about "orbit utilization" quantitatively, it is obviously necessary to have a measure for that term. In fact, depending on circumstance, more than one measure may be meaningful. Once we have a measure, the question arises as to how "good" a particular orbit environment is, in terms of that measure. This implies having a reference with respect to which we can assess any particular realization of the set of factors whose goodness we wish to determine. Thus, the notion of efficiency of orbit utilization naturally arises. This subject is discussed in Section 5.11. In some sense, this section could have preceded all of the others, in fact could have been the first section of the preceding chapter. However, at too early a stage, the discussion could have been digressive. In any case, since the subject is central, though somewhat abstract, the reader may find it useful at any point.

A question of the greatest significance concerns the capacity (in terms of some measure) of the orbit, for the assessment of the degree of "congestion" certainly rests on the gap between capacity and demand. The capacity of the orbit is not unique, for it depends upon a set of constraints, such as the technical standards to be assumed. However, all previous attempts at calculating the capacity have agreed on the general conclusion, that the capacity is "large." In Section 5.12, we present one such capacity calculation, both to illustrate the calculational approach as well as to impart some feeling for the actual numbers.

Finally, in section 5.13 we take up briefly a number of "miscellaneous" topics which are difficult to analyze mathematically, such as operational considerations, or which do not seem to warrant at this stage a full exposition. However, the reader will find fuller accounts in the cited references.

5.1 Pair-wise Consideration of Non-Homogeneous Carriers

Consider two mutually interfering carriers each associated with a different geostationary satellite network, which we identify as network 1 or 2. The quantities of interest will be subscripted, according to the notational convention of Section 2.1. For simplicity of exposition we shall deal only with the downlink, but an exactly parallel development applies to the uplink. Further for convenience we take both signals to be analog and assume the relevant performance measure against both noise and interference is a suitably defined SNR.

Thus, using the earth station antenna sidelobe model of (4-23) we have

$$\Delta \theta_{21}^{\beta 2} = \frac{(S/N)_{12}}{R_{12}} \frac{A \, P_{st2} \, G_{st21} \, L_{st21}}{P_{st1} \, G_{st11} \, G_{gr11} \, L_{st11} \, Y_{st21}} \tag{5-1}$$

as the necessary spacing to satisfy the permitted amount of interference noise generated into carrier 1 from carrier 2, this noise being defined through the signal-to-interference noise ratio $(S/N)_{12}$. Similarly, for network 2 one has

$$\Delta \theta_{12}^{\beta 1} = \frac{(S/N)_{21}}{R_{21}} \frac{A \, P_{st1} \, G_{st12} \, L_{st12}}{P_{st2} \, G_{st22} \, G_{gr22} \, L_{st22} \, Y_{st12}} \tag{5-2}$$

as the necessary spacing to satisfy its interference noise budget. Evidently, in order for the allowable interference noise not to be exceeded in either network the actual spacing must be at least equal to the larger of the two computed from (5-1) and (5-2). For arbitrary implementations it may happen that one angle is much larger than the other, $\Delta\theta_{ij} \gg \Delta\theta_{ji}$. There then exists what could be regarded as an inefficiency in the sense that the interference noise budget of the network requiring the smaller spacing is not entirely used. It would seem that this leads to wasted potential capacity for that network, or that somehow full use of the orbit resource is not being made. Different points of view are possible in this contest. For one thing, the inequality $\Delta\theta_{ij} \gg \Delta\theta_{ji}$ may be of no practical significance if the absolute value of $\Delta\theta_{ij}$ is itself of reasonably small magnitude. This inequality is sometimes attributed to the very fact that there is non-homogeneity, implying that the extent of the disparity between the two angles should be assessed in relation to the angular separation that would exist in a homogeneous environment. This presents a problem because there are two homoeneous systems which could serve as the reference. One is where all networks are of type 1, and in this case we label the necessary separation by $\Delta\theta_{11}$; the second reference system is where all networks are of type 2 and for this situation we denote the necessary spacing $\Delta\theta_{22}$. The question is: to which of these two homogeneous separations is it proper to compare the non-homogeneous angles and what conclusions is it fair to make? The fact is that any "answer" to this question must to some extent be subjective. It would appear sensible, however, to say that if the larger of $(\Delta\theta_{12}, \Delta\theta_{21})$ is not substantially larger than the larger of $(\Delta\theta_{11}, \Delta\theta_{22})$, then any disparity between $\Delta\theta_{12}, \Delta\theta_{21}$ is not properly attributable to non-homogeneity as such. On the other hand, if the larger of $(\Delta\theta_{12}, \Delta\theta_{21})$ is considerably larger than the larger of $(\Delta\theta_{11}, \Delta\theta_{22})$, then it would seem appropriate to ascribe an inefficiency to the existence of non-homogeneity. In Section 5.6, we shall show some computations which will illustrate both of the possibilities just discussed. In Section 5.3 we discuss some of the means which have been proposed for limiting non-homogeneity and, thus, preventing to some extent inordinately large satellite spacings.

A major problem in the latter connection is exactly in what manner can one limit the degree of nonhomogeneity. Every parameter on the right-hand side of both (5-1) and (5-2) potentially contributes to non-homogeneity. For example, if both networks are identical in every respect but, say, earth station antenna size, and if $G_{gr11} \triangleq G_0 >$ $G_{gr22} \triangleq G$, then

$$\left(\frac{\Delta\theta}{\Delta\theta_0}\right) \propto \left(\frac{G_0}{G}\right)^{1/\beta} \tag{5-3}$$

where $\Delta\theta_0$ is the spacing required when $G = G_0$. By taking one parameter at a time, as in (5-3), one could establish a series of non-homogeneous spacing factors. It would clearly be not only impractical from a regulatory standpoint, but also unacceptably and unnecessarily constraining to system designers to specify limits on every relevant parameter.

In one attempt to meet these difficulties [20,21] it was found useful to define four parameters which are functions of the implementation parameters but do not explicitly specify them, and which (approximately) embody the notion of non-homogeneity in a fairly general way. This notion rests on the basic observation that non-homogeneity between two networks is a direct consequence of the interference-producing capability of one and the susceptibility to interference of the other. These two aspects of network interaction can be characterized by power densities. The parameters which have been defined are the following:*

P_u = the maximum uplink *e.i.r.p.* per unit bandwidth in the direction of the geostationary orbit radiated at an angle θ to the axis of the main beam of the earth station antenna;

S_u = the uplink sensitivity, defined as the minimum interference power flux density (PFD) at the geostationary satellite orbit which corresponds to the recommended maximum single entry of interference in a channel;

P_d = the maximum *PFD* produced at the earth's surface by the satellite emissions;

S_d = the downlink sensitivity, defined as the minimum interference *PFD* at the earth's surface, arriving at an angle θ to the direction of the wanted signal, which corresponds to the recommended maximum single entry of interference in a channel.

Parameters P_u and P_d reflect the potential of a network to cause

---

*These parameters have been labeled A,B,C,D, respectively, in the CCIR literature, but the symbology used here is somewhat more suggestive.

interference while S_u and S_d are indicative of the susceptibility of a network to interference. Thus, in principle, constraining the values of these parameters within certain bounds will control the degree of non-homogeneity. The implications of this approach will be considered in 5.3.1.

It should be pointed out that the *Radio Regulations* already constrain P_d in different frequency bands. However, the purpose there is to protect sensitive terrestrial radio-relay receivers from undue interference, rather than optimizing the use of the geostationary orbit. The corresponding limitation on satellite e.i.r.p. is taken up in Section 5.8.

5.2 The Concept of Isolation: Examples of Pair-Wise Non-Homogeneity

Because of the many ways in which mutually interfering networks can differ from one another, it is helpful to try to capture in a simple way the relative degree of non-homogeneity between two *given* networks. Comparing the emission (P_u, P_d) and sensitivity (S_u, S_d) parameters of two networks would give such an indication. A more readily intuitive measure of non-homogeneity is a quantity which has been called internetwork isolation [22, 23]. The term is closely related to its usual connotation; i.e., the "physical" isolation required between two networks to satisfy the necessary carrier-to-interference ratio. This isolation stems, for example, from antenna discrimination or polarization isolation, rather than signal structure, the latter being implicit in the required CIR. Isolation is not necessarily symmetrical between two networks.

An expression for isolation can be obtained as follows. Ignoring slant range differences and isolation mechanisms other than antenna gains, and focusing on network 1 as the wanted one, the equations of Section 4.1 give us for the uplink and downlink carrier-to-interference ratios

$$\left(\frac{C}{I}\right)_{s12} = \frac{P_{gt11}\, G_{gt11}\, G_{sr11}}{P_{gt21}\, G_{gt21}\, G_{sr12}} \tag{5-4}$$

$$\left(\frac{C}{I}\right)_{g12} = \frac{P_{st11}\, G_{st11}\, G_{gr11}}{P_{st21}\, G_{st21}\, G_{gr12}} \tag{5-5}$$

define the following quantities:

$$\sigma_u = \frac{G_{gt22}}{G_{gt11}} \; ; \sigma_d = \frac{G_{gr22}}{G_{gr11}}$$

$$\gamma_u = \frac{G_{sr22} \, T_{s2}}{G_{sr11} \, T_{s1}} \; ; \gamma_d = \frac{G_{st22} \, T_{g2}}{G_{st11} \, T_{g1}}$$

where T_s and T_g are, respectively, the satellite and earth station receiver noise temperatures. Further, let

$$(C/T)_{s1} = \frac{P_{gt11} \, G_{gt11} \, G_{sr11}}{T_{s1}} \; ; (C/T)_{s2} = \frac{P_{gt22} \, G_{gt22} \, G_{sr22}}{T_{s2}}$$

$$(C/T)_{g1} = \frac{P_{st11} \, G_{st11} \, G_{gr11}}{T_{g1}} \; ; (C/T)_{g2} = \frac{P_{st22} \, G_{st22} \, G_{gr22}}{T_{g2}}$$

Thus,

$$\frac{P_{gt22}}{P_{gt11}} = \frac{(C/T)_{s2}}{(C/T)_{s1}} \, (\sigma_u \, \gamma_u)^{-1}; \frac{P_{st22}}{P_{st11}} = \frac{(C/T)_{g2}}{(C/T)_{g1}} \, (\sigma_d \, \gamma_d)^{-1}$$

Some further definitions are useful:

$$(C/T)^{-1}_1 = (C/T)^{-1}_{s1} + (C/T)^{-1}_{g2}$$

$$(C/T)^{-1}_1 = (C/T)^{-1}_{s1} + (C/T)^{-1}_{g2}$$

$$(C/T)_{g2}/(C/T)_{s2} \triangleq \rho_2$$

$$[(C/T)_{s2}/(C/T)_{s1}] \cdot [(C/T)_{61}/(C/T)_{g2}] \triangleq R$$

Recalling that

$$(C/T)^{-1}_{12} = (C/T)^{-1}_{s12} + (C/T)^{-1}_{g12}$$

a sequence of manipulations yields

$$\left(\frac{C}{I}\right)_{12} \frac{(C/T)_2}{(C/T)_1} \frac{1+\rho_2}{1+\rho_2 R} = \left\{ \frac{R}{\sigma_u \, \gamma_u} \frac{G_{gt21} \, G_{sr12}}{G_{gt11} \, G_{sr11}} \right.$$
$$\left. + \frac{1}{\sigma_d \, \gamma_d} \frac{G_{st21} \, G_{gr12}}{G_{st11} \, G_{gr11}} \right\}^{-1} \quad (5\text{-}6)$$

which is not very intuitive as it stands. However, it will be made so by a few additional assumptions. If we label the RHS of (5-6) by d_{12}, we have

$$\left(\frac{C}{I}\right)_{12} \frac{(C/T)_2}{(C/T)_1} \frac{1+\rho_2}{1+\rho_2 R} = d_{12}$$

and the quantity d_{12} is specifically what has been defined as internetwork isolation [22, 23]. In the particular instance that $R = 1$ and T_{s1} = T_{s2}, T_{g1} = T_{g2}, and the ratio of uplink to downlink power in the two networks is identical, we have

$$\left(\frac{C}{I}\right)_{12} C_{21} = d_{12} \qquad (5\text{-}7)$$

where C_{21} is the ratio of power in the interfering to that in the wanted network. In this instance, we see that d_{12} is truly an "isolation" in the usual interpretation of that term. The required isolation is that necessary to meet the required CIR. With $R = 1$, Weiss [22, 23] has tabulated

$$d_{12,\ req} = (C/I)_{12,\ req} \frac{(C/T)_2}{(C/T)_1} \delta_A \qquad (5\text{-}8)$$

for a variety of mutually interfering carrier pairs having typical implementations in the 4/6 GHz band. The "aggregation" factor σ_A in (5-8) is a multiplying factor which assumes that when the interfering signal's bandwidth is much smaller than that of the wanted one, as many interfering carriers are assumed as are necessary to fill up the bandwidth of the interfered-with signal. The results, which originally appeared in [22] and later in [23], are shown here in Table 5-1. This chart gives a good idea of the interference asymmetry; i.e., non-homogeneity, that may be found in practice.

5.3 Methods of Limiting Non-Homogeneity

We have seen that substantial non-homogeneity can exist, the result of which is that the spacing required between two networks can be substantially larger than either would require in the presence of identical interfering networks. In order to mitigate the effect of such incompatibilities, proposals have been made to constrain the degree of non-homogeneity between networks. Basic to these methods is the definition of non-homogeneity through the emission and sensitivity parameters defined previously.

5.3.1 Constraints on the Emission and Sensitivity Parameters

One means of reducing non-homogeneity which has been proposed is to set definite values on the emission and sensitivity parameters P_u, P_d, S_u, and S_d. The goal, ideally, would be to have compatible pairings (P_u, S_u) and (P_d, S_d) meaning that for networks designed to meet these values the satellite spacing at which the interference allowance is met would be reasonably small.

The form of the specifications which have been suggested for the parameters in question is of the following kind:

$$10 \log P_u \le A \text{ - } 25 \log \theta \ (dBW/4 \ kHz)$$

$$10 \log S_u \ge B \ (dBW/m^2/4 \ kHz)$$

$$10 \log P_d \le C \ (dBW/m^2/4 \ kHz)$$

$$10 \log S_d \ge D + 25 \log \theta \ (dBW/m^2/4 \ kHz)$$

As mentioned above, P_u and S_u, jointly, and P_d and S_d jointly imply a satellite spacing which can be deduced from Figure 5-1. For illustration, two horizontal lines are drawn for hypothetical values of B and C. Intersection of the B line with a chosen A line determines a spacing, as does the intersection of the C line with some D line.

If some specific numerical set A, B, C, D were chosen as a regulatory standard, what would the implications be on satellite networks? Clearly, the intended design of only a few networks would happen to

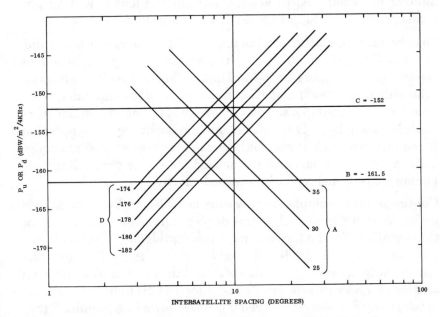

Figure 5-1 Relationship Between Emission and Sensitivity Parameters

adhere to that specific set. Networks which had intended to use larger emission densities would have to accept lower performance against thermal noise, and those with higher sensitivities than the stipulated values would suffer greater interference noise.

Thus, it can be seen that limiting non-homogeneity by setting definite values for the emission and sensitivity power densities has significant implications on system design or performance. To illustrate these implications, suppose it were desired to constrain the non-homogeneity so that spacings no greater than 10° were required to satisfy the interference budget. This would be satisfied if, say, D = -180 and C≤-155. This would imply that for networks with downlink sensitivities less than -180, more interference would be suffered. On the other hand, what kind of constraint does C = -155 impose? To answer this question we need to specifically relate C to the system design. The details are given in Appendix F, and the results shown in Table 5-2 for a range of FDM/FM telephoning carriers operating at 4GHz, the specific properties of which are also given in that Appendix. Each column applies to a different value of earth station G/T. Thus, one can see that if C = -155 were taken as a standard, either a number of "small" earth station implementations would be precluded, or they would have to accept lower performance.

To overcome this difficulty one might adopt a more relaxed standard, say C = -150, in which case the associated non-homogeneous separation would increase to 16°. Though this is rather large from an orbit utilization standpoint, it should be pointed out that this specific angle is still linked to a sensitivity of -180. If some connection were made between the PFD of a signal and the sensitivity of signals that it can interfere with, it may be possible to regain smaller spacings. This in fact is the basis for the second method of controlling non-homogeneity, which will be discussed shortly.

Continuing the example, we now consider the uplink situation. Logically, we would impose the same degree of non-homogeneity as for the downlink. For the 10° spacing, a compatible pair of numbers is A = 30 and B = -158. Table 5-3, derived in Appendix F, shows the relationship between that value of A and the earth station antenna size for various values of satellite G/T for a 6 GHz uplink. For the set of telephony carriers considered it is also shown in Appendix F that the median value of uplink sensitivity occures for $(G/T)_s \approx -4dB$. Hence the use of higher satellite G/T (implying "spot" beams) permits the use of small earth station antennas, but the uplink sensitivity would be greater, thus resulting in more interference than the standard. This latter effect, however, may be ameliorated by the fact that spot beam satellites would likely be subject to fewer significant interference entries.

We have seen that power flux densities are convenient quantities for characterizing non-homogeneity. However, as regulatory tools there may be considerable difficulty in applying them as controls on non-homogeneity without unduly penalizing classes of systems.

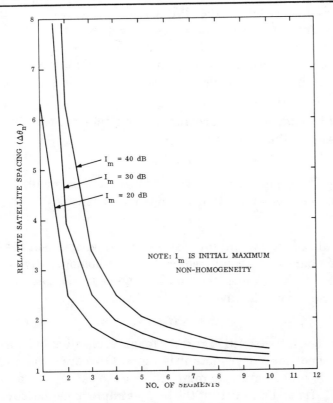

Figure 5-2 Illustration of Effect of Band Segmenting on Satellite Spacing

5.3.2 Band Segmentation

As already alluded to, one way in which the method above may become usable is by segregating networks in categories of relative non-homogeneity, a technique referred to as band segmenting [24, 25]. That is, a frequency band would be divided into n segments. In each segment the degree of non-homogeneity is constrained not to exceed a certain amount, which might not necessarily be identical in all segments. That is, P_u/S_u and/or P_d/S_d are upper bounded. Let I_m bet the maximum initial ratio of these quantities over the whole band. There are different possible ways of reducing this maximum non-homogeneity, depending in practice upon the actual valves of P_u, P_d, S_u and S_d that exist. In one idealized conception, P_u and S_u could be continuously variable over the band, but with a fixed ratio at any frequency. Of course, this could be realized only in the unlikely event that the actual values could be so arranged. What is more likely to be practical is that sub-bands could be defined so that within each the ratio of maximum P_u to minimum S_u would be approximately equal, but the absolute values would vary from one sub-band to another.

The satellite spacing, $\Delta\theta$, associated with non-homogeneity I_m can be expressed in the form [25]

$$\Delta\theta \propto I_m^{0.4}$$

assuming CCIR sidelobes. Hence, if after band segmentation the maximum non-homogeneity is I'_m, the new spacing is

$$\Delta\theta' \propto I_m'^{0.4}$$

In an idealized case we might assume that if we segment the band into n sub-bands, we could reduce the non-homogeneity in each sub-band to $(1/n)$ of the initial non-homogeneity dB, i.e., $I'_m = I^{1/n}_m$. In that case, we would have

$$\Delta\theta_n \propto (I_m)^{0.4/n} \tag{5-9}$$

Equation (5-9) is plotted in Figure 5-2 for several values of I_m, assuming the proportionality constant is unity. Of course, the absolute values of P_u and S_u must be matched so as to satisfy the assumed value of I'_m. It can be seen that the largest improvement occurs with just two band segments. This improvement also applies if we have n segments in each of which the non-homogeneity does not exceed $\sqrt{I_m}$.

5.4 On the Distribution of Interference Entries

Limiting the possible range of non-homogeneity is one way to prevent, or at least control, the inefficiency in orbit utilization that may result from it. However, this will inevitably impose some implementation constraints. Some means are available for lessening the negative effects of non-homogeniety without specifically regulating implementations. These means might be collectively put in the category of "optimization," that is of optimally distributing the interference into networks sharing a common part of the orbit. One technique in this category is based on the relative positioning of satellites, and is discussed in the next section. Another consideration, discussed here, involves the relative amount (distribution) of interference noise permitted to be caused by the various carriers interfering with any wanted carrier.

By international agreement the total amount of interference (either post-detection or predetection) permitted into a wanted signal of given type may not exceed a certain value, which we shall refer to as total interference criterion (TIC); also by agreement the interference permitted to be caused by any single interferer (or "entry") may not exceed another (obviously smaller) value which we refer to as single entry criterion (SEC). The question at hand is does there exist an "optimal" ratio $v = SEC/TIC$ from the orbit utilization standpoint? Unfortunately, except under highly structural conditions, the answer cannot be affirmative, even if one could readily define optimality. It will be seen as rather obvious that either extreme, $v \approx 0$ or $v \approx 1$ is grossly inefficient under any definition of optimality, so a reasonable value probably lies "somewhere" in the middle. In fact, one can go further and argue that to allow for uncertainties in the orbital environment the SEC ought to be flexible and thus maximally accommodate non-homogeneity. Consider, for example, the situation where the SEC is satisfied for one or perhaps two entries but the TIC is not fully used up. In this case presumably greater oribt utilization would be available if one of the interfering satellites were moved closer, thus using up the interference margin. (This of course assumes that there is also interference margin on the "other" satellites.) Thus, even though the single entry maximum were exceeded the total interference limit would still be observed.

Since from the point of view of any interfered-with network it is only the total interference that matters, it might be asked why any single entry criterion is required at all. There are two reasons for this. One is that if the SEC were uncontrolled and a single entry were to use up most of the TIC it would generally make it very difficult to accommodate other satellites in the same part of the orbit, and could well preclude the accommodation in that part of the orbit of requirements that might arise in the future. Further, from a practical standpoint the current international *(Radio Regulations)* procedures call for bilateral coordination when a new network is to be implemented. Clearly such a process must rest on a definite SEC.

On the other hand, the SEC should not be too small, for otherwise the required satellite separation to meet it (for co-coverage or adjacent-area coverage) could be forced to be larger than would be deemed satisfactory from the orbit efficiency perspective. But a single entry criterion does not bring with it the obligation to an interfering network to ensure that the maximum total interference into the interfered-with network is not exceeded.

Thus, a good single entry criterion would optimally permit the orbit utilization efficiency associated with the total interference criterion to be closely approximated in practice, and it would imply with high probability that the total interference criterion would not be exceeded. These desiderata would be guaranteed if the interference environment in which a network existed were completely predictable. In that case it might be possible to mitigate the inefficiencies that might result from substantial non-homogeneity, by judiciously choosing the single entry criterion. Some observers have argued that this should be done by making the SEC spacing-dependent and coverage-area-dependent on the premise that good orbit utilization requires that widely separated satellites should not induce significant interference into one another, and similarly for satellites whose coverage areas are sufficiently separated. This may, however, impose significant restrictions on some networks.

One sufficiently structured environment in which we can readily study the relationship between the SEC and the TIC is the homogeneous model. If we denote by I_S the largest single entry and the total interference by I_T, it follows from Section 4.2 that

$$I_S/I_T = \{Z\,(\beta,\,N)\}^{-1}$$

As an example, let $\beta = 2.5$, whence $I_S/I_T \approx 0.4$. If $SEC < 0.4\ TIC$ it is clear that when the SEC is satisfied, the TIC will not be completely used up, implying a loss of orbit efficiency. If $SEC > 0.4\ TIC$ then when the SEC is met, the TIC will be exceeded. Thus, the optimum condition is when both the SEC and TIC are simultaneously satisfied, a statement which applies generally to non-homogeneous orbit utilization.

In practice, of course, the orbit environment is neither as structured nor predictable as the homogeneous one. The effect of uncertainty in the environment can be appreciated to some extent by postulating a homogeneous situation where β is uncertain. Suppose, for example, that an SEC has been derived on the basis of assuming $\beta = 2.5$, but that the actual β turns out to be larger. Thus, the total interference into a network is given by

$$I_T = TIC\frac{Z\,(\beta,\,N)}{Z\,(2.5,\,N)}$$

which, for $\beta > 2.5$, clearly is less than the TIC. There is thus an inefficiency which may be characterized by

$$\frac{\Delta\theta'}{\Delta\theta} = \left\{\frac{Z\,(2.5,\,N)}{Z\,(\beta,\,N)}\right\}^{1/\beta}$$

where $\Delta\theta$ is the spacing that satisfies the SEC and $\Delta\theta'$ is the spacing that would meet the TIC.

The object of this example is to highlight the fact that uncertainties in future implementations will inevitably make any SEC less than optimal, and further strengthens the case for a certain amount of flexibility in the value of the SEC.* A good deal of study has been made on the relationship between the SEC and TIC, based on various models of orbit occupancy [26, 27, 28, 29, 30] and on the way in which interference is likely to accumulate taking into account statistical factors. In some of these studies a mix of global and limited

---

*It may be mentioned in passing that computer optimizations of the type described in Section 5.6 have generally shown better orbit utilization when the TIC alone is the controlling criterion than when both the TIC and the SEC are operative.

coverage (i.e., spot beam) networks is assumed, while in others the orbit is occupied strictly with limited coverage satellites. Statistical considerations arise when considering likely frequency plans in any satellite *vis-a-vis* any other interfering satellite. The arrangement of carrier types will vary from one network to another so that a given carrier will encounter a range of interferers.

The net result of these studies is that the total interference induced into any carrier will rarely exceed by more than 2-3 dB that generated by the largest single entry. This fact should be taken into consideration when regulations specifying the SEC and TIC are developed.

5.5 Consideration of a Limited Non-Homogeneity: Clustering

Although the unconstrained non-homogeneous environment cannot be generally formulated, special cases may be so amenable. We consider one such here which is useful for two reasons. First, it will lead to the idea of "clustering" which can provide helpful guidance in arranging satellites in practice [31]. Secondly, even for as simple a non-homogeneity as we will be considering, it will illustrate the significant increase in analytical complexity that accompanies any departure from homogeneity and thus impart to the reader an appreciation for the difficulties involved in the general case.

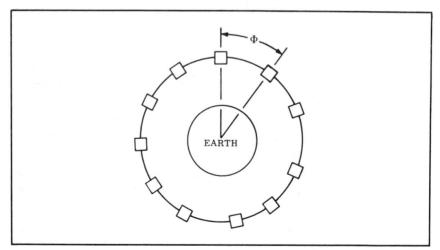

Figure 5-3 Homogeneous Deployment — Type *A* Satellite

The non-homogeneous environment we are considering is specifically one in which only two distinct types, a and b, of network exist. (For notational clarity we shall usually denote them by underlining: a, b). We shall assume that the satellite antennas of both types are global coverage, or at least common-area coverage.

Consider first the situation where the orbit is occupied only by type a satellites which are all identical, as illustrated in Figure 2-3. This, of course, is the homogeneous model, and will be referred to here as the Hom a case. For a given set of implementation and modulation parameters, a spacing ϕ is required to meet the interference noise limitation. Similarly, we can consider the orbit to be populated by type b satellites only, which we refer to as the Hom b case, and for this case let ψ be the satellite spacing necessary to meet interference noise constraints.

We now consider how the two types of satellite can share the orbit. One method, termed large group development (LGD) is the simplest extension of homogeneous development and is illustrated in Figure 5-4. As the figure shows, half the orbit is occupied by type a and the other half by type b satellites. Except for edge effects at the interface,

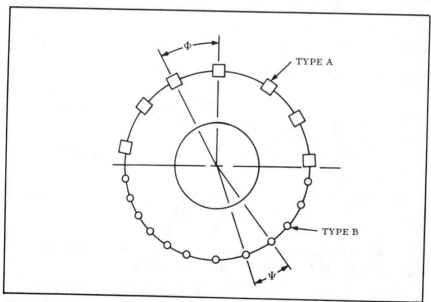

Figure 5-4 Large Group Deployment (LGD)

each half of the orbit behaves essentially as it would in the homogeneous situation, but with only 180° instead of 360° of available arc. If we measure the efficiency of orbit utilization by the number of satellites in orbit, we have for this case

$$n \ (LGD) \approx \frac{180}{\phi} + \frac{180}{\psi} = \frac{180}{\phi}(1 + \omega) \qquad (5\text{-}10)$$

where $\omega \triangleq \phi/\psi$ is the relative "inherent" utilization of the two types of network.

The LGD is of course not very realistic because one can expect both types to exist within a relatively small arc. In order to satisfy this objective we consider next a scheme which is also an extension of the homogeneous case, in which type $\underline{a}$ and type $\underline{b}$ are alternated along the orbit. We call this the single alternating deployment (SAD) method, and it is illustrated in Figure 5-5. The spacing between type $\underline{a}$ satellites is now ϕ' and that between type b satellites is ψ'. Symmetry obviously dictates that $\phi' = \psi'$ but evidently ϕ' is not necessarily ϕ and ψ' not ψ. In fact, by imposing the necessary condition that the effective* carrier-to-interference ratio or the interference noise must remain the same, regardless of the deployment method, one can

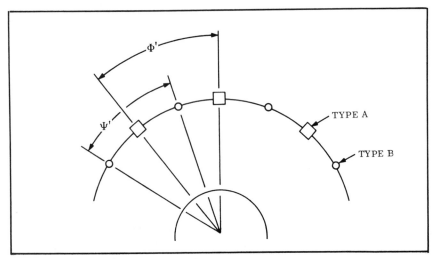

Figure 5-5 Single Alternating Deployment (SAG)

---

*The effective carrier-to-interference ratio takes into account the relative susceptibilities to interference of the signals in the two types of network.

arrive at a relationship between old and new spacings so that this requirement is met, namely:

$$\phi' \geq \phi \, [1 + \frac{w_{ba}}{r}(2^{\beta} - 1)]^{1/\beta} \qquad (5\text{-}11)$$

$$\psi' \geq \psi \, [1 + rw_{ab} \, (2^{\beta} - 1)]^{1/\beta} \qquad (5\text{-}12)$$

which are special cases of the general relations developed in Appendix G, and

r = ratio or eirp of type <u>a</u> to that of type <u>b</u> satellite;

w_{ba} = ratio of protection ratio or receiver transfer characteristic when type <u>a</u> interferes with type <u>a</u>, to that when type <u>b</u> interferes with type <u>a</u>;

w_{ab} = ratio of protection ratio or receiver transfer characteristic when type <u>b</u> interferes with type <u>b</u>, to that when type <u>a</u> interferes with type <u>b</u>;

β = as before, the earth station antenna sidelobe decay exponent, but in the context of this particular section we have used for analytical convenience the alternative sidelobe model of Equation 4-40a.

It may be noted that the minimum spacing occurs when both (5-11) and (5-12) are simultaneously satisfied with equality. Since $\phi' = \psi'$, this will occur (by taking the ratio of these two equations) when

$$\frac{\phi}{\psi} = \omega = \left\{ \frac{1 + rw_{ab} \, (2^{b} - 1)}{1 + rw_{ba} \, (2^{b} - 1) \, /r} \right\}^{1/\beta} \qquad (5\text{-}13)$$

Of course, this equality will hold only if r, β, w_{ab} and w_{ba} have certain specific values. The orbit utilization for this scheme in terms of number of satellites, remembering that $\phi' = \psi'$, is simply

$$n \, (SAD) = \frac{360}{\phi'} + \frac{360}{\psi'} = \frac{720}{\phi} \, \frac{\phi}{\phi'} \qquad (5\text{-}14)$$

The last deployment scheme considered is a generalization of the SAD and is termed the cluster alternating deployment (CAD) mode. It is illustrated in Figure 5-6 which shows that between every pair of type a satellites, spaced ϕ'' degrees apart, there are n type <u>b</u> satellites spaced ψ'' degrees from each other. In general, $\psi'' \neq \phi''/(n + 1)$, that is the <u>b</u> satellites are not evenly spaced between the <u>a</u> types but are "bunched" up near the middle. In the figure the two type $\underline{b}$ satellites at either end of the $\underline{b}$ cluster; i.e., closest to the type $\underline{a}$ satellites, are termed the "flank" satellites and labeled b_1; the adjoining $\underline{b}$ satellites, labeled b_2, are termed "second" satellites. If one denotes $(C/I)_{b_1}$ and $(C/I)_{b_2}$ as the CIR for the flank and second satellites, respectively, we should logically require

$$(C/I)_{b_1} = (C/I)_{b_2} \tag{5-15}$$

since the same performance is required in both. When this requirement is satisfied it does not mean that the interference noise is equalized in all $\underline{b}$ satellites; for this to be true we would need

$$(C/I)_{b_1} = (C/I)_{b_2} = (C/I)_{b_3} = \ldots (C/I)_{b_{u12}} \tag{5-16}$$

However, if (5-15) is satisfied the CIR in the other $\underline{b}$ satellites will be still higher. It would be possible to examine the CAD under the constraint in (5-16) but this over-complicates matters. We shall adopt (5-15) as our basic constraint, which imposes a condition on the ratio ψ''/ϕ'', say $(\psi''/\phi'')_0$. We also require that the effective CIR (or interference noise) remain the same whatever the deployment. This imposes conditions on ϕ''/ϕ and ψ''/ψ. Manipulations of these conditions subject to the constraint on $(\psi''/\phi'')_0$ then leads to numerical solutions for ϕ''/ϕ and ψ''/ψ. The mathematical details are some-

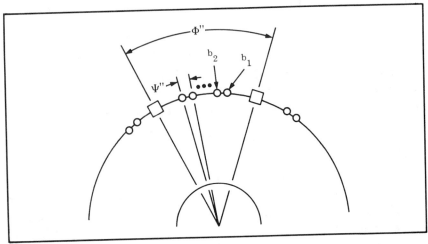

Figure 5-6 Cluster Alternating Deployment (CAD)

what involved and are given in Appendix G. The utilization of this scheme is given by

$$n \ (CAD) = \frac{360}{\phi''} + n \ \frac{360}{\phi''} = \frac{360}{\phi} \ \frac{\phi}{\phi''} \ (1 + n) \qquad (5\text{-}17)$$

The deployment schemes given could be generalized still further. For example, one could postulate clusters of m type $\underline{a}$ satellites and between these clusters would be clusters of n type $\underline{b}$ satellites. Or, one could make the spacings between the type $\underline{b}$ satellites non-uniform. We shall not, however, consider these extensions, as they do not materially shed additional light on the present discussion.

The relative efficiency of the schemes considered is best elucidated by numerical example. For that purpose we take the simple case where $w_{ab} = w_{ba} = 1$, and assume only that there is an *e.i.r.p.* difference r between the two types. Figure 5-7 shows the utilization in numbers of satellites as a function of n for various *e.i.r.p.* ratios *(10 log r)* assuming the initial (homogeneous) spacings are such that $\phi/\psi = \omega = 3$, and for $\beta = 2.5$.

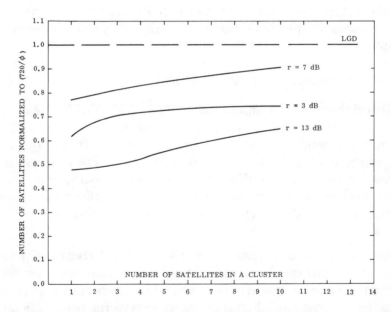

Figure 5-7 Orbit Utilization as a Function of Cluster Deployment and Power Unbalance

The ordinate is given by (5-10), (5-14), or (5-17) according to the deployment mode considered, normalized with respect to the factor $(720/\phi)$. Note that in computing the latter two equations the factor (ϕ/ϕ') or (ϕ/ϕ'') appears; the value of these ratios is controlled by the most restrictive of the inequality conditions. For example, in computing (5-14) one must first see whether ϕ' or ψ' is required to be larger; the smaller of the two must then be set equal to the larger.

Note that the point for $n = 1$ corresponds to the SAD mode. The figure shows that the improvement that can be obtained by clustering more than one low power satellite between two higher power satellites. Note, too, the different behavior as a function of the power unbalance, in particular the fact that there is a specific unbalance which maximizes orbit utilization. In practice, the causality is reversed. That is, for a given power unbalance there is an initial set of parameters $(\omega, w_{ab}, \text{ and } w_{ba})$ that is optimum.

The main lesson to be learned here is that judicious relative positioning of satellites in a non-homogeneous environment can be beneficial in minimizing average intersatellite spacing. The general principle that applies in establishing a satellite ordering is to minimize the number of interfaces between significantly non-homogeneous satellites.

5.6 Orbit Utilization Computer Programs

Beyond the understanding that analysis can provide it is of only limited usefulness in practical situations where the assumptions that render analysis tractable cannot be justified. If the term "computerized analysis" may be allowed, it is a practical necessity to handle an arbitrary configuration of satellites. Computer programs have been developed for this purpose, and here we describe in broadbrush fashion their capabilities and the general methodology involved.

Two types of programs are extant, by custom referred to as "analysis" and "synthesis" programs. Analysis programs are intended to examine a more or less arbitrary ensemble of networks to determine interference compatibility; thermal noise performance, while incidental in this context, is normally also calculated since margin with respect to noise might be taken advantage of to allow somewhat more interference noise than the accepted limit. To accomplish this objective, a program performs a repeated pair-wise analysis, computing for every carrier the interference noise (or carrier-to-interference ratio) induced in it by every other carrier whose bandwidth overlaps its own. In effect such a program mechanizes the equations of Section 4.1, which have been written in rather general form precisely for this purpose. No geometrical approximations need be made since all angles and distances can be computed straightforwardly using spherical trigonometry and vector analysis. Thus, one needs as inputs to the program (for every carrier), earth station coordinates, satellite coordinates, antenna boresight pointing vectors, antenna patterns, *e.i.r.p.*, and modulation characteristics. In addition one needs stored in the program either in equation form or table look-up the receiver transfer characteristic for all possible combinations of interfering carriers, or the protection ratio, or (for digital signals) a method of taking account of the interference in computing the BER. Generally speaking this type of information is the weak link, i.e., is either approximate, unavailable, or too complex to recalculate for every carrier pair. In the latter two instances one often equates the interference to thermal noise of the same power.

In the first analysis program to be developed [15d], three types of carriers could be accepted: FDM/FM telephony, FM television, and multilevel PSK. Two reference antenna patterns for the earth station antenna were available, the standard CCIR pattern or a modified Rice pattern [32]. For the satellite antenna circular spot beams of any beamwidth or elliptical spot beams with arbitrary geometrical axial ratios could be used. In addition a cross-polarized pattern could be invoked for the study of orthogonally polarized transmissions. This type of analysis program is in widespread use and its capabilities upgraded in some of the versions in various ways, for example to handle shaped beam patterns or a larger range of signal types.

As a particular example of the use of this type of program, computations have been made showing the interference noise or carrier-to-interference ratio as a function of satellite spacing for three types of mutually interfering carriers, the specific characteristics of which are shown on Table 5-4. There are thus nine pair-wise combinations when each carrier is considered both as a wanted and an unwanted signal. The result of the computations is shown in Figures 5-8 and

5-9. These figures also show the effect of orthogonal polarization on the mutually interfering carriers. The first number next to each curve indicates the wanted signal and the second number the unwanted signal; the numbering is the same as in Table 5-4. Computational time for these figures is on the order of seconds.

Although an analysis program would typically be used to examine a specific set of networks, it can be seen that a program of this type can be of more general deductive value. The results given, for example, provide general insight into the effect in spacing of non-homogeneity. These results were alluded to in Section 5.1. We see (using the notation of that section) that $\Delta\theta_{11} \approx \Delta\theta_{22} \approx \Delta\theta_{12} \gg \Delta\theta_{21}$, so that in

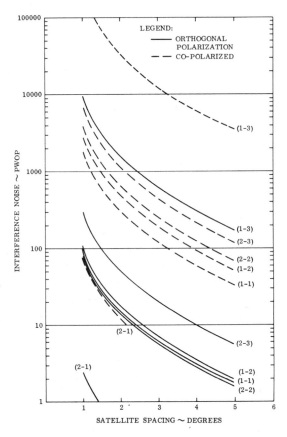

Figure 5-8 Interference *vs* Orbital Spacing for Various Carrier Combinations—Fixed Satellite Service (FDM Telephony)

Source: Ref. [25] and [42].

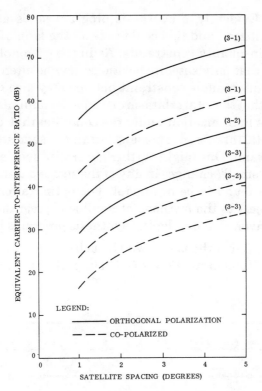

Figure 5-9 Interference *vs* Orbital Spacing for Various Carrier Combinations—Fixed Satellite Service (FM Television)

spite of the last inequality we would not say that this carrier pair causes inefficient orbit utilization. On the other hand, it can be seen that $\Delta\theta_{31} \ll \Delta\theta_{11} \approx \ll \Delta\theta_{33} \approx \theta_{13}$, which indicates a definite in compatibility that results in an inordinately large spacing requirements in order to protect carrier 1.

If the result of examining a set of networks reveals that one or more carriers receives more interference than the maximum permissible, one can use the analysis program to find the conditions which would eliminate that situation. One way to do this would be to modify initial satellite positions and then rerun the program. This may have to be repeated several times before finding a set of satellite positions for which interference conditions are satisfied. The iterated use of an analysis program, in a fashion similar to that just described, in effect constitutes a "synthesis" program when the iterations are automated. Evidently, a synthesis program needs an internally-built stop criterion and this could take on the form of one or more equality or inequality constraints. As in the expample above, one may decree that only satellite positions may be altered subject to, say, an elevation angle constraint but one may also construct the program so that the total orbital arc occupied is minimized. One may perform this minimization under the condition that both a single entry and total interference noise constraint be satisfied, or only the latter constraint. One might further generalize such a program by permitting "small" changes in any of the signal parameters if such changes are found to be beneficial. The optimization problem at hand is subject to the methods of nonlinear programming which have been invoked in developing synthesis programs [33, 34].

An example [33] of the use to which such a program can be put is capsuled in Table 5-5 and Figure 5-10. The problem is the assignment

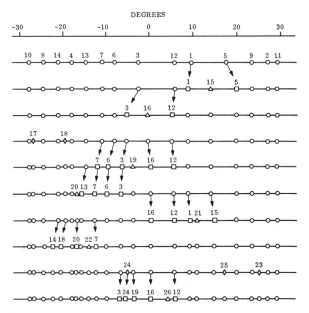

O EXISTING SATELLITES   □ SATELLITES REQUIRED TO BE RELOCATED

△ NEWLY ENTERED SATELLITES   ◊ NEWLY ENTERED SATELLITES WITHOUT RELOCATION OF ADJACENT SATELLITES

Figure 5-10 Satellite Arrangement for the Dynamic Orbital Position Assignment

of satellite positions to a set of satellites that are implemented sequentially in time. The table shows the relevant geometrical parameters. It is assumed that initially satellites 1 to 14 are in orbit in the orbit locations shown on the first line of Figure 5-10. Succeeding lines show the sequential introduction of new satellites which can sometimes be done without relocating existing ones. Thus, a systhesis program can be a powerful tool in finding optimum satellite arrangements. However, the computing time can become very significant especially if other factors, such as beam shapes and coordinated traffic plans, are to be optimized as well.

5.7 Narrow-Beam Satellite Antennas: Frequency Re-use at the Satellite

Narrow-beam (or "spot" beams) satellite antennas present one of the greatest potential tools for significantly improving the utilization of the geostationary orbit. By "narrow-beam" we mean a beam whose width in any plane (peripendicular to the direction of propagation) as seen from the gestationary orbit is substantially less than the earth's disk ($\approx 18°$). To the extent that independent earthward beams can be transmitted from the same position in orbit, this

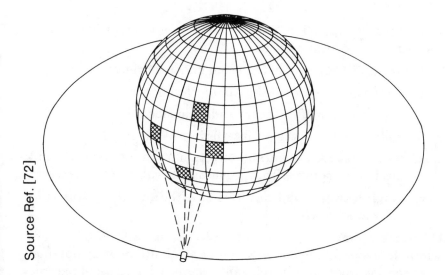

Source Ref. [72]

Figure 5-11 Illustration of Frequency Re-use at the Satellite through Narrow-Beam Satellite Antennas

constitutes the same number of frequency re-uses. The potential for frequency re-use at the satellite depends on the size of a typical coverage area and upon the required areal separation necessary to achieve the requisite inter-beam isolation. The latter, in turn, is a function of the achievable satellite antenna sidelobe pattern. Figure 5-11 gives a visual feeling for the situation at hand. Note that the frequency re-use could be achieved either by a single satellite or several independent satellites. Since the angle subtended at the satellite by most service areas of interest, including even large countries, is only on the order of 1° to 3° in the largest dimension (exceptionally, the USA subtends approximately a 5° × 3.5° solid angle), one can see the significant frequency re-use potentially available. It will be noted that the full potential of frequency re-use will be realized only by providing total interconnectivity among the various beams.

To place the preceding ideas on a more quantitative footing, consider the carrier-to-interference ratio ("interbeam" interference) at a point within any of the coverage areas due to sidelobe radiation from other beams.

$$\left(\frac{C}{I}\right)^{-1} = \frac{\Sigma \, P_i \, G_i \, (\alpha_i)}{P_o \, G_o \, (\alpha_o)} \tag{5-18}$$

where the summation is taken over all interfer beams originating at the same orbital location, and

P_o = transmitted power of the beam under consideration

P_i = transmitted power of the ith interfering beam

G_o = gain function of the beam under consideration

G_i = gain function of the ith interfering beam

α_o = angle between pointing direction of wanted beam and arbitrary point within coverage area (wanted receiver)

α_i = angle between pointing direction of the ith bean and direction of wanted receiver.

Of course, normally α_o is equal to or less than half the -3dB beamwidth. In general, the angles α_i can be straightforwardly obtained from the dot product of the relevant vectors. Given a set of parameters defined in (5-18) it is a straightforward matter to compute the resulting CIR. The reverse process, however, is considerably more involved. That is, for a specified CIR one may ask how many frequency re-uses are potentially available. The CIR specification is not sufficient in itself to bound the problem. One must in addition define the coverage areas as well as the satellite antenna radiation patterns.

Because of the complexity of the problem, simplifying assumptions are usually made to render it tractable. In particular it is usually taken that the projected coverage areas as viewed from the satellite are of equal size.* Under this assumption, and assuming the beam gain function to be of the form (4-40a), Fuenzalida [35, 36] has investigated the potential for frequency re-use as a function of the sidelobe decay exponent and the CIR. However, strictly from the re-use potential viewpoint a simpler approach is possible.

Consider Figure 5-12 which shows a hexagonal tiling of the earth's disk as viewed from orbit; this geometry gives the densest satellite beam packing [37]. The maximum number of frequency re-uses per satellite, for two orthogonal polarizations, is

$$N_R = \frac{2\Omega}{\phi}$$

where Ω is the solid angle subtended by the earth (0.07 steradian) and ϕ is the smallest solid angle subtended at the satellite such that the requisite interbeam isolation is attained. If we label the area covered by ϕ the "coverage" area we need to distinguish it from the "service" area, that part within ϕ intended for service from a single beam, normally defined by the 3dB beamwidth. Thus, for co-frequency beams we can write the beam center separation Δc as

$$\Delta c = K\Omega_b \qquad (5-19)$$

where Ω_b is the individual (3dB) beamwidth and K is a factor which ensures that when the separation is as in (5-19) the beams are "independent," i.e., the requisite CIR exists. One then has

$$\phi = \frac{\sqrt{3'}}{2}(K\Omega_b)^2$$

---

*This assumption does not conform to reality, in general, not just because realistic coverage areas are unequal, but even for equal such areas the curvature of the earth requires different beamwidths, especially for the more extreme latitudes.

and using the rule-of-thumb

$$\Omega_b = 1.2 \, (\lambda/D_s)$$

where λ is the wavelength and D_s the antenna diameter. Combining the previous equations

$$N_R = (D_S/\lambda)^2 \, (2.985K)^{-2} \tag{5-20}$$

If there are N_S satellites in orbit, the implication of Figure 5-12 is that they all have the same antenna diameter, whence the total area of all satellite antenna apertures is

$$A = (\pi/4) \, D_5^2 \, N_5$$

The number of total frequency uses available from orbit is thus [37]

$$N_T = N_R \, N_S = (A/\lambda^2)/(7K^2) \tag{5-21}$$

Equation (5-21) is plotted in Figure 5-13 as a function of the normalized total aperture using the value $K = 2.2$. This value of K should yield interbeam isolation of at least 30 dB. Figure 5-14 shows, for fixed aperture, the effect of varying K, which is the simplest way of characterizing sidelobe behavior. N_T may be taken as the appropriate measure of orbit utilization in this context.

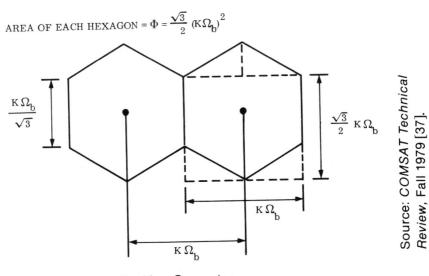

AREA OF EACH HEXAGON $= \Phi = \dfrac{\sqrt{3}}{2} \, (K\Omega_b)^2$

$\dfrac{K\Omega_b}{\sqrt{3}}$

$\dfrac{\sqrt{3}}{2} \, K\Omega_b$

$K\Omega_b$

$K\Omega_b$

Source: *COMSAT Technical Review*, Fall 1979 [37].

Figure 5-12 Beam Packing Geometry

Source: COMSAT *Technical Review,* Fall 1979 [37].

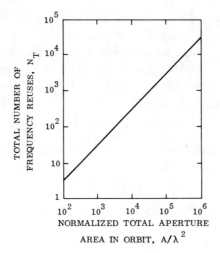

Figure 5-13 Frequency Re-uses as a Function of Total Aperture

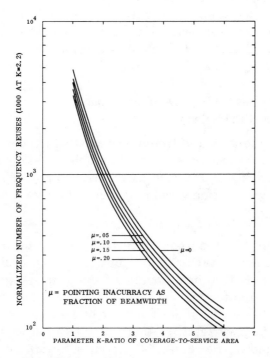

Figure 5-14 Effect of Beam Sidelobe Characteristics and Pointing Accuracy on Frequency Re-use.

The preceding equations were derived on the implicit assumption of perfect antenna pointing. The effect of inaccurate pointing is readily deduced for the simple model we have been considering. In particular, we may suppose that the pointing errors have the effect of sweeping a larger area, corresponding to an effective increase in the beamwidth. Thus, we set

$$\Omega_6' = (1 + \mu)\,\Omega_b$$

where μ is the pointing error as a fraction of the beamwidth. It is easy to see that (5-21) then extends to

$$N_T = (A/\lambda^2)/[7K^2\,(1 + \mu)^2] \qquad (5\text{-}22)$$

which is shown in Figure 5-14 for several values of μ. It can be seen that even for a seemingly small inaccuracy of 10% the orbit utilization potential is reduced to about 80% of its maximum value, so that efforts should be made to minimize μ. In recognition of the importance of pointing accuracy the WARC 79 stipulated that the pointing accuracy of satellite narrow-beam antennas should be no worse than 10% of the half-power beamwidth relative to the nominal pointing direction, or 0.3 degree relative to that direction, whichever is greater [3b].

5.7.1 Techniques for Improving the Capability of Frequency Re-use at the Satellite

A number of factors and techniques exist which can increase the capability of frequency re-use at the satellite, taking into account practical considerations such as the irregularity of real coverage areas. Here we mention very briefly several relevant items.

a) Practical Antenna Patterns

As can be appreciated from Figure 5-14, the narrow beam satellite antenna sidelobes have a significant impact on orbit utilization. There are many technical factors affecting the design and performance of such antennas and these are discussed by the CCIR [38] where also the status of development by various organizations is summarized. As with earth station antennas a good deal of attention has been paid to the subject of reference patterns, to be used for interference calculations in lieu of actual patterns. However, there does not get seem to have evolved a model with general validity or acceptance. One such pattern for the fixed satellite service has been tentatively evolved [38] while another adopted at the 1977 WARC is widely used as a broadcasting standard [3e]. The latter pattern is less conservative than the FSS reference pattern, but as can be seen from Figure 5-15 it is still conservative in comparison to achievable performance with controlled aperture illumination taper.

b) Scanning Spot Beams

Up to now it has been implicit that the spot beams pointed permanently in a given direction. However, it has been shown [39 and references therein] that in certain situations scanning spot beams can make optimal use of satellite power. In particular, in areas where there would be low traffic density, dedicated spot beams would be under-utilized and hence represent uneconomical satellite design. In such cases scanning spot beams could be used to advantage. A side benefit would be the reduction of interference since at any one time there would be less beams than service areas.

c) Shaped Beams

Another possibility to increasing the orbit utilization potential of

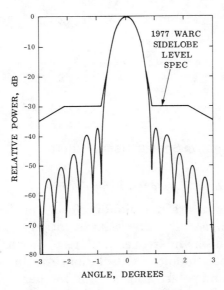

Figure 5-15 Satellite Antenna Reference Pattern and Computed Pattern with Tapered Aperture Illumination

narrow beam antennas is to shape the beams to conform closely to the contour of a coverage area. Such shaped beams may be particularly useful where the angular separation between two separate coverage areas is fairly small. An illustrative computer-generated shaped beam is shown in Figure 5-16 [40]. Typically, such beams require relatively large feed horn clusters and a larger aperture ($\sim$ 2-3 times) than would be required for a simple (elliptical) pattern. A possible drawback, however, is that the flexibility of satellite repositioning may be significantly curtailed unless the pattern can be made reconfigurable. This subject is revisited in Chapter 8, Section 8.3.1., within the context of nulling, or interference cancellation, and a beam shaping technique that is amenable to reconfiguration is briefly discussed.

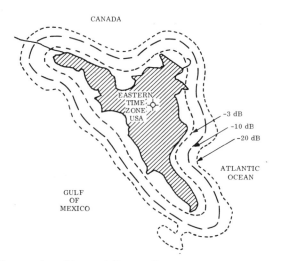

Figure 5-16 Illustrative Shaped Beam Pattern

d) Polarization and Frequency Interleaving

Where service areas are unavoidably closer than would be required to achieve the necessary interbeam isolation, a judicious assignment of polarization and frequency plans to the various beams may provide the needed additional isolation. An illustrative arrangement of beams and polarization *(P)* and frequency *(F)* assignments is shown in Figure 5-17 [35], where P_1 and P_2 represent orthogonal polarizations and F_1 and F_2 stand for two interleaved frequency plans. As was pointed out earlier, however, the latter measure is likely to provide significant assistance only for analog signals. The extent of orbit utilization increase is shown in **Figure** 5-18 which for given P and F improvements show the possible increase in beam density (γ) relative to the case where both $P = F = 0$ dB [35].

Source: COMSAT Corp. Technical Memo, CL-49-69 [35]; AIAA Paper #70-442 [36].

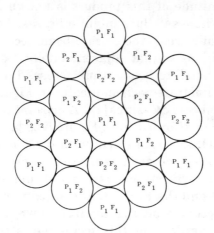

Figure 5-17 Arrangement of Polarizations and Frequency Plans for Earthward-Directed Beams

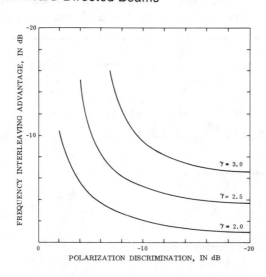

Figure 5-18 Resulting Increase in the Beam Density *vs* Polarization Discrimination and Interleaving Advantage

e) Crossed-Beam Arrangements

The interference seen at an earth station is proportional to the product of the earth station antenna gain in the direction of an interfering satellite and the gain of the interfering satellite antenna in the direction of the wanted earth station. A technique which can reduce the magnitude of this product is known as crossed-beam geometry [41]. Where satellites might otherwise be co-longitudinal with their coverage areas, the crossed-beam technique rearranges the satellite positions so that adjacent satellites do not illuminate adjacent areas. Satellite beams thus "cross" the beams of other satellites, as is illustrated schematically in Figure 5-19, where n_c indicates the number of crossings [42]. A more intuitive appreciation of the technique is afforded from Figure 5-20 [43] which shows a hypothetical satellite arrangement with $n_c = 0$ and $n_c = 2$. For the situation illustrated, the orbit utilization improvement has been computed for an earth station $D/\lambda = 100$ and satellite antenna beamwidth of 2° and the FSS satellite antenna reference pattern. The resulting effect on orbit utilization is shown on Figure 5-21 [43]. A possible disadvantage of this technique is that the satellite antenna squint angles are smaller than when $n_c = 0$, thus increasing the sensitivity to pointing inaccuracy; and the elevation angles are smaller, thus increasing the path loss and atmospheric attenuation.

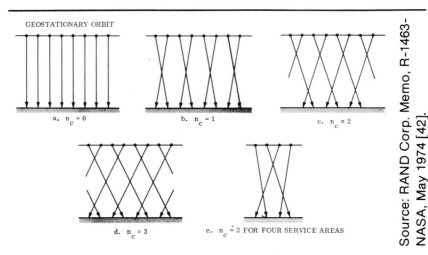

GEOSTATIONARY ORBIT

a. $n_c = 0$ b. $n_c = 1$ c. $n_c = 2$

d. $n_c = 3$ e. $n_c = 2$ FOR FOUR SERVICE AREAS

Source: RAND Corp. Memo, R-1463-NASA, May 1974 [42].

Figure 5-19 Schematic Examples of Crossed-Beam Geometry

However, these may be compensated for by a satellite gain increase which is possible (though not taken advantage of in the example) as a result of the longitudinal offset. This latter possibility is discussed next.

f) Satellite Gain Increase as a Function of Longitudinal Offset

A prescribed area on the earth subtends a solid angle from the geostationary orbit that depends upon its shape and location. In

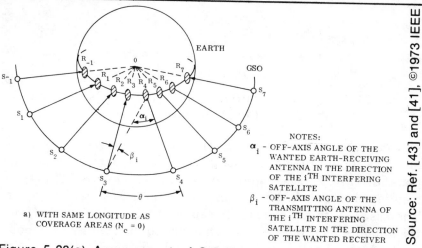

NOTES:
α_i - OFF-AXIS ANGLE OF THE WANTED EARTH-RECEIVING ANTENNA IN THE DIRECTION OF THE iTH INTERFERING SATELLITE

β_i - OFF-AXIS ANGLE OF THE TRANSMITTING ANTENNA OF THE iTH INTERFERING SATELLITE IN THE DIRECTION OF THE WANTED RECEIVER

a) WITH SAME LONGITUDE AS COVERAGE AREAS (N_c = 0)

Source: Ref. [43] and [41], ©1973 IEEE

Figure 5-20(a) Arrangement of Satellites with Same Longitude as Coverage Areas

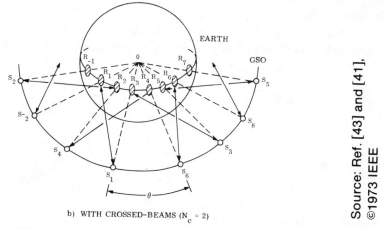

b) WITH CROSSED-BEAMS (N_c = 2)

Source: Ref. [43] and [41], ©1973 IEEE

Figure 5-20(b) Arrangement of Satellites with Crossed-Beams

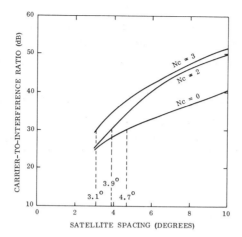

Source: Ref. [43] and [41],©1973 IEEE

Figure 5-21 An Example of the Reduction in Minimum Satellite Spacing by the Crossed-Beam Technique

particular, an area of given size subtends a progressively smaller angle as the latitude or the longitude relative to the sub-satellite point increase. This is graphically illustrated in Figure 5-22 [44]. As mentioned earlier, the longitudinal offset increases the slant range. The loss due to that factor is shown in the figure and subtracted from the antenna gain increase, which is due to the smaller subtended angle. As can be seen, rather substantial net gain is available. Not shown is the additional loss due to atmospheric/precipitation attenuation. This specifically is related to the elevation angle, but may be quite mild in certain climatic regions. Thus, the use of crossed-beam geometry and the consequent antenna gain increase made possible thereby can jointly prove to be very useful.

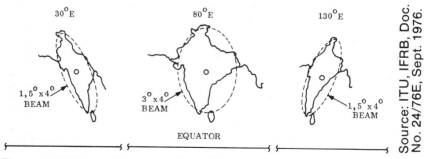

Source: ITU, IFRB, Doc. No. 24/76E, Sept. 1976.

Figure 5-22 Beam Coverage from Three Different Satellite Positions

5.8 Power Density Constraints

The *Radio Regulations* contain a number of restrictions [3c, 3d] on the power density or power flux density (power per unit bandwidth and unit area) of emissions, the purpose being to provide inter-service or intra-service protection but not specifically focused on improving orbit utilization in the sense of Section 5.3. The most ubiquitous of these regulations affecting geostationary satellite networks are the ones which limit the power flux density (PFD) at the earth's surface [3d]. This imposes a direct limitation on the satellite *e.i.r.p.*, depending upon the signal and modulation characteristics. We have seen earlier [see, e.g., Figures 4-21, 4-27] that, especially for digital systems, greater potential orbit capacity is available as the CNR increases. Now, the *e.i.r.p.* is not the sole determinant of the CNR, but for an earth station of fixed character-istics *(G/T)* the CNR is proportional to the *e.i.r.p.*. Thus, the PFD restrictions may impose an ultimate limit on the orbit capacity, though the more practical effect is likely to be a constraint on the minimum size of earth station antennas or to impose a requirement for energy dispersal, i.e., artifically spread the signal power over a larger bandwidth. In the next two subsections we examine the satel-lite *e.i.r.p.* limitations implied by PFD constraints for FDM/FM telephony and PSK signals, respectively.

5.8.1 Satellite EIRP Limitations due to Power Flux Density Constraints for FDM/FM Telephony

The constraint on the power flux density, *F,* at the surface of the earth due to satellite emissions can be expressed in the form

$$10 \, log \, F \le D + K \, (\delta), \, dBW/m^2/B \qquad (5\text{-}23)$$

where the parameter D depends on the frequency band of interest, $K(\delta)$ is a function of the angle of arrival, *(δ = 0* corresponds to grazing incidence), and B is a reference bandwidth which also depends on the frequency in question. Currently, $B = 4kHz$ is the norm for fre-quencies less than 12.75 GHz, and $B = 1MHz$ is standard for frequen-cies greater than 17.7 GHz. It is expected that the smaller reference bandwidth will be limiting because the power in spectral lines remains unaffected by that bandwidth. In calculations we will therefore assume $B = 4kHz$.

The power flux density in the "worst" 4kHz band in the δ direction is given by

$$F = \frac{P_{4,\,m}}{4\,\pi d^2\,(\delta)} \qquad (5\text{-}24)$$

where $d(\delta)$ is the slant range (meters) and $P_{4,\,m}$ is the maximum power in any 4kHz band. the determination of $P_{4,\,m}$ depends on the power spectral density (PSD) of the signal, a relatively complex function of the modulation parameters. However, for certain limiting cases the PSD is relatively straight forward to obtain. These limiting cases are when the *rms* modulation index M_0 is "large," i.e., $M_1 \gtrsim 1$, and when it is "small," i.e., $M_0 \ll 1$. Under the first condition the PSD approaches a continuous Gaussian-shaped spectrum irrespective of the baseband spectral shape, hence independent of the form of pre-emphasis network. On the other hand, for medium or small values of M_0 the modulated carrier PSD has a complicated shape that depends

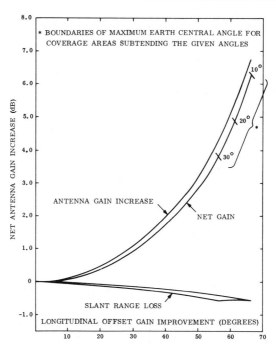

Source: ITU, IFRB, Doc. No. 24/76E, Sept. 1976.

Figure 5-23 Potential Gain Improvement *vs* Displacement of Satellite from Center of Coverage Area

on the baseband spectrum, and hence on the type of pre-emphasis. However, for these conditions a significant spectral line appears at the carrier frequency and we shall assume this component above determines $P_{4, m}$.

In Appendix H we will show that for the limiting cases in question, (5-23) translates into the following limitations on the satellite *e.i.r.p.*, measured in dB;

$$eirp \leq 15.2 + D + K(\delta) + 20 \log d(\delta)$$
$$+ 10 \log M_0 + 20 \log d(\delta) \tag{5-25}$$

$$eirp \leq 11 + D + K(\delta) + 20 \log d(\delta)$$
$$+ 4.34 M_0^2 [0.4/\varepsilon + 1.15 + 0.368(1 + \varepsilon + \varepsilon^2)] \tag{5-26}$$

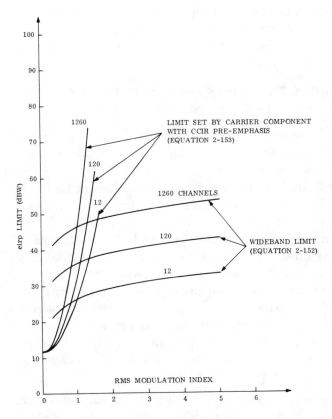

Figure 5-24 Satellite EIRP Limits at Zero Elevation

for M_0 large and small, respectively, where n is the number of channels, ε is the ratio of lowest to highest baseband frequency, and in (5-26) CCIR pre-emphasis has been assumed.

To illustrate the implications of (5-25) and (5-26), Figure 5-24 shows the *e.i.r.p.* constraint determined therefrom for $K(\delta) = 0$, $D = -152$, and n = *12*, *120*, and *1260* channels, using values of ε in Table H-1 of Appendix H. The actual curve near the intersection of the two types would likely effect a smooth transition. at low modulation index, the PFD constraint may limit the system designer's options, or impose the need for energy dispersal. The low modulation index condition may also occur in nominally large index systems when the loading is light since the modulation index is a function of the loading. Equation (5-26) can also be applied to this latter situation by choosing the worst likely baseband arrangement. As the loading varies, so would the necessary amount of energy dispersal. the CCIR [45] has reported on possible methods of energy dispersal applicable to this adaptive requirement.

5.8.2 Satellite EIRP Limitations due to Power Flux Density Constraints for Digital (MCPSK) Signals

As with the previous case of telephony signals we concentrate here on the smaller of the reference bandwidths, i.e., 4kHz. We expect this to impose the most stringent limitations, which occur in the presence of spectral lines. As before, the essential element for the purpose at hand is $P_{4, m}$ which depends further on the modulated carrier PSD. The latter is a function of the phase-modulating waveform and of the input symbol probability distribution. In the particular case of rectangular signalling pulses, which we assume here, the PSD is composed of a continuous part and a single spectral componenet at the carrier frequency. It is not possible to give general quantitative results in a compact fashion. Hence, for illustrative purposes, which should be generally representative, we present results only for the case $M = 4$ *(QPSK)* and for the situation where three of the four symbols are equiprobable with probability of occurrence $\frac{1}{3}[1\text{-}p]$ and the fourth symbol has probability p.

Under these conditions, calling upon the results of Appendix I, we find the *e.i.r.p.* limitation implied by the continuous portion of the spectrum above is given by (in dBW),

$$eirp \leq D + K(\delta) + 20 \log d(\delta) + 20 \log (R/4000)$$
$$- 10 \log \left\{ 1 - \left(\frac{4p\text{-}1}{3} \right)^2 \right\} \tag{5-27}$$

where R is the symbol rate. For the carrier spike, the corresponding equation is

$$eirp \leq D + K(\delta) + 20 \log d(\delta) - 10 \log \left(\frac{4p\text{-}1}{3} \right)^2 + 11 \tag{5-28}$$

For the case studied previously, $\delta = 0$, $K(\delta) = 0$, $D = -152$, Equation (5-27) is bounded in the "worst" case by

$$eirp \leq 0.4 - 10 \log (R/4000) \tag{5-29}$$

which occurs in the statistically balanced situation, $p = 0.25$. Thus, the *e.i.r.p.* is limited in this instance by the multiple of 4000 which characterizes the data rate. But of course the *e.i.r.p.* is required to be proportional to the data rate, all other things being fixed. Thus, Equation (5-29) represents an equal constraint for any data rate.

The *e.i.r.p.* constraint set by the carrier spike is shown graphically in Figure 5-25. For moderate statistical imbalance and low data rates, this constraint does not pose a problem. However, even for moderate imbalance the carrier spike will constrain the *e.i.r.p.* for higher data rates. For example, suppose $p = 0.35$; from Figure 5-25 we have *e.i.r.p.* $\leq 29\,dBW$. From Equation (5-29), *e.i.r.p.* $\leq 29\,dBW$ for $R \approx 3 \times 10^6$. Thus for data rates greater than about *3M baud/s* the carrier spike would prevent the *e.i.r.p.* from increasing, such an increase being required to support higher data rates.

A statistical imbalance among the data symbols can be expected to exist. For example, if an analog baseband is sampled and quantized an imbalance will inevitably result if the baseband doe not have a uniform probability density. The results of this section show that designers of wideband data transmission systems will have to pay careful attention to mitigating the carrier spike, either (for example) by baseband compression or energy dispersal. Methods for accomplishing the latter, with digital signals, are also discussed in [45].

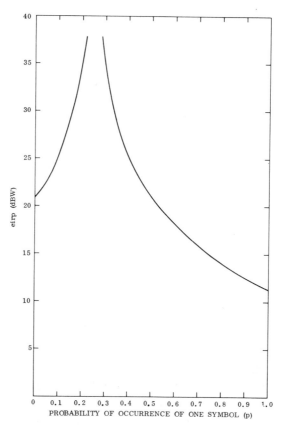

Figure 5-25 EIRP Limitation for 4-Phase PSK as a Function of Symbol Probability Imbalance

5.9 The Use of Frequency Bands for Both Uplink and Downlink

Frequency bands have been typically allocated for one direction of transmission, either uplink or downlink. However, on a limited basis the most recent revisions to the *Table of Frequency Allocations* permits bidirectional use of certain bands, i.e., both uplink and downlink are permitted. This mode of operation has both advantages and disadvantages in a very general context. However, from the orbit utilization viewpoint it can bring about appreciable improvement, as will now be demonstrated.

Consider Figure 5-26 which shows two satellites using frequency bands in the reverse sense, i.e., the uplink frequency for satellite 1 is the downlink frequency for satellite 2 and *vice versa*. Considering satellite 1 as the wanted one the (uplink) CIR is given by

$$\left(\frac{C}{I}\right)^{(r)}_{s12} = \frac{P_{gt11}\, G_{gt11}\, G_{sr11}\, L_{gt11}}{P_{st21}\, G_{st21}\, G_{sr12}\, L_{st21}} \tag{5-30}$$

where we use the convention of Section 4.1 and assume co-channel operation and no polarization isolation, and the superscript implies reversed frequency operation.

From the figure, it is seen that the off-axis antenna gains can be written as

$$G_{st21} = G_{st2}\,(\propto + \eta);\ G_{sr1}\,(\propto + \delta)$$

In other words $(\propto + \eta)$ is the angle off boresight of the interfering antenna towards the wanted satellite, and $(\propto + \delta)$ is the angle off boresight of the wanted receiving antenna in the direction of the interfering satellite. The angles δ and η are the angles between the subsatellite and beam pointing directions. Even for rather large angular separations θ these gains must be relatively small. For example, if $\theta = 120°$, $\propto = 30°$ and even for global coverage beams 30° is about four half-beamwidths. The inter-satellite distance is given by

$$d_{12} = h\,(1 + \Delta)\,\sqrt{2\,(1 - \cos\theta)}$$

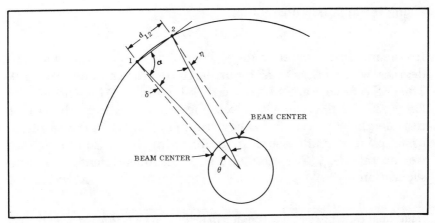

Figure 5-26 Geostationary Orbit Geometry for Reverse-Frequency Assignments

where $\Delta = R_e/h$, R_e is the Earth's radius, and h is the geostationary orbit altitude.

The interference at the earth station, which is due to the uplink of satellite 2, cannot be given a simple general representation, and moreover is unrelated to orbital parameters. It depends on the proximity of the two earth stations, the terrain, and isolation mechanisms such as site shielding. As a simple approximation we assume as in [46] that this interference is equal to that produced in the satellite.*

Consider now the "quasi-homogeneous" model wherein co-frequency satellites and reverse-frequency satellites alternate in the orbit at spacing $\Delta\theta$. We assume homogeneity in the sense that all earth station powers are equal to one another, all earth earth station antenna diameters and sidelobe patterns are identical, all satellite transmit powers are the same, and all satellite antenna characteristics are identical. In conjunction with the latter the homogeneous model stipulates earth coverage (or beamwidth coverage of all relevant earth stations), an assumption we shall continue here. It is clear that this assumption underbounds the orbit utilization potential of reversed-frequency assignments since the ratio of off-axis angles $(\approx\propto)$ to 3dB beamwidth is thereby minimized; equivalently the isolation due to satellite antenna gain is less than for a beam with more restricted coverage.

Using the approximate relation

$$G_{\mathrm{m}} \approx 27{,}000/\theta_3^2$$

for the maximum antenna gain, where θ_3 is the 3dB beamwidth in degrees, we find $G_m \approx 20dB$ for an earth coverage antenna $(\theta_3 \approx 17)$. The FSS reference antenna pattern [38] shows that $G_m/G(\theta) = .01$ for $2.6\ (\theta_3/2) < \theta \le 6.3\ (\theta_3/2)$, which encompasses all but the fewest near-neighbor reverse-frequency stallites. Thus, as a conservative assumption we shall assume in the following that in Equation (5-30) we can take $G_{sr11}/G_{sr12} = 100$ and $G_{st21} = 1$, independent of satellite separation.

---

*Mathematically, this will make this interference source a function of satellite spacing, which it is not. However, it is a convenient artifice. We shall thus, after computing the total interference into the satellite, double that amount.

Thus, the total CIR for reverse-frequency contributions to a given network can be approximated by*

$$\left\{(C/I)^{-1}\right\}^{(r)} = \frac{(P_s/P_g)\,(\lambda_u/\lambda_d)^2}{69\,G_r} \sum_{k\,odd} [(1 - \cos k\,\Delta\theta)]^{-1}$$

where G_r is the earth station on-axis receive gain. But,

$$2\,(1 - \cos k\,\Delta\theta) = 4\sin^2(k\,\Delta\theta/2) \approx (k\,\Delta\theta)^2$$

where the last approximation overestimates the interference. Converting to degrees, we have

$$\left\{(C/I)^{-1}\right\}^{(r)} = \frac{(P_s/P_g)(\lambda_u/\lambda_d)^2}{69\,G_r} \sum_{k\,odd} (k\,\Delta\theta \cdot \pi/180)^{-2}$$

$$\approx \frac{112\,(P_s/P_g)\,(\lambda_u/\lambda_d)^2}{G_r(\Delta\theta)^2} \tag{5-31}$$

The co-frequency CIR is given by Equation (4-30):

$$\left\{(C/I)^{-1}\right\}^{(c)} = [1 + (\lambda_u/\lambda_d)^2]\frac{A\,Z\,(\beta,\,N)}{G_r(\Delta\theta)^\beta}$$

since co-frequency satellites are spaced at multiples of $2\Delta\theta$. Hence the total CIR is

$$(C/I)^{-1} = \frac{112\,(P_s/P_g)\,(\lambda_u/\lambda_d)^2}{G_r(\Delta\theta)^2} = [1 + (\lambda_u/\lambda_d)^2]\frac{A\,Z\,(\beta,\,N)}{G_r(2\Delta\theta_0)^\beta} \tag{5-32}$$

For co-frequency operation alone (the "standard" homogeneous case), we have

$$(C/I)_0^{-1} = [1 + (\lambda_u/\lambda_d)^2]\frac{A\,Z\,(\beta,\,N)}{G_r(\Delta\theta_0)^\beta} \tag{5-33}$$

Therefore, setting (5-32) equal to (5-33) yields the desired relationship between the spacings in the two situations:

$$(\Delta\theta/\Delta\theta_0)^\beta = 2^{-\beta} + (\Delta\theta)^{\beta-2}\frac{112\,(P_s/P_g)}{A\,Z\,(\beta,\,N)\,[1 + (\lambda_d/\lambda_u)^2]} \tag{5-34}$$

We may expect $(P_s/P_g) < 0.1$; hence for the CCIR pattern $(A \approx 1585)$, and assuming $\lambda_u/\lambda_d = 1.5$, the coefficient of $(\Delta\theta)^{\beta-2}$ on the right-hand side of Equation (5-17) is less than 8×10^{-4}. For any resonable spacing, therefore, the second term on the RHS of Equation (5-34) is negligi-

---

*Including a factor of 2 for earth station interference, as previously discussed.

ble. Hence, $(\Delta\theta/\Delta\theta_0) \approx 0.5$, which shows the significant improvement in orbit utilization that might be possible by allowing the use of reverse-frequency assignments. The 2:1 improvement just shown is specifically dependent on the model that was used and may differ for other assumptions, e.g., with narrow-beam antennas.

The equations derived assumed that reversed-frequency interfering satellites were in one another's antenna sidelobes. This is clearly violated when the orbital separation nears 180°, in which case the full main beam gain of the interfering satellite is seen by the wanted satellite. This will impose certain limitations on the receiving gain of antennas toward the earth's limb, or on the interfering power in that direction, as described in [46]. Another liability of reverse-frequency operation is increased interference into terrestrial systems and additional coordination constraints.

5.10 Propagation Factors

In earlier discussions we have not been explicit regarding the actual frequencies over which particular uplink and downlink bands are defined. In fact the actual frequencies used have a significant bearing on a network's design and operation, owing to the frequency dependence of propagation conditions, which have a substantial effect at frequencies about 10GHz and above. The occurrence of rain, in particular, can induce significant attenuation of the signal for fractions of the time that are small but operationally not insignificant. The influence of rain on satellite communications at high frequencies has long been appreciated and has been extensively discussed (see, e.g., [39, 54, 55, 56, 57, 58]) from the phenomenological and statistical standpoints, the ultimate aim being to assist the system designer so as to correctly provide for a given outage probability. However, beyond a recognition of its fundamental role [59], the effect of rain on orbit utilization has not been evaluated in a general way, and indeed it is difficult to do so.

Nevertheless, we can give a preliminary indication, through a simplified example [60] of how rain attenuation can have a detrimental effect on orbit utilization. Consider a single interfering satellite at $\Delta\theta°$ from the wanted one and make the assumption that the occurence of rain simultaneously over the wanted transmitting and receiving earth stations has negligible probability. Further assume conservatively that when the desired uplink is attenuated, the uplink interferer is not. Therefore, we need to protect the link against either uplink or downlink rain attenuation. In these two instances we have, respectively, for the carrier-to-noise plus interference, denoted C/NPI,*

$$(C/NPI)^{-1} = \alpha_u \, (\rho_u^{-1} + \eta_u^{-1}) + \rho_d^{-1} + \eta_d^{-1}$$

$$(C/NPI)^{-1} = \alpha_d \, \gamma \, \rho_d^{-1} + \rho_u^{-1} + \eta_u^{-1} + \eta_d^{-1}$$

where

$\alpha_u \, (\alpha_d)$ = uplink (downlink) attenuation (≥ 1) corresponding to the desired outage probability,

$\rho_u \, (\rho_d)$ = clear weather uplink (downlink) carrier-to-noise ratio,

$\eta_u \, (\eta_d)$ = clear weather uplink (downlink) carrier-to-noise ratio,

γ = increase in receiver noise temperature.

It should be noted that the downlink CIR is unaffected by the propagation conditions since it can safely be assumed that both the wanted and interfering signals pass through the same rain cell. It is also assumed that gain control in the satellite is used to maintain the radiated power constraint.

Equating the right-hand sides of the two previous equations (with $\gamma = 1$) permits us to solve for ρ_u:

$$\rho_u^{-1} = \frac{(\alpha_u + \alpha_d - 2) \, (C/NPI)^{-1} \, \mu}{(\alpha_u \, \alpha_d - 1) \, (1 + \delta + \mu) \, (\alpha_u + \alpha_d - 2) \, \delta \, \mu}$$

where

$\delta = \eta_u / \eta_d$

$\mu = (\eta_u^{-1} + \eta_d^{-1}) / (\rho_u^{-1} + \rho_d^{-1})$

Since for identical networks we have

$$\rho_u = G_{0, u} \, 10^{-3.2} \, (\Delta\theta)^{2.5}$$

for the CCIR antenna pattern, where $G_{0, u}$ is the uplink transmit antenna gain, we can relate spacing to the propagation conditions. The result is shown in Figure 5-27 (taken from [60]) for the particular case *(C/NPI)* = *15db*, μ = *10dB*; also, the 30/20 GHz band was assumed, implying δ = 2.25 and $\alpha_u \approx \alpha_d^{2.3}$

---

*It may be noted that C/NPI is an adequate measure of performance for purposes of the example; it essentially treats the interference as equivalent noise.

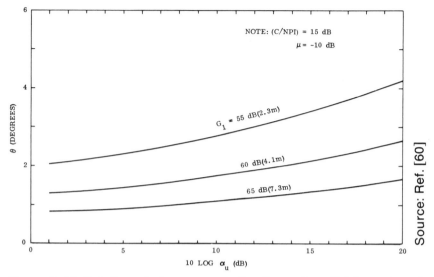

Figure 5-27 Relations between up-path attenuation (*10 log x*ᵤ) and Satellite Spacing (*θ*) (Values in parentheses show the diameter of 30 Ghz antenna of 60% efficiency)

The deficiency with this example is that it does not present a complete picture, which would be given by the total distributions as a function of the margins. This is the indicated direction for future work.

Still, the message is unequivocal that rain attenuation requires a larger spacing than would otherwise be the case with no attenuation. Actually, there is an additional degrading effect not reflected in this example, namely that due to the depolarization induced by precipitation [54]. This phenomenon will evidently reduce the isolation between nominally orthogonally polarized waves, although this degradation can be mitigated at the cost of implementation complexity [e.g., 61]. Compensation of depolarization is discussed at greater length in Chapter 8.

The attenuation due to rain can be more or less neutralized both as far as its effect on orbit utilization and its impact on satellite design are concerned. This is achieved by using site diversity on the earth, i.e., using two earth stations sufficiently widely separated that the probability of rain over both sites at the same time is negligible.

Although there are obvious economic consequences, these should at least be partially compensated by the savings gained by not having to build a satellite as high powered as would have been necessary without site diversity. However, there is an additional operational problem, namely the fact that the diversity stations must be connected by a diversity interconnection link (DIL) so that the signals arriving at the two earth stations can be appropriately combined.

Another method to partially compensate for rain, again with operational/economic implication, is to use adaptive uplink power control, i.e., increase the uplink power in proportion to the attenuation [62, 63, 64]. Actually, this approach can take one of two forms: in a single carrier-per-transponder mode which normally operates at saturation, uplink power control can compensate only for uplink fading; but in the case where many (small) carriers share a transponder in a more or less linear mode, uplink power control can compensate to some extent for downlink attenuation. In the latter instance, however, more interference into the uplink of other networks will result. A related method which may be useful when the satellite connects many dispersed earth terminals is downlink adaptive power control (or equivalently adaptive error correction coding). Clearly, no advantage can be expected if every carrier destination is protected, but the fact that these destinations are widely separated makes it unlikely that rain of sufficient intensity will be occurring at more than one location. Thus, a common reserve of power (or time elements in a TDMA system) adaptively assigned to a downlink fading carrier may be a relatively efficient method of compensating for rain. There is, however, one identifiable drawback, namely that when a carrier's power is elevated it will during that interval induce additional interference into the earth station receivers of other networks where rain is not occurring.

Thus, without applying specific means of compensation, rain at frequencies greater than 10 GHz will have a negative effect on orbit utilization. There are, however, technical means for mitigating the effects of rain and while these may have a net cost, in some sense, to the system designer, they may nevertheless be considered preferrable to building sufficient margin into the network; and at the same time these means will also tend to have a beneficial effect on orbit utilization.

5.11 Measures of Orbit Utilization: Criteria of Efficiency

In the preceding sections of this chapter we have used various quantities at different times to represent what was meant by orbit utilization. These quantities were intuitively pleasing, or at least plausible, in the context: e.g., number of satellites per unit arc, information rate per unit bandwidth and unit arc, or carrier-to-interference ratio. Certainly, as far as the first two* are concerned, one would expect any reasonable measure of orbit utilization** to increase monotinically with these quantities, all else being fixed. But neither of these quantities is a sufficiently encompassing measure for all situations of interest. The first says nothing about the capacity of individual satellites, while the second does not constrain either the individual satellite capacity nor the number of satellites. Clearly, in practice it is important that the satellite capacity be cost-effective to the operator, and it is also important to accommodate in orbit as many satellites as there are separate entities wishing to operate satellites. Furthermore, there are inevitable operational and cost considerations not explicitly reflected in the measures mentioned. In the most general sense, a measure might be symbolically expressed as

$$M = F(K_t, K_s, K_c) \qquad (5\text{-}35)$$

where $F(\cdot)$ is a suitable functional, K_t is the set of relevant technical factors, K_s is a set of factors reflecting social utility, and K_c embodies cost factors which could include development costs, maintenance costs, revenues, and others.

In order to make the concept of orbit utilization measure most useful it would be highly desirable if a single measure were to be appropriate for all frequency bands and all services and all applications. Unfortunately one cannot expect this to be the case. In other words the preferred form of $F(\cdot)$ will generally depend on the user class. A less abstract formulation than (5-46) which may still possess sufficient generality to be widely applicable emerges from the fundamental assumption that the essential purpose of the orbit/spectrum

---

*The third, carrier-to-interference ratio, is obviously related to satellite spacing, hence to number of satellites per unit arc.

**By a "measure" of orbit utilization we mean a suitably defined function of the variables (or parameters) of interest, the implication being that it is desirable to maximize that measure.

resource is the transmission of information. Thus it seems natural that a meaningful measure of orbit utilization would reflect, as a minimum, the amount of information transmitted, R, relative to the amount of the resource used up. One would also expect such a measure to be proportional to social utility, however, defined, and to be inversely proportional to the cost of providing the service. One may thus consider as reasonably comprehensive the measure

$$M = \frac{R \cdot (social\ utility)^{\propto}}{(\theta)^{\nu}\ (W)^{\mu}\ (cost)^{\delta}} \tag{5-36}$$

where

$$\theta = orbital\ arc\ used$$
$$W = occupied\ bandwidth$$

and the exponents $\propto$, ν, μ, and δ reflect the relative importance of the elements in question, a judgment which must be somewhat subjective and which may well depend on the service and frequency band being examined. The difficulty with (5-47), aside from the question of relative weights, is how to quantify social utility and cost. The former is inherently quantitative and the latter, while quantitative in nature, is not unambiguous and furthermore is time-varying. If the concept of measure is to be truly useful it would seem imperative to place it on a firmly quantitative basis. This leads us to delete social utility and cost and propose the more restricted measure

$$M = \frac{R}{(\theta)^{\nu}\ (W)^{\mu}} \tag{5-37}$$

which rests on strictly technical factors which can be well defined. Note that if we were computing M for two different sets of implementation parameters but for the same basic service, the social utility factor would be common and therefore cancel in the comparison. So, for such situations, the social utility factor was redundant in the first place. The cost factor, however, was not. Consider, for example, the case where θ is made to decrease strictly through a uniform increase in earth station antenna size. The measure increases but, clearly, so does the cost no matter how figured. Thus, one must bear this in mind when using a strictly technical measure such as (5-48). This measure can be reduced to special cases that may be particularly relevant under some conditions:

a) total information rate $(v = \mu = 0)$;

b) information rate per unit bandwidth $(v = 0, \mu = 1)$;

c) information rate per unit arc $(v = 1, \mu = 0)$;

d) information rate per unit arc and bandwidth $(v = \mu = 1)$.

The latter measure, in particular, we have found convenient to use as a general criterion. The equality of exponents is a convenience which does not affect any comparison (and avoids a value judgment) when we keep the bandwidth and arc constant.

One basic purpose of a measure of orbit utilization is to provide guidance on the relative effectiveness (from the orbit utilization standpoint) of various technical designs or features. Ideally, one would like to be able to evaluate this property for single networks. This is not possible, however, because a measure of orbit utilization intrinsically involves at least two satellites. Thus, to assess the orbit utilization potential of a prospective design one must define an environment. There are many possible ways to do this, among them the following:

a) Define another network with a single satellite, identical to the subject satellite considered;

b) Define a set of networks identical to the one being considered, i.e., the homogeneous model;

c) Define a "standard" reference network which would serve as the basis for calculation with any network being considered;

d) Define a non-homogeneous but tractable set of other networks to provide a more realistic context;

e) Use the "actual" environment in which the network under consideration might be found.

all of these methods have one or more difficulties associated with them, some of which are discussed in [70]. However, the homogeneous model is perhaps the least ambiguous and involves the most straightforward computation of any method.* Perhaps most important (and this is related to ambiguity) is that since all networks are identical, one can isolate the effect of changes of single parameters and thus obtain a clearer idea of the significance of these parameters.

In the present context, the notion of efficiency naturally arises. In fact, once a particular measure is adopted, the value of the measure computed for two sets of conditions provides an indication of the relative efficency of orbit utilization. But we do not necessarily know in either case the true efficiency in the standard interpretation of the term, i.e., as a ratio of an actual quantity with respect to the maximum value that quantity could ever have.

For an appropriately defined environment and for our usual measure (information rate per unit bandwidth and arc) the ultimate capacity of the orbit is deducible from information-theoretic considerations, which lead to unsurpassable bounds that can therefore be used as references for an efficiency calculation. For example, given a set of satellites $\{S_j\}$ each with power constraint P_j and antenna radiation pattern G_j, and given a set of associated earth stations with specified parameters such that the C/NPI at each earth station is ρ_j, and the bandwidth, equal to W, is common to all, then the orbit capacity is bounded by $\Sigma W \log_2(1 + \rho_j)$, or the normalized bound $\theta^{-1}\Sigma \log_2(1 + \rho_j)$ $bps/Hz/deg$, where θ is the orbital arc subtended by all the satellites considered.* The information rates achieved by the actual designs can be legitimately compared to such a bound. For a homogeneous environment we have in fact found such a bound in Section 4.13.4, and plotted in Figure 4-37, where the intersatellite spacing was itself optimized to yield the best ratio of satellite capacity to occupied arc.

In the preceding discussion the expression for the ultimate capacity rested on a set of specific assumptions. These assumptions may correspond to an existing configuration of satellites, or they may correspond to a tentative, yet-to-be implemented group of satellites.

---

*Method (a) is the simplest in this respect but method (b) is only trivially more complex.
*For simplicity we ignore the uplink in this discussion.

In the latter case, there may be a still greater capacity (both actual and ultimate) achievable by a more intelligent use of the resources available, subject to the same constraints. The simplest example would be a re-ordering of the satellite positions to avoid overly non-homogeneous neighbors, where that might have been the case. Another example might be a judicious shaping of the antenna pattern of one or more of the satellites, subject to aperture constraints. Extending this line of thought, Acampora [71] has found an upper bound on the capacity of a single frequency re-use satellite providing communication between N distributed earth stations. The bound specifies the N-dimensional region wherein it is possible to transmit information with arbitrarily high reliability at rate R, to earth station 1, rate R_2 to earth station 2, ---, rate R_N to earth station N, subject to a constraint on satellite power, bandwidth, and aperture size. This bound is all-inclusive, and no TDMA, FDMA, nor SSMA scheme, antenna pattern shaping, or scanning beam concept can provide better performance. This capacity region does indeed, then, specify the ultimate capacity for this single satellite. For a number of satellites of the type at given orbital positions the orbit capacity can be found straightforwardly if the inter-satellite interference is treated like thermal nise and used to correspondingly reduce the CNR in the single satellite case.

To summarize this section, there are a number of possible meaningful measures of orbit utilization. However, only those measures involving technical quantities appear amenable to quantification, and of these perhaps the most generally useful and comprehensive is the information rate per degree of orbit and unit bandwidth. The homogeneous model provides a context in which such a measure can be computed without conceptual difficulties, and further permits study of the effect on orbit utilization (as embodied in this measure or other, related simpler measures such as average intersatellite spacing) of individual technical factors. A measure can provide quantitative comparison of two different orbit-use techniques. However, an estimation of the absolute efficiency of any actual or proposed set of satellite networks with respect to orbit utilization requires knowledge of the theoretically maximum utilization possible. The theoretical bound is obtainable, under the given constraints, from information-theoretic considerations. The knowledge of such a bound is useful because it may indicate, where the efficiency is very low, that a reconsideration of a proposed manner of orbit usage would be in order.

5.12 On the Capacity of the Geostationary Orbit: An Example

A recurrent theme in the subject of geostationary orbit utilization concerns the "capacity" of the orbit, not in the ultimate sense discussed in Section 5.11, but, as it is customarily used in this context, in the sense of what is practical and cost-effectively achievable at any point in time. The question is crucial from a regulatory standpoint, because if the capacity is seen to be insufficient to meet projected demand (it being assumed that capacity and demand are estimated for the same period) then it is clear that regulatory agencies must take an active role and somehow distribute parcels of the resource to those wishing to make use of it. On the other hand, if the capacity estimate significantly exceeds the demand forecast, regulation can be much more passive, letting matters evolve as they will, perhaps only instituting certain minimum technical standards to prevent gross waste of the resource.

In the preceding sections of this chapter we have largely dealt with normalized measures. Here, we will present an example calculation for the orbit capacity in absolute terms. The process of the example, as well as the resulting numbers, will be instructive. A number of capacity estimates [39, 72, 73, 74] have been made over time, for different frequency bands, type of service, and assumptions. The capacity estimate must obviously rest on a large number of assumptions, and it is equally clear that the usefulness of the estimate is in direct proportion to the correspondence between these assumptions and actuality. Although this cannot be known for certain before the fact, certain general trends are clear and can be used for guidance.

Our capacity example is based on [39, 72]. We limit ourselves to the fixed satellite service (FSS). As mentioned above, the time element has an important bearing on the capacity estimate because the steady progress of technology inevitably makes the interpretation of what is practical and cost-effective a function of the time. Hence two capacity estimates will actually be developed, one applicable roughly to the current time frame (1980-1985) and one to the near future (1995-2000). The technical assumptions will be consistent with these time frames and, in particular, the frequency bands taken into use will be assumed to be below 14.5 GHz in the first instance and below 31 GHz in the second. We shall refer to the two time frames as Case I and Case II, respectively.

5.12.1 Service Area Model

Perhaps the most central part of the model lies in the coverage assumptions. In the earlier days of communication satellites, connections were exclusively international. However, a strong trend is emerging for "domestic" applications, implying restricted area coverage. While the future will undoubtedly see a mix of international and domestic networks, we will restrict ourselves in this example to the latter type, both for simplicity and because these will likely be much more numerous than the other type.

Thus, the coverage model is as follows. The portion of the earth's surface visible from the orbit is divided into equal-area zones, corresponding to the coverage area of individual satellite antenna beams, illustrated in Figure 5-28. The orbit is similarly divided into equal-length segments. The zones are divided into four groups labeled A, B, C, D. From each location in the orbit marked by a particular letter, one satellite or a group of satellites communicates to all of the areas on the earth marked by the same letter that can be seen from that point on the orbit. In this way each satellite serves several areas and each area is served by several satellites. The first is made possible by separating the areas served from the same orbital location and the second is enabled by separating satellites along the orbit. Since those coverage areas that lie over oceans are not realistic service areas, those areas are excluded from consideration in the capacity calculations; the resulting model is shown in Figure 5-30.

Based on this geometrical model we can roughly calculate the number of frequency re-uses. Let the area of each zone be equal to D^2, where D will be referred to as the zone dimension. Since about half the earth's surface is visible from any point on the orbit, the number of zones visible from that point is half the spherical surface divided by the zone area. The number of reuses from that point, n_R, is one-fourth of the total, by the assumptions of the model. Hence

$$n_R = \frac{\pi}{2} \ (R/D)^2$$

where R, the earth's radius, is about 4000 (statute) miles. The total number of re-uses is therefore

$$N_R = \frac{360}{\Delta\theta} \cdot \frac{\pi}{2} \ (R/D)^2 \qquad (5\text{-}38)$$

where $\Delta\theta$ is the inter-satellite spacing. Equation (5-49) does not yet account for land areas only, nor does it account for areas in the polar regions invisible from the orbit or for a strip around the periphery of the earth from which the elevation angles would be too small to be usable in practice. Table 5-6 shows for some values of D and $\Delta\theta$ the number of frequency reuses computed from (5-49) in the third column; a corrected number in the fourth column, based on a minimum elevation angle of 10 degrees; and in the fifth column the number of re-uses over land areas, on the basis that oceans cover approximately two thirds of the earth's area. This last column forms the basis for the capacity calculations though a further reduction will be taken to account for other factors as discussed later.

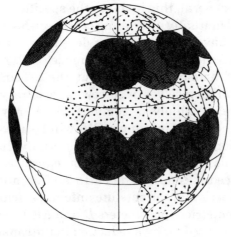

Source Ref. [72]

Figure 5-28 Orbit Capacity Model

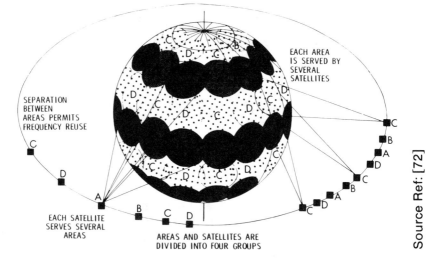

Figure 5-29   Modified Orbit Capacity Model (Land Area Coverage Only)

5.12.2 Internetwork Interference

Implicit in the capacity calculations must be the requirement that the permissible interference levels are not exceeded. This requirement is met by choosing a compatible set of geometrical and network parameters. In general, this involves the specific modulation characteristics of all mutually interfering carriers, which further implies that we would have to make rather detailed assumptions in this regard for each frequency re-use in the model. This not only necessitates a tedious amount of detail, but the assumptions thereby invoked would in themselves be somewhat arbitrary. The approach taken, therefore, is simply to compute the carrier-to-interference ratio and assume that a minimum CIR will be adequate to protect any transmission. If we further assume initially that all networks are homogeneous, except for modulation characteristics, the only assumptions needed concern the earth station antenna size and sidelobe characteristics, the satellite antenna sidelobe characteristics and the geometrical parameters D and $\Delta\theta$. These idealizations will later be accounted for by a reduction in the capacity that would result from them.

We shall assume $D = 2000$ and $\Delta\theta = 5$ for Case I and $D = 1000$ and $\Delta\theta = 5$ for Case II. We further assume earth station antenna size of 5 meters with CCIR sidelobes and satellite antenna sidelobe pattern as given by the WARC-BS *Final Acts* [3e]; a graph of this pattern appears in Figure 5-15. Three types of interference can be identified. One involves satellites at the same orbit location as the desired satellite but which serve other service zones. Interference is governed in this case by the satellite antenna sidelobe levels. The second case involves cross-polarized transmissions, principally over the same path as the desired transmission. This interference is also determined mainly by satellite antenna characteristics. The third interference mode involves satellites which serve the same or adjacent geographic areas as the desired satellite but which are not at the same orbital position. This interference is controlled by the earth station antenna sidelobe levels and the intersatellite spacing.

The first of these cases is the most significant and the most complex to determine. Taking into account the fact that the satellite antenna beamwidth is a function of the zone location for fixed zone dimension, and for the moment assuming frequency reuse over the full earth, it can be determined that for the worst earth terminal location in the worst zone $C/I = 21\ dB$,* this arising from the sum of 12 entries from other networks. The second interference mode is characterized by isolation of at least 33 dB,* using the cross-polarized pattern in [3e]. In the third case, one finds $C/I \approx 31.7\ dB$* for the 5 meter antenna in the 6/4 GHz band. More isolation would be available with lower sidelobes, larger antennas, and higher frequencies.

An overall $C/I \approx 17.4\ dB$ for homogeneous transmissions results from the combination of 21, 33, and 31.7 dB in each direction of transmission. However, these numbers arose from consideration of full earth frequency re-use. Since only about one-third of the earth's surface is land, as mentioned before, we can assume on the average a 5 dB improvement from this fact. Furthermore, only half the theoretical number of frequency reuses will be taken, as discussed in the next subsection, yielding a further 3 dB improvement. Thus the expected C/I corresponding to the assumptions in the capacity calcualtion for the Case II model is on the order of 23 dB. The value for

---

*These values of C/I apply roughly to either uplink or downlink.

Case I will be somewhat higher due to the smaller number of possible interference entries. These values of CIR are compatible with a practical interference allowance.

Interference between satellites at the same orbit location, the most serious source of interference, can be reduced by displacing the interfering satellites slightly. For example, assume that the satellites associated with a given orbit location are divided into two groups located at $\pm$ 0.35° from the nominal location, and that the zones served from that location are divided into corresponding groups. The effect of this procedure is to improve the worst-case interference situation by moving the interfering satellite 0.7° away from the desired satellite. Thus, if a 5-meter earth station antenna is used at 4 GHz, it will provide 6 dB of added isolation.* The penalty associated with this procedure is that the spacing between some satellites in adjacent orbit locations is decreased from 5° to 4.3°, which may necessitate a small increase in earth station antenna diameter. It is evident that interference could be further reduced by increasing the amount of offset. It is also possible that a small increase in nominal satellite spacing, e.g., from 5° to 5.5°, would yield a net improvement in orbit capacity by permitting larger offsets.

5.12.3 Orbit Capacity Calculation

Having determined the total number of frequency re-uses that result in interference compatibility, we can determine the orbit capacity by associating with each re-use a certain communications capacity. Telephone and TV channels provide a convenient and practical measure of capacity. Table 5-7 summarizes the capacity values associated with various multiple access and signal processing techniques. The first entries in each classification are representative of current practice, and the succeeding entries represent more advanced techniques that have been demonstrated theoretically or experimentally, some of which are already in use in some networks [17, 18, 19, 75, 76, 77]. Using an average, or "representative", value per 40 MHz, times the number of 40 MHz bands in the allocated bandwidth, times the number of frequency reuses, yields the capacity which is shown in Table 5-8.

---

*This was assumed in arriving at the 21 dB C/I value given earlier.

In arriving at Table 5-8 the following general rules were applied: (a) for Case I only frequency bands below 14.5 GHz were used, and for Case II only bands below 31 GHz; (b) the use of orthogonal polarization is assumed only for the bands below 14.5 GHz; (c) the theoretical number of land-area frequency re-uses (right-most column of Table 5-6) is reduced by 50% in order to allow for non-homogeneity, non-uniformity of coverage areas, and other factors which may prevent the idealized conditions assumed in the model from being realized in practice.

The orbit capacity calculations are clearly based on a great simplification of the way that orbit usage develops in actuality.

Interpretation of the results should therefore take not of the following:

(a) The existence of global coverage networks is ignored. Such networks, while adding capacity of their own, in fact reduce the possibilities for frequency re-use and consequently tend to reduce orbit capacity;

(b) In practice orbit locations for satellites are not selected on the basis of an overall plan which maximizes compatibility with future systems; an optimum arrangement is therefore not assured;

(c) Different networks will have coverage areas of various sizes and shapes, which will tend to reduce the possible number of frequency re-uses from a given orbit location if only simple beam patterns are used;

(d) Satellite antenna sidelobe levels in the FSS are not now stipulated either in the *Radio Regulations* nor by a CCIR *Recommendation;* the FSS reference pattern described in [38] is in fact not as good as the pattern assumed in these calculations;

(e) For various reasons it will not generally be feasible to upgrade the performance of older satellite networks to the levels assumed for Case II; thus, in the latter time-frame one is likely to see a mix of new and older technology;

(f) Dual polarization operation may be feasible in the 18/30 GHz bands, which would increase capacity for Case II by 50%;

(g) Operations may commence in the bands above 30 GHz during the Case II time-frame, which would also increase capacity;

(h) The capacity attributable to a given geographic zone will be greater for zones near the equator and less than average for zones at high latitudes;

(i) Networks will not be homogeneous either in the terminal parameters or in the signal/modulation characteristics.

The capacity calcualtions compensate to some extent for the effects of (a) - (d) and (i) by reducing the number of frequency re-uses by half. Item (e) was accounted for by using transponder capacity values about half of what has already been demonstrated to be possible. On the other hand, the possibilities for increased capacity outlined in items (f) and (g) were not counted, nor was the sidelobe level improvement anticipated in the future used to raise the Case II capacity.

5.12.4 Conclusions

In developing the capacity model our purpose has not been so much to obtain a specific value for the orbit capacity — for surely there is no single such number — as to establish whether the capacity, roughly measured, appears to be adequate relative to demand in the foreseeable future. The answer seems to be definitely positive. The orbit capacity predicted by the model is extremely large, and in this respect all capacity estimates [39, 72, 73] are in general agreement. A reasonable amount of care was taken to be conservative by decreasing capacity to account for certain idealizations and by not invoking all of the technical advances known to be possible. Even so, the conclusion as to the adequacy of orbit capacity does not appear to be critically dependent on the assumptions made, except perhaps that the majority of future networks will use limited-coverage antenna beams. This conclusion is bolstered by comparing the computed capacities of 7450 and 47300 equivalent 40 MHz transponders (Case I and Case II, respectively) to the number now in worldwide use, on the order of 500, or to the projected demand of about 10,000 [see Chapter 1] in the Case II time frame.

5.13 Miscellaneous Technical/Operational Factors

In this section we consider only briefly a number of factors which have or may have an impact on orbit utilization, but which appear as yet somewhat difficult to quantify in a general way. Further studies are required before their full significance is appreciated. These factors also have a strong operational/economic flavor, which by nature complicates their evaluation.

5.13.1 Intersatellite Links

Intersatellite links (ISL) provide a connection between gestationary satellites,* and from this simple fact emerges their beneficial properties. Since the purpose of a gestationary satellite system is eventually to permit communication between two points on the earth, the inclusion of an ISL transforms a network into a three-hop one instead of what is typically a two-hop configuration. However, in some cases the connection between two points on the earth via a two-hop network may be physically impossible, or may require undesirably low elevation angles. Even when these constraints do not apply, it is clear that much more satellite position flexibility is afforded by an ISL, especially when the satellites employ spot beams. Following these lines of thought, we may then identify the following potential benefits of intersatellite links [47]:

a) to offer an alternative to the provision of additional earth station antennas in a multinetwork system with increasing connectivity requirements;

b) to provide interconnection of widely separated earth stations which otherwise could only be served through "double-hops", i.e., two two-hop links via a relaying earth station. Here there is a system trade in cost/complexity between an ISL and a double hop;

c) to alleviate satellite orbit position constraints in multinetwork systems for the purpose of improving operating performance (e.g., location of satellites so as to maximize earth station antenna elevation angles) and, as an extension, to alleviate orbit crowding

---

*Intersatellite links may also be used with non-gestationary satellites, of course, but we do not discuss these since they are outside the intended scope of this book.

problems where this positioning flexibility permits satellites to be placed in orbital arcs that are less congested than the ones where they might otherwise have to be without an ISL;

d) to permit consolidation of the coverage areas of separate-coverage networks (e.g., consolidation of coverages of different-coverage multibeam satellites);

e) to permit international program exchange between two or more broadcasting-satellite networks;

f) to allow the resources of individual satellites to be shared, thereby resulting in more efficient use of the communications capacity of the total satellite system.

Among the foregoing, item (c) is most directly related to orbit utilization, *per se,* though this effect appears difficult to quantify.

A number of frequency bands have been allocated for use by the intersatellite service. Aside from sharing considerations with other services which may use these frequencies (these considerations are discussed in [Reference 48]), a question of interest concerns the frequency re-use potential in the intersatellite service, for that potential has a bearing on the ultimate utility of this service. This question has been considered in some detail [47], and the following tentative conclusions have been reached:

a) the frequency re-use potential for geostationary intersatellite links is high and may allow on the order of several hundred intersatellite links to be accommodated with acceptable interference levels;

b) the major factors which affect the frequency re-use potential are the antenna sidelobe discrimination and the orbit ellipticity. It is found that high sidelobe discrimination and low orbit ellipticities improve the re-use potential;

c) very short intersatellite links with co-directional frequency assignments, such as might be of particular interest to early users of this service, cannot overlap. The smallest length of similar intersatellite links for which overlapping is possible depends on antenna discrimination and orbit ellipticity — the higher the former and the lower the latter, the smaller is the link length for which overlapping is possible;

d) the use of several intersatellite link sets of sufficiently different link length allows frequency re-use to be increased to a multiple of that achievable with a single set of short identical links.

Possible system configurations involving ISLs and system design/ tradeoff considerations are discussed at some length in [49, 50].

5.13.2 Pairing of Up/Down Frequency Bands

For any space service a number of frequency bands have been allocated for the uplink and downlink of satellite networks. In general, a system designer has the freedom to choose uplink and downlink bands among the allowed possibilities, as the constraints or desiderata on the system may dictate. While there has been a certain amount of undirected standardization — certain up/down bands tend to be paired, such as 6/4 GHz and 14/11 GHz — there is at present no requirement for specific band pairings. The question arises whether it would be beneficial from the orbit utilization standpoint to have standard up/down frequency band pairings, or, conversely, whether it is detrimental to orbit utilization if a variety of such pairings are used.

A simple example [51] suggests that orbit inefficiency could result from the use of mixed band pairings. Consider the situation where frequency bands U_1 and U_2 are available for uplinks and bands D_1 and D_2 for the downlink. If (U_1, D_1) and (U_2, D_2) are always used as pairings, then satellites using either pair can be independently placed in orbit as closely as the permissible interference criteria will allow. Figure 5-30 illustrates an orbit arrangement in which we may suppose the satellites using different band pairs are co-located and require the same spacing between other satellites using the same band pair. If most systems adopt these pairings as standard, but now a satellite using a "non-standard" pairing (U_1, D_2) is introduced, as shown in the lower part of the figure, this will preclude other networks wishing to use either standard pairing from operating in the arc where before two satellites could be accommodated. It would, of course, be possible to co-locate another non-standard satellite using (U_2, D_1) with the one using (U_1, D_2). However, this implies that an operator of the last satellite introduced wished to use the pairing (U_2, D), which might not conform to his preferred design. Thus, if a large measure of uncoordinated band pairings were to develop in

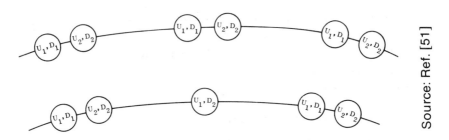

Figure 5-30 Illustration of Effect of Mixed Frequency Band Pairing
on Orbit Utilization

more or less random fashion, it would be difficult to coordinate both satellite locations and band pairings so as to preserve orbit utilization efficiency. Even where standard band pairing is observed, orbit inefficiency can arise from the use of non-standard translation frequencies. Consider the case where a satellite occupies only a part of the bandwidth of the allocated frequency band. If the designer chooses to pair the top part of the uplink with the lower part of the downlink band in order, say, to simplify implementation then this will "sterilize" the whole of both bands in that part of the orbit for operators using the normal translation frequencies.

Some specific band pairings have been suggested [51]. However, there are difficulties with establishing a rigid band pairing pattern globally. There may be legitimate reasons for using a non-standard combination of up/down frequency bands, for example to overcome a particular interference problem or meet a particular service requirement. Further, for a satellite using more than one pair of bands, cross-strapping within the satellite would then presumably not be permitted, even if this interconnection were desired to provide a special service or were needed to overcome a failure in one of the links. Other factors that would militate against a rigid pairing concept are:

Only a few of the frequency bands allocated to the fixed satellite service have been taken into extensive use, and the best ways of using the others have yet to be studied thoroughly;

There are differences between the fixed satellite service in the three ITU regions;

In certain parts of the spectrum, the bandwidth allocated for earth-to-space links is not equal to that allocated in the other direction, so that one uplink band may have to be paired with more than one downlink band.

Therefore, while it seems in principle desirable to establish an orderly pattern of frequency band pairings and translation frequencies, additional studies and operational experience are necessary before the practicality or feasibility of such standardization can be ascertained.

5.13.3 Multiple Band/Service Satellites

A problem that is in a sense an extension of the one in the preceding subsection concerns the impact on orbit utilization of using more than one pair of up/down bands on the same satellite, or, more generally, M uplink and N uplink bands to provide connections in L different services to which these bands have been individually or jointly allocated. We shall refer to these as multiple band/service (MBS) satellites [52, 53].

A possible negative effect on orbit utilization can be envisaged from the use of MBS satellites. This possibility stems from the fact that considerable differences are likely to exist in the characteristics of the links in the different bands/services in question, and consequently the satellite spacing required to satisfy the interference criteria may vary significantly from one link to another. If the latter is true, this represents a waste in orbit utilization in the sense that the spacing protecting the most sensitive link, i.e., the largest one, must prevail. The other links would then be operating with less interference than permitted. If these links had been on separate stations, these stations could then have been placed closer together in the orbit. Hence, potential orbit utilization would have thereby been lost by implementing links for different bands/services on the same satellite.

A further detrimental effect to orbit utilization can arise if the different links have associated with them substantially different coverage areas, and hence service arcs. For then, the flexibility of orbit positioning which might have existed for some of the links would be limited by the most constraining link.

The coverage liability, if it exists, cannot easily be mitigated. However, it is possible by proper choice of transmission parameters to equalize the interference/spacing differential between the different links. But in order to be effective this must be consciously incorporated into the design at an early stage. Consider, for example, a satellite using two band pairs P/Q and X/Y, and suppose nominal designs have been completed so that at frequencies P/Q the thermal noise and interference budgets are just satisfied (at a given spacing), but at frequencies X/Y there is interference margin. By changing the modulation index, say, so that noise-plus-interference jointly satisfy the total performance requirement, one can increase the information rate thus raising the orbit utilization. This implies that the accepted interference limits would be exceeded, but in fact the net effect would be beneficial not only from the viewpoint of orbit utilization, but also from that of the system itself.

The preceding considerations are further complicated if different satellites use different sets of band pairs. Here it would appear that the liabilities associated with both the use of multiple band pairs and non-standardized pairings would become manifest, but it is not clear how they would combine.

While there are potential problems with MBS satellites, these must be weighed against the operational and economic advantages that provide an incentive for their use. In fact, an increasing number of satellite networks exist or are being planned with a capability to operate in more than one frequency band pair, either within the fixed-satellite service or in two or more services including the fixed-satellite service. For example, beginning with INTELSAT V, INTELSAT spacecraft will provide fixed-satellite services in both the 6/4 GHz and 14/11 GHz bands. the ARABSAT and INSAT systems will use the 6/4 GHz band for fixed-satellite services and the 2.6 GHz band for broadcasting satellite service. In addition, INSAT will operate in the meteorological satellite service in still another frequency band.

5.13.4 Processing Satellites

While the term "processing" satellite does not yet seem to have a commonly accepted definition, we shall use it here to connote any satellite which does more than simple frequency translation. Therefore, we include switching as well as demodulation, detection, remodulation, and baseband processing. Processing at the satellite is sure to become important in future applications, and in some forms is under active development, if not imminently to be put into practice. The scanning beam technique mentioned in Section 5.7.1 is one example of "processing," and the multiple band operation discussed in the preceding subsection is another example, provided there is cross-strapping between bands.

Given the variety of meanings that can be given to "processing," it is difficult to make definite statements concerning its effect on orbit utilization without constructing a formal model, and this may be a fruitful area for future work. However, without such a model it is still possible to make an assessment, and in general it would appear that processing at the satellite should have a beneficial effect on orbit utilization.

This conclusion can be arrived at from consideration of two examples of processing. The first, switching in the satellite, does not in itself affect the interference environment, but it is a necessary function to permit the full potential of frequency re-use at the satellite to be realized. If switching were not available, then only ground locations within the same beam could communicate with one another.

This would represent an inefficiency if the capacity of that beam/ bandwidth would not be fully utilized by users within the beam coverage area. Switching at the satellite permits the interconnection of users from the different coverage areas, and this will tend to increase the traffic "throughput" at a given orbit location.

On the other hand, demodulation/remodulation need not in itself affect the traffic capacity, but it will generally have an influence on interference. This can be illustrated in the case of digital transmission, where a bit stream is detected in the satellite and remodulated on a downlink carrier. For this example, we assume that the interference can be treated as thermal noise, as far as its effect on detection performance is concerned. Further for simplicity assume that we are dealing with binary PSK and a distortionless system. Hence Equation (4-88) applies, with $m = 1$, $M = 2$, so that the bit error rate (BER) can be expressed as

$$P_e = \tfrac{1}{2} \, erfc \left(\sqrt{2 \, (C/NPI)} \right) \qquad (5\text{-}39)$$

The carrier-to-noise plus interference ratio, (C/NPI), in the case of the transponding link, is given by

$$\left(\frac{C}{NPI} \right)^{-1} = \left(\frac{C}{N} \right)^{-1} + \left(\frac{C}{I} \right)^{-1}$$

$$= \left(\frac{C}{N} \right)_u^{-1} + \left(\frac{C}{N} \right)_d^{-1} + \left(\frac{C}{I} \right)_u^{-1} + \left(\frac{C}{I} \right)_d^{-1} \qquad (5\text{-}40)$$

where subscripts u and d denote uplink and downlink, respectively.

In the regenerative (demod/remod) link a bit error rate is associated with each path. The uplink BER and downlink BER can be respectively written as

$$P_{eu} = \tfrac{1}{2} \, erfc \left(\sqrt{2 \, (C/NPI)_u} \right) \qquad (5\text{-}41a)$$

$$P_{ed} = \tfrac{1}{2} \, erfc \left(\sqrt{2 \, (C/NPI)_d} \right) \qquad (5\text{-}41b)$$

where the uplink and downlink (C/NPI) are given by

$$\left(\frac{C}{NPI} \right)_u^{-1} = \left(\frac{C}{N} \right)_u^{-1} + \left(\frac{C}{I} \right)_u^{-1} \qquad (5\text{-}42)$$

$$\left(\frac{C}{NPI} \right)_d^{-1} = \left(\frac{C}{N} \right)_d^{-1} + \left(\frac{C}{I} \right)_d^{-1} \qquad (5\text{-}43)$$

The same symbols in (5-41) to (5-43) do not necessarily have the same value since the design and environment in the two situations need not be identical. To highlight this distinction we first make the following definitions:

$$\left. \begin{array}{l} (C/N)_u = \alpha_u \, (C/I)_u \\ (C/N)_d = \alpha_d \, (C/I)_d \end{array} \right\} \; \textit{Transponder} \qquad (5\text{-}44)$$

$$\left. \begin{array}{l} (C/N)_u = a_u \, (C/I)_u \\ (C/N)_d = a_d \, (C/I)_d \end{array} \right\} \; \textit{Regenerative} \qquad (5\text{-}45)$$

To continue, we shall now assume a homogeneous orbit environment, for which we can write [see Chapter 4]

$$\left(\frac{C}{I}\right)_u = K_u\, \Delta\theta^\beta; \quad \left(\frac{C}{I}\right)_d = K_d\, \Delta\theta^\beta \tag{5-42}$$

If we let $\Delta\theta_1$ be the spacing in the transponder case and $\Delta\theta_2$ the spacing in the regenerative repeater case, a few simple manipulations using (5-40), (5-41), and (5-42) give rise to

$$\left(\frac{C}{NPI}\right) = \Delta\theta_1^\beta \left\{ \left(\frac{1+\alpha_u}{\alpha_u}\right)\frac{1}{K_u} \right.$$
$$\left. + \left(\frac{1+\alpha_d}{\alpha_d}\right)\frac{1}{K_d} \right\}^{-1}; \textit{Transponder} \tag{5-47}$$

$$\left(\frac{C}{NPI}\right)_u = \Delta\theta_2^\beta \left\{ \left(\frac{1+a_u}{a_u}\right)\frac{1}{K_u} \right\}^{-1}$$
$$; \textit{Regenerative}$$
$$\left(\frac{C}{NPI}\right)_d = \Delta\theta_2^\beta \left\{ \left(\frac{1+a_d}{a_d}\right)\frac{1}{K_d} \right\}^{-1} \tag{5-48}$$

Realizing that in the latter case the BER is given by

$$P_e = P_{eu} + P_{ed} \tag{5-49}$$

a constraint is set which must be adhered to while manipulating the other variables. Thus, substituting (5-43) into (5-36) and (5-44) into (5-38), subject to the condition (5-45), yields a relationship between $\Delta\theta_1$ and $\Delta\theta_2$ as a function of the link parameters. This relationship is easy to compute, but not straightforwardly interpretable from the equations themselves. However, a simple case will illustrate the procedure. For this case we assume, first of all, that

$$\frac{1+a_u}{a_u}\,\frac{1}{K_u} = \frac{1+a_d}{a_d}\,\frac{1}{K_d} = \rho^{-1}$$

which balances the uplink and downlink in the regenerative case and is the most efficient from the spacing aspect. Next, we also assume $\alpha_u = a_u$ and $\alpha_d = a_d$. The result of these assumptions is

$$P_e = \tfrac{1}{2}\,erfc\ \left\{ \sqrt{\rho\,(\Delta\theta_1)^\beta} \right\}\ ; \textit{Transponder}$$
$$P_{eu} = P_{ed} = \tfrac{1}{2}\,P_e = \tfrac{1}{2}\,erfc\ \left\{ \sqrt{2\rho\,(\Delta\theta_2)^\beta} \right\}\ ; \textit{Regenerative}$$

For $P_e = 10^{-6}$ $\rho(\Delta\theta_1)^\beta = 11.30$, and $2\rho(\Delta\theta_2)^\beta = 11.97$, whence $\Delta\theta_2/\Delta\theta_1 = $

0.775 for $\beta = 2.5$. Thus, the spacing in the regenerative case is reduced to about ¾ of that in the transponder case. However, since the spacing is closer in the former case, the assumption $\alpha_u = a_u$ and $\alpha_d = a_d$ implies that the C/N is higher by the inverse ratio, i.e., by about 1db. In actuality, the P_e *vs.* C/N relationship is likely to be worse (shallower slope) in the transponder case, due to the cumulative effect of uplink and downlink distortions, than for either link in the regenerative case. Thus, the 1dB C/N penalty for the latter in our example would probably not exist. The net conclusion, therefore, is that processing of the type just examined can have a beneficial effect on orbit utilization.

5.13.5 Quasi-Stationary Orbits

The geostationary satellite orbit is normally visualized as a circle, so that its angular size is merely 360°. This size could be effectively increased by employing orbits having a fixed mean longitude with a moderately small north-south excursion. Two or perhaps more satellites could then effectively use the same longitude (and frequencies) but be sufficiently angularly separated by proper phasing along the quasi-stationary orbit. Several possibilities have been pointed out for "longitude re-use" [65, 66, 67, and references in 59]; however, all of them will substantially increase the cost of the space and ground segments in relation to that of a standard gestationary satellite network. Commercial exploitation of these orbits is not expected until saturation of the geostationary orbit.

5.13.6 Operational and Techno-Economic Considerations

Given a more or less arbitrary assemblage of geostationary satellite networks, it is conceptually straightforward to optimize orbit utilization by arranging their order in orbit, properly shaping and orienting their coverage beams, and optimally juxtaposing carrier frequency assignments. Or given a requirement for a number of networks and associated traffic-carrying capacity and connectivity matrix, it is always possible to accommodate these networks by suitable specification of the technical parameters. In either instance the freedom implied can be expected to exceed the constraints on cost or flexibility likely to exist in practice. Here we consider some of these practical limitations.

a) Permissible Level of Interference

Generally speaking the higher the permissible interference level the closer satellites can be placed in orbit, hence the greater the number of satellites that can be accommodated. (Recall from Section 4.13.3, however, that increasing the interference indefinitely does not keep on increasing the orbit capacity in all circumstances.) This, however, will typically require that the individual satellite capacity be decreased, with an obvious economic penalty. Thus, since orbit efficiency and economic viability indicate opposing trends, a proper compromise must be struck to achieve a reasonable level for both.

b) Technical Parameters and Subsystems

Orbit utilization could be increased through improvements in the technical parameters or subsystems that affect it. For example, reduction in the sidelobe levels of antennas or more accurate pointing would be cases in point. While the history of technology has evidenced that improvements evolve continually, there is generally an economic burden associated with newer technology at least until such time that is has become generally commercially available ("state-of-the-art"). In addition, new technology cannot be applied instantaneously, as existing systems (which may be difficult and expensive to modify) must continue to function over their useful lives. However, it may not be unreasonable to adopt increasingly stringent technical standards to new systems as technology evolves in the course of time. In this regard the notion of time-phased introduction of orbit-conservation measures has begun to surface in the CCIR [68].

c) Satellite Replacement [69]

As just mentioned, technical modifications are difficult to apply to existing networks, impossible at this juncture to the space segment. Current satellite lifetimes are about seven years and are likely to increase to ten years in the near future. It is expected that an entity having a system providing service would have firm plans for replacement satellites well before the expected end of the existing system's life so that the service could be continued. It is also a factor that significant growth in the amount of service being provided is generally experienced by FSS systems and replacement satellites are often built with greater capabilities than the satellites that they replace. This usually necessitates significant changes in some of the characteristics of the satellite replaced which could affect the planning or coordination process, and technical improvements tend naturally to be applied.

For example, the INTELSAT IV satellites provided 12 transponders in global coverage beams in each of the ocean regions served by INTELSAT. Where the traffic growth was such as to saturate an INTELSAT IV satellite, expansion was accomplished in two ways: (1) use of multiple satellites, and (2) replacement of the INTELSAT IV satellite by an INTELSAT IV-A satellite which used East and West hemispheric beam antennas to achieve dual frequency reuse with a total of 20 transponders. As a result of the same growth force, the INTELSAT IV and IV-A satellites are being replaced by INTELSAT V satellites which add new frequency bands (14/11 GHz) and introduce the use of dual polarization on each of the hemispheric beams to achieve four-fold frequency re-use.

Other examples are provided by recent U.S. domestic communications satellites which have employed cross-polarization techniques to double the number of channels provided in the 6/4 GHz bands compared to earlier or first generation models.

In this context the regulatory environment has a significant bearing. It is clear that, for a given orbit location, the network characteristics which had to be coordinated introduced significant changes that may not have been possible in an overly constrained system of allotment. The alternative of adding satellites would, of course, also not be possible if a system of allotment were used that did not include such additional orbit locations. Historically, the most effective method of increasing FSS capabilities to meet expanding demands has been the introduction and utilization of advancing technology. Thus, any contemplated system of planning or coordination should continue to make provisions for growth of system capability by appropriate use of new technological developments.

d) Satellite Position Flexibility [69, 70]

As mentioned above, if a set of satellites were given at a fixed point in time, it would be possible to specify their relative positions for maximum orbit utilization. However, in practice networks are introduced sequentially, and the orbit location of any one satellite cannot be optimized with respect to those of others which have yet to be planned for. The idealized situation can be approached to some extent, however, if operational satellites can be relocated in orbit so that the set existing at any one time can be mutually placed in relative positions more nearly approximating those which could have been designed if all satellites had been introduced simultaneously. However, there are limitations to the degree of flexibility in the relocation of satellites.

For geostationary satellites, the range of acceptable orbital positions capable of providing specified coverages, i.e., the service arc, depends on the longitudinal extent and latitude of the coverage areas and on propagation factors that limit the minimum acceptable elevation angle within the coverage areas. Satisfactory domestic coverage of a country near the equator can normally be achieved from a wide range of orbital positions, whereas coverage of the same-sized country located in the higher latitudes can be provided from within only a relatively small service arc. In addition, some satellite services which require global or hemispherical coverage would desire to maximize the coverage of each satellite in their network. The maximum practical longitudinal extent of a geosynchronous satellite is approximately 150 degrees (±75 degrees) [see also Fig. B-4, Appendix B]. This type of requirement severely restricts the service arc.

In any planning activity, special consideration should be given to those networks whose requirements can only be satisfied within limited service arcs. Conversely, where the useful service arc for a particular system or country is relatively large, it may be feasible to incorporate some flexibility in the location of some satellite(s) by providing alternate orbital slots coupled with the ability to change orbital positions.

A factor which limits satellite position flexibility is that associated with the antenna pattern as it appears on the earth. If the "footprint" has been achieved through a combination of actual antenna patterns and skewing associated with the orbit position itself, then the satellite may have limited freedom to be moved. Further, if the beam has been shaped by use of a multifeed array, and reconfiguration is not possible, then repositioning may be even more restricted.

However, while the arc over which the satellite may be repositioned may be significantly less than the service arc, it may still be on the order of a few tens of degrees. In this latter case, a possible solution would be to have steerable antennas which would permit some orbit position changing without requiring reconfiguration of the satellite antenna feed system. This problem, however, is intensified when the satellite has a multiplicity of spot beams.

Another factor that should be taken into account in connection with satellite repositioning is the ability of the earth stations in the network to repoint their antennas to new satellite positions. As a minimum, it means an interruption of operating service, although this can be kept to a short interval if the system has originally been designed with this contingency in mind. At worst, it may entail major interruptions of service. This would be the case only if the antennas require modifications of their support foundations, adjustments in polarization orientations of their feeds, and other similar considerations. Site selection could be made more difficult, however, since it must be chosen to provide visibility to any potential satellite location, and in a similar vein coordination may be more difficult.

Another potential problem, which could be overcome with labor cost, occurs when the ground segment includes a large number of unattended earth stations.

Finally, there is a penalty incurred by the loss of fuel required to reposition the satellite. The fuel expended would be a function of the speed at which the repositioning takes place, and could have an impact on the useful life of the satellite. There is thus a trade-off between the length of service disruption and the number of relocations that could take place without unduly affecting satellite life.

e) Transmission Planning [69]

To some extent, transmission planning has been done on a relatively ad-hoc basis with little standardization of bandwidths and powers among different systems and with little concern to the position of spectrum utilized by any particular combination of transmissions. In some of the coordination activities that have taken place, one of the techniques used to solve the more critical problems was to do some alignment of compatible transmissions between the networks of different Administrations or domestic satellite communications systems. Eventually, such alignment tends to have a "domino" effect. When two parties have agreed to a particular alignment, a third party may have to use a transmission plan similar to that of the two existing systems to achieve compatibility. This presents no problem so long as the magnitude and nature of the services permit this type of alignment. However, as additional systems are planned, it may be found that such alignments become increasingly difficult.

From the operator's point of view, the need for and the desire to retain complete flexibility of frequency planning is very high. As the satellites become larger, they carry many more channels when they go into service, and if they are replacements, they are likely to be immediately operating at 75% of their total capacity. A large amount of traffic is already fixed in one pattern, and it is not easy to introduce a brand new set of frequency assignments in a new satellite. There is pressure to retain as many of the existing assignments as possible. To some extent, these assignments become fixed, thus avoiding impacting other systems which might have already coordinated with these assignments.

It is also true that, as satellites get bigger, additional capacity is obtained by increasing the number of frequency reuses on a single satellite. The resultant self-interference problem can be greatly reduced by adjusting frequency assignments for internal compatibility. However, if the system has been coordinated on the basis of frequency assignments with another system, the intersystem interference problem must also be reassessed if there is an attempt to minimize intrasystem interference.

f) Satellite Orientation and Antenna Beam Pointing

An increasingly important factor in the effective utilization of the geostationary orbit is the accuracy with which satellite station-keeping and antenna beam pointing can be controlled. WARC 79 requires that new systems in the fixed satellite service have a longitudinal station-keeping capacity of $\pm 0.1°$. Section 4.7 shows that the benefit of improved station-keeping may have reached the point of diminishing returns at this level.

Latitude stationkeeping and yaw axis control are also important parameters in satellite stationkeeping and orientation, but limits have not been established or recommended to date because of the complexity in evaluating the effect of these parameters on intersystem interference. WARC 79 tightened the requirement on beam-pointing accuracy of satellite antennas to be capable of being maintained within either ten percent of the half-power beamwidth or 0.3 degree relative to the nominal pointing direction, whichever is greater. The impact of pointing tolerance has been considered in Section 5.7.

Any further reduction in tolerances on the foregoing satellite parameters in order to increase orbit capacity by reducing satellite orbital separations should take into account the associated penalties in increased design complexity, operational sensitivity and costs.

g) System Planning for Growth

In many, if not most, cases an initial operational system (network) is only a partial realization of a planned capability or of a capability that evolves in a direction that was not entirely foreseen. To the extent that they can be predicted or anticipated, it is important, both for the system owner/operator and from the standpoint of orbit utilization, to factor in future plans when implementing the initial capability. There may otherwise be detrimental consequences from both points of view. We have in effect already considered such implications in the discussions under c), d), and e) above. For example, earth station site selection should be made with a view not only of compatible locations for the "current" satellite, but also for other satellites that might be used to expand the system.

TABLE 5-1

Distribution of Isolation Requirements (in dB).

d_{high}	d_{low}	Δd	Analog (FM)					Digital		
			TV	24 ch	252 ch	972 ch	1332 ch	1.6 Mb	60 Mb	SCPC
50.8	35.4	15.4	X							●
43.5	28.4	15.1	X	O						
43.0	25.9	17.1		O			X			
42.8	33.0	9.8					X			O
42.6	26.3	16.3			O		X			
40.8	34.1	6.7				O	X			
40.6	31.7	8.9	O				X			
40.6	31.2	9.4					X		O	
40.3	—	—					Δ			
38.3	32.5	5.8					X		O	
37.0	29.4	7.6	X		O					
36.9	25.6	11.3		O		X				
36.1	26.8	9.3			O	X				
35.6	31.4	4.2						O	X	
34.8	29.2	5.6	O						X	
34.6	32.3	2.3	O				X			
34.6	30.8	3.8					X	O		
34.6	—	—					Δ			
34.5	—	—	Δ					X		
33.7	32.5	1.2	O							
33.6	30.1	3.5						O		X
33.2	31.2	2.0							O	X
33.0	27.2	5.8			O				X	
32.9	32.1	0.8				O				X
32.8	32.3	0.5				O			X	
32.5	—	—						Δ		
32.5	—	—							Δ	
32.3	26.1	2.5	O						X	
32.2	27.7	4.5			O					X
31.7	25.3	6.4	O	X						
31.7	26.6	5.1	O							X
31.5	29.5	2.0	O					X		
30.4	30.4	0.0			X=O			X=O		
30.1	—	—		Δ						
29.2	—	—	Δ							
28.5	—	—								Δ

The isolation d_{high} is that required to protect transmission of the type O from those of the type X. The isolation d_{low} indicates the reverse case; Δ denotes indentical transmission types.

Sources: Ref. [22] and [23], © 1980 IEEE.

TABLE 5-2

POWER FLUX DENSITY (P_d) REQUIRED TO MEET SERVICE
REQUIREMENTS ON 4 GHz DOWN-LINK ($dBW/m^2/4\,kHz$)

	(G/T)$_e$	40.7 dB	31.7 dB	25.7 dB	19.7 dB
CARRIER SIZE	24	-179.8	-170.8	-164.8	-158.8
	60	-171.0	-162.0	-156.0	-150.0
	72	-179.3	-170.3	-164.3	-158.3
	132	-170.2	-161.2	-155.2	-149.2
	192	-171.1	-162.1	-156.2	-150.2
	252	-171.5	-162.5	-156.5	-150.5
	432	-168.5	-159.5	-153.5	-147.5
	612	-167.8	-158.8	-152.8	-146.8
	792	-167.5	-158.5	-152.5	-146.5
	972	-174.4	-165.4	-152.4	-135.4
	1092	-169.2	-160.2	-154.2	-148.2

TABLE 5-3

MINIMUM EARTH STATION ANTENNA DIAMETER(m)
REQUIRED TO MEET 6 GHz UP-LINK OFF-BEAM EIRP (P_u) LIMITS

		$P_u = 30 - 25\log$ (dBW/4 kHz)					
	(G/T)s	5	0	-5	-10	-15	-20
CARRIER SIZE	24	0.46	0.83	1.47	2.62	4.65	8.27
	60	1.3	2.3	4.1	7.3	13.0	23.0
	72	0.06	1.5	2.7	4.8	8.6	15.2
	132	1.4	2.5	4.4	7.9	14.1	25.0
	192	1.28	2.3	4.1	7.2	12.8	22.8
	252	1.21	2.15	3.8	6.8	12.1	21.5
	432	1.71	3.0	5.4	9.6	17.1	30.4
	612	1.85	3.3	5.9	10.4	18.5	32.9
	792	1.92	3.4	6.0	10.8	19.2	34.1
	972	0.87	1.54	2.7	4.9	8.7	15.4
	1092	1.58	2.8	5.0	8.9	15.8	28.0

TABLE 5-4

CARRIER CHARACTERISTICS

(for example of Figures 5-8 and 5-9)

Carrier No.	1	2	3
Type of Service	Thin-route FDM Telephony (12 ch.)	Heavy-route FDM Telephony (960 ch.)	TV Distribution
Ground Antenna Dia (M) Uplink Downlink	8.5 29.0	29.0 29.0	10.7 10.7
Frequency (GHz) Uplink Downlink	5.985 3.760	5.985 3.760	5.985 3.760
Modulation Index	4.22	1.10	3.5
RF Bandwidth (MHz)	2.0	36.	36.
RF Power (Watts) Uplink Downlink	9. .03	126. 4.	1000. 4.
Antenna On-axis Gain (dB) Uplink Transmit Uplink Receive Downlink Transmit Downlink Receive	50.5 30.0 29.5 58.5	62.0 30.0 29.5 58.5	52.5 30.0 29.5 50.0

TABLE 5-5

ID of Coverage	3 db beam width (degrees)	Beam Center Location		Service Arc	
		Relative Latitude (degs.)	Relative Longitude (degs.)	From (degs.)	To (degs.)
1	2.0	61.0	9.5	- 2.5	30.0
2	2.0	62.5	27.0	10.0	30.0
3	2.0	53.5	-2.5	-15.0	5.0
4	1.5	65.0	-18.0	-30.0	5.0
5	1.5	49.0	17.5	5.0	30.0
6	1.0	40.0	-8.0	-20.0	5.0
7	1.0	49.5	6.0	-20.0	0.0
8	1.0	53.0	-7.5	-30.0	-10.0
9	1.0	51.5	4.5	0.0	30.0
10	1.0	52.5	6.0	-30.0	10.0
11	1.0	47.0	8.0	-10.0	30.0
12	2.0	47.5	2.5	-5.0	20.0
13	1.0	56.5	9.0	-25.0	10.0
14	2.0	40.5	-3.5	-30.0	0.0
15	1.0	47.5	8.5	0.0	25.0
16	1.5	52.5	12.5	-10.0	15.0
17	1.0	41.5	20.0	-30.0	30.0
18	1.0	47.5	20.0	-30.0	30.0
19	2.0	42.0	12.5	-15.0	15.0
20	1.0	42.5	25.5	-30.0	30.0
21	1.5	46.0	25.0	0.0	30.0
22	1.5	35.0	38.0	-30.0	30.0
23	3.0	24.0	45.0	-30.0	30.0
24	1.0	33.5	36.0	-30.0	30.0
25	1.5	33.0	43.0	-30.0	30.0
26	1.0	32.0	37.0	-30.0	30.0

Sources: Ref. [33]

TABLE 5-6

FREQUENCY RE-USE SUMMARY FOR CAPACITY MODEL

Zone Dimension (Miles)	Satellite Spacing (Degrees)	Frequency Re-Uses Per Band — Upper Bound	Frequency Re-Uses Per Band — Full Earth (Elevation $\chi \geq 10°$)	Frequency Re-Uses Per Band — Land Areas Only
2000	10	226	150	50
2000	5	452	300	100
1000	10	905	600	200
1000	5	1809	1200	4000

TABLE 5-7

SATELLITE TRANSPONDER COMMUNICATIONS CAPACITY

SERVICE	OPERATING MODE	CHANNEL CAPACITY PER 40 MHz TRANSPONDER
ANALOG TELEPHONE	SCPC/FDMA FDM/FDMA FDM/FDMA-COMPANDED FDM/FDMA-COMPANDED- SPEECH INTERPOLATION	1200 500 1000 1600

DIGITAL TELEPHONE	PCM SCPC/FDMA
	
	ΔMOD SCPC/FDMA
	PCM/TDMA
	ΔMOD/TDMA
	PCM/TDMA-SPEECH
	  INTERPOLATION
	ΔMOD/TDMA-SPEECH
	  INTERPOLATION
		800
		
		1600
		1000
		2000
		2000
		
		4000

ANALOG TELEVISION	"SMALL" EARTH TERMINALS
	"LARGE" EARTH TERMINALS
	"SMALL" EARTH TERMINALS-
	  MULTIPLEXED
	"LARGE" EARTH TERMINALS-
	  MULTIPLEXED
		1
		2
		2
		
		4

DIGITAL TELEVISION	DATA COMPRESSION
	  (BROADCAST QUALITY)
	DATA COMPRESSION
	  (TELECONFERENCE QUALITY)
		3
		
		10

Sources: Ref. [72]

Note: Since the initial preparation of this table, further progress has been made in the techniques for increasing transponder utilization. Perhaps the most significant of these is companded single-sideband (CSSB) transmission for voice traffic (see, e.g., Brown, *et al.*, in the Select Bibliography). It is estimated that with linear transponders as many as 6,000 channels could be supported with $5m$ earth-station antennas.

TABLE 5-8

ORBIT COMMUNICATION CAPACITY DETERMINATION

	Coverage Zone Dimension (Miles)	Satellite Spacing (Deg.)	(1), (2) Number of Frequency Reuses	(3) Allocated Bandwidth Below 14.5 GHz (MHz)	Allocated Bandwidth Above 14.5 GHz MHz)	(4) Total Effective Bandwidth (MHz)
CASE I (1980 - 1985)	2000	5	50	2980	-	5960
CASE II (1995 - 2000)	1000	5	200	2980	3400	9460
	Equivalent 40 MHz Transponders	Telephone Channels Per Transponder	Television Channels Per Transponder	Total Telephone Channels	Total Television Channels
CASE I (1980 - 1985)	149	750	1	5.5×10^6	7×10^3
CASE II (1995 - 2000)	236.5	2000	3	91×10^6	1.4×10^5

(1) Based on coverage of non-ocean areas only.
(2) 50% of theoretical value, to allow for non-uniformity of network parameters.
(3) Where allocations are to a single Region, one-third of the allocation is counted.
(4) Includes effect of dual polarization below 14.5 GHz.

REFERENCES/PART TWO

1. CCIR: XIVth Plenary Assembly, Kyoto, Japan, Vol. IV, *Rec. 465-1* (Reference earth station radiation pattern for use in coordination and interference assessment in the frequency range from 2 to about 10 GHz) Geneva, 1978.

2. CCIR: XIVth Plenary Assembly, Kyoto, Japan, Vol. IV, *Report 390-3* (Earth station antennae for fixed satellite service), Geneva, 1978.

3. ITU, *Final Acts of the World Administrative Radio Conference*, Geneva, 1979.

    (a) Article N27, Section III.

    (b) Article N27, Section IV.

    (c) Article N25.

    (d) Article N26.

    (e) Appendix 30.

4. CCIR: XIVth Plenary Assembly, Kyoto, Japan, Vol. IV, *Report 391-3* (Radiation diagrams of antennae for earth stations in the fixed satellite service for use in interference studies and the determination of a design objective), Geneva, 1978.

5. Harris, A.B., B. Claydon, and K.M. Keen, "Reducing the Side Lobes of Earth-Station Antennas," *Conf. Rcd.*, Vol. 1, pp. 6.6.1 - 6.6.5, International Conference Communication (ICC '79), Boston, MA, June 10-14, 1979.

6. CCIR: XIVth Plenary Assembly, Kyoto, Japan, Vol. IV, *Report 555-1* (Discrimination by means of orthogonal circular and linear polarizations), Geneva, 1978.

7. CCIR: XIVth Plenary Assembly, Kyoto, Japan, Vol. IV, *Report 556-1* (Factors affecting station-keeping of geostationary satellites of the fixed satellite service), Geneva, 1978.

8. Takahashi, K., "Collision Between Satellites in Stationary Orbits," *IEEE Trans. Aerosp. & Electronic Syst.*, Vol. AES-

17, No. 4, pp. 591-596, July 1981. (Additions and Corrections, Sept. 1981.)

9. CCIR: XIVth Plenary Assembly, Kyoto, Japan, Vol. IV, *Rec. 353-3* (Allowable noise power in the hypothetical reference circuit for frequency-division multiplex telephony in the fixed satellite service), Geneva, 1978.

10. CCIR: XIVth Plenary Assembly, Kyoto, Japan, Vol. IV, *Rec. 356-4* (Maximum allowable values of interference from line-of-site radio-relay systems in a telephone channel of a system in the fixed satellite service employing frequency modulation, when the same frequency bands are shared by both systems), Geneva, 1978.

11. Giger, A.J., and J.G. Chaffe, "The FM Modulator with Negative Feedback," *B.S.T.J.*, July 1963, Part I.

12. Hedinger, R.A., and M.C. Jeruchim, "On the Relationship Between Geostationary Orbit Capacity and the Interference Allowance," (publication pending).

13. CCIR: XIVth Plenary Assembly, Kyoto, Japan, Vol. IV, *Rec. 466-2* (Maximum permissible level of interference in a telephone channel of a geostationary satellite network in the fixed satellite service employing frequency modulation with frequency-division multiplex, caused by other networks of this service), Geneva, 1978.

14. Lindsey, W.C., and M.K. Simon, "Telecommunication Systems Engineering," Prentice-Hall Book Co., Englewood-Cliffs, N.J., 1973.

15. Jeruchim, M.C., and D.A. Kane, G.E. Co. Doc. No. 70SD4293, Vol. IV, 31 December 1970. (NTIS: PB-203-389.)

16. CCIR: XIVth Plenary Assembly, Kyoto, Japan, Vol. IV, *Rec. 522*, (Allowable bit error rates at the output of the hypothetical reference circuit for systems in the fixed satellite using pulse-code modulation for telephony), Geneva, 1978.

17. Skevington, R.C., "Compandors for FDM/FM Voice Traffic," *Conf. Rcd.*, Vol. 1, pp. 12.5.1 - 12.5.5, Nat'l Telecommun. Conf., (NTC '79), Washington, D.C., Nov. 1979.

18.   Szarvas, G.G., and H.G. Suyderhoud, "Enhancement of FDM-FM Satellite Capacity by Use of Compandors," COM-SAT Tech. Rev., Vol. II, No. 1, Spring 1981, pp. 1-58.

19.   Welti, G.R., and R.K. Kwan, "Comparison of Signal Processing Techniques for Satellite Telephony," Conf. Rcd., National Telecommunication Conference (NTC '77), Los Angeles, CA, Dec. 1977.

20.   CCIR: Study Groups Doc. USSG 4/19 (USA), "Concerning Emission and Sensitivity Constraints on Earth and Space Stations of Geostationary Networks in the Fixed-Satellite Service," 6 May 1977.

21.   CCIR: IWP 4/1 Report, Tokyo, June 1978.

22.   CCIR: Doc. IWP 4/1-2 (USA), "The Concept of Internetwork Isolation for the Assessment of Potential Interference Between Geostationary Networks of the Fixed-Satellite Service," Contribution to IWP 4/1 Meeting, Tokyo, 24 March 1978.

23.   Weiss, H.J., "Relating to the Efficiency of Utilization of the Geostationary Orbit/Spectrum in the Fixed-Satellite Service," Proc. IEEE, Vol., 68, No. 12, Dec. 1980, pp. 1484-1496.

24.   CCIR: Study Groups Doc. IWP 4/1-7 (USA), "A Method of Reducing Inhomogeneity Within the Fixed Satellite Service Bands," 2 April 1980.

25.   Long, W.G., Jr., "Reduction in Inhomogeneity Within the Fixed-Satellite Service Frequency Bands," Conf. Rcd., NTC '78, Birmingham, AL, Dec. 1978.

26.   CCIR: Study Groups Doc. IWP 4/1/615 (USA), "Single Entry Interference Criterion," Tokyo, 1978.

27.   CCIR: Study Groups Doc. IWP 4/1/616 (USA), "Evaluation of Interference among Communications Satellite Networks Sharing the Same Orbit/Frequency Space," Tokyo, 1978.

28.   CCIR: Study Groups Doc. IWP 4/1/617 (USA), "A Study on the Characteristics of Interference in FM Satellite Systems," Tokyo, 1978.

29.   CCIR: Study Groups Doc. IWP 4/1-3 (USA), "Factors Affecting the Accumulation of Interference in a Network of the Fixed-Satellite Service Due to Individual Interference Contributions from Other Such Networks," Tokyo, 10 Feb. 1978.

30.   CCIR: Study Groups Doc. USSG 4/6 (USA), "The Relationship Between the Single Entry Criterion and the Maximum Total Internetwork Interference," 1 April 1980.

31.   CCIR: Special Joint Meeting, Doc. M/80, "Deployment of Dissimilar Satellites in the Geo-Stationary Satellite Orbit," 16 Nov. 1970.

32.   Rice, P.L., et al., "Idealized Pencil-Beam Antenna Patterns for Use in Interference Studies," IEEE Trans. Commun. Tech., Feb. 1970.

33.   CCIR: Doc. IWP 4/1-944, IWP 4/1 Meeting, 28 May 1981.

34.   Ito, Y., T. Mizuno, and T. Muratani, "Effective Utilization of Geostationary Orbit Through Optimization," *IEEE Trans. Commun.,* Vol. COM-27, No. 10, pp. 1551-1558, Oct. 1979.

35.   Fuenzalida, J.C., "Reuse of the Frequency Spectrum at the Satellite," COMSAT Corp. Tech. Memo CL-49-69, 28 Nov. 1969.

36.   Fuenzalida, J.C., and E. Podraczky, "Reuse of the Frequency Spectrum at the Satellite," AIAA Paper No. 70-442, AIAA 3rd Communications Satellite Systems Conf., Los Angeles, CA, April 6-8, 1970.

37.   Welti, G.R., "Frequency Reuse Limits for the Geostationary Orbit," COMSAT Tech. Rev., Vol. 9, No. 2B, Fall 1979, pp. 723-730.

38.   CCIR: XIVth Plenary Assembly, Kyoto, Japan, Vol. IV, *Report 558-1* (Satellite antenna patterns in the fixed satellite service), Geneva, 1978.

39.   Reudink, D.O., A.S. Acampora, and Y.-S. Yeh, "The Transmission Capacity of Multibeam Communication Satellites," *Proc. IEEE,* Vol. 69, No. 2, Feb. 1981, pp. 209-225.

40.   Brunstein, S., CCIR Information Document — JPL Contoured Beam Antenna Computer Generated Pattern, 21 April 1975.

41.   Matsushita M. and J. Majima, "Efficient utilization of orbit/frequency for satellite broadcasting," *IEEE Trans., Aerosp. Electron. Syst.,* Vol. AES-9, pp. 2-10, Jan. 1973.

42.   Reinhart, E.E. "Orbit-spectrum sharing between the fixed-satellite and broadcasting services with applications to 12 GHz domestic system," RAND Corp., Santa Monica, CA, R-1463-NASA, May 1974.

43.   CCIR: Japanese contribution to IWP 4/1 Meeting, "Efficient Utilization of Orbit/Spectrum by Use of Crossed-Beam Arrangement of Satellite Systems," 3-8 June 1975.

44.   ITU: Seminar on Frequency Management and Use of the Frequency Spectrum, Organized by the International Frequency Registration Board (IFRB); Doc. No. 24/76-E, Sept. 1976.

45.   CCIR: XIVth Plenary Assembly, Kyoto, Japan, Vol. IV, *Report 384-3* (Energy dispersal in the fixed satellite service), Geneva, 1978.

46.   CCIR: XIVth Plenary Assembly, Kyoto, Japan, Vol. IV, *Report 557* (The use of frequency bands allocated to the fixed satellite service for both the up-path and down-path of geostationary-satellite systems), Geneva, 1978.

47.   Arnold, H.W., D.C. Cox, H.H. Hoffman, and R.P. Leek, "Measurements and Prediction of the Polarization — Dependent Properties of Rain and Ice Depolarization," *IEEE Trans. Commun.,* Vol. COM-29, No. 5, pp. 710-715.

48.   Arnold, H.W., D.C. Cox, and A.J. Rustako, Jr., "Rain Attenuation at 10-30 GHz Along Earth-Space Paths: Elevation Angle, Frequency, Seasonal, and Diurnal Effects," *IEEE Trans. Commun.,* Vol. COM-29, No. 5, pp. 716-721, May 1981.

49.   Castel, R.E., and C.W. Bostian, "Combining the Effects of Rain-Induced Attenuation and Depolarization in Digital Satellite Systems," *IEEE Trans. Aerosp. & Electronic Syst.,* Vol. AES-15, No. 2, pp. 299-301, March 1979.

50. Crane, R.K., "Prediction of Attenuation by Rain," *IEEE Trans. Commun.*, Vol. COM-28, No. 9, Sept. 1980, pp. 1717-1733.

51. Lin, S.H., H.J. Bergmann, and M.V. Pursley, "Rain Attenuation on Earth-Satellite Paths — A Summary of 10-Year Experiments and Studies," *B.S.T.J.*, Vol. 59, No. 2, Feb. 1980, pp. 183-228.

52. CCIR: XIVth Plenary Assembly, Kyoto, Japan, Vol. IV, *Report 453-2* (Technical factors influencing the efficiency of use of the geostationary satellite orbit by radiocommunication satellites sharing the same frequency bands. General Summary), Geneva, 1978.

53. CCIR: Japanese contribution to IWP 4/1 Meeting, "Orbit Spacing of Geostationary Fixed-Satellites Operating at Frequencies Above 10 GHz," 3-8 June 1975.

54. Lee, L.-S., "The Feasibility of Two One-Parameter Polarization Control Methods in Satellite Communications," *IEEE Trans. Commun.*, Vol. COM-29, No. 5, pp. 735-743, May 1981.

55. Ince, A.N., D.W. Brown, and J.A. Midgley, "Power Control Algorithms for Satellite Communications Systems," *IEEE Trans. Commun.*, Vol. COM-24, No. 2, Feb. 1976.

56. Lyons, R.G., "A Statistical Analysis of Transit Power Control to Compensate up- and down-link Fading in an FDMA Satellite Communication System," *IEEE Trans. Commun.*, Vol. COM-24, No. 6, June 1976.

57. Maseng, T., and P.M. Bakken, "A Stochastic Dymanic Model of Rain Attenuation," *IEEE Trans. Commun.*, Vol. COM-29, No. 5, pp. 660-669, May 1981.

58. CCIR: XIVth Plenary Assembly, Kyoto, Japan, Vol. IV, *Report 711* (Criteria of efficiency of use of the geostationary-satellite orbit), Geneva, 1978.

59. Acampora, A.S., "The Ultimate Capacity of Frequency-Reuse Communication Satellites," *B.S.T.J.*, Vol. 59, No. 7, Sept. 1980, pp. 1089-1122.

60. CCIR: Study Groups Doc. IWP 4/1/808 (USA), "Communication Capacity of the Geostationary Satellite Orbit," 30 June 1980.

61. CCIR: Study Groups Doc. IWP 4/1/J-3, Japanese contribution to IWP 4/1 meeting, "The Capacity of the Geostationary Satellite Orbit," 18-29 May 1981.

62. Weinberger, H.L., "Communication Satellite Spectrum Conservation Through Advanced Technology," Fifth Intern. Wroclaw Symp. on Electromagnetic Compatibility, Wroclaw, Poland, Sept. 1980.

63. Abbott, L., "Transmission of Four Simultaneous Television Programs via a Single Satellite Channel," *SMPTE Journal,* Feb. 1979.

64. Campanella, S.J., "Digital Speech Interpolation," *COMSAT Tech. Rev.,* Vol. 6, No. 1, Spring 1976, pp. 127-158.

65. Kaneko, H., and T. Ishiguro, "Digital Transmission of Broadcast Television with Reduced Bit Rate," *Conf. Rcd,* Vol. 3, pp. 41.4.1 - 41.4.6, Nat'l Telecomun. Conf. (NTC 77), Los Angeles, CA, Dec. 1977.

66. CCIR: Study Groups Doc. USSG 4/1 (USA) "Some Considerations Regarding Frequence Reuse in the Intersatellite Service," 10 Jan. 1980.

67. CCIR: UK Contribution to IWP 4/1 Meeting, "Provisional CCIR Technical Report for WARC-84; Elements of Chapter 18 Involving Inter-Satellite Links," 18-29 May 1981.

68. CCIR: XIVth Plenary Assembly, Kyoto, Japan, Vol. IV, *Report 451-2* (Factors affecting the system design and the selection of frequencies for inter-satellite links of the fixed satellite service), Geneva, 1978.

70. CCIR: UK Contribution to IWP 4/1 Meeting, "Provisional CCIR Technical Report for WARC-84; Chapter 8: the Inter-Satellite Service," 18-29 May 1981.

71. CCIR: Study Group Doc. for IWP 4/1 (UK Contribution), "Pairing of Up and Down Path FX-SAT Frequency Bands as a Means of Improving Efficiency of Use of the Geostationary Orbit," 11 March 1981.

72.   CCIR: Japanese contribution to IWP 4/1 Meeting, "Orbit/ Spectrum Utilization Efficiency of a Multi-Band Satellite System," 3-8 June 1975.

       CCIR: Study Groups Doc. IWP-4/1-US5 (USA), "The Effect of Multiple Band/Service Satellites on Orbit and Spectrum Utilization," April 1975.

73.   Ballard, A.M., "Rosette Constellations of Earth Satellites," *IEEE Trans. Aerosp. & Electronic Sys.*, Vol. 16, No. 5, Sept. 1980, pp. 656-673.

74.   Wadsworth, D.V.Z., "Longitude-Reuse Plan Doubles Communication Satellite Capacity of Geostationary Arc," *Proc. AIAA 8th Communications Satellite Systems Conf.*, 1980.

75.   Wang, J.H.C., and E.D. Davis, "The Quasi-Stationary Satellite Orbit," Federal Communications Commission, *Report No. RS 75-01*, Feb., 1975.

76.   CCIR: UK Contribution to IWP 4/1 Meeting, "Draft Element for Provisional CCIR Technical Report for WARC-84; Chapter 2. Technical Principles of Efficient Orbit/Spectrum Utilization," 18-29 May 1981.

77.   CCIR: USA Contribution to IWP 4/1 Meeting, "Operational and Technical Issues in the Fixed-Satellite Service," 18-29 May 1981.

78.   CCIR: Canadian Contribution to IWP 4/1 Meeting, "The Potential Flexibility in Re-positioning of Communication Satellites in the Geostationary Orbit," 3-8 June 1975.

Chapter Six

THE PERFORMANCE OF ANALOG SIGNALS IN AN INTERFERENCE ENVIRONMENT

6.0 Introduction

While there are many factors relevant to orbit utilization, they all revolve around one central and inescapable causation; namely, interference. If it were not for the presence of interference the subject of orbit utilization would not exist in a communication sense. (It might exist eventually in a spatial sense, when collision probabilities become large enough, but that is another subject matter. See, e.g., Ref. [1]). Interference can be controlled or attenuated in various ways but it cannot be eliminated. Its existence is as physically fundamental as that of noise in electronic devices, and stems from the fact that it is not possible to confine radiated energy to within a prescribed volume; stated another way, an antenna which illuminates only an intended area of reception, and nowhere else, is physically unrealizable. The amount of interference permitted to be induced by one network into another is at the core of the orbit utilization problem. That amount is measured either at the input to the wanted signal's receiver, or at its output, depending upon the specifics of the case. In the most general terms we seek to evaluate an expression of the form

$$Q = f(S_w, D_w; S_I, D_I) \qquad (6\text{-}1)$$

where

Q = wanted signal quality, e.g., SNR and error probability;

S = the set of parameters specifying the modulation characteristics; e.g., signal type, modulation index, and baseband bandwidth; subscript w and I refer to wanted and interfering signals, respectively, the latter of which can be arbitrary in number;

D = the set of network design particulars, for example the link parameters such as *e.i.r.p.*, frequency, antenna size; and the details of filtering, multiple access, amplification, detection; subscript w and I refer to wanted and interfering signals, respectively.

There are two distinct and separable aspects to the evaluation of (6-1). The first is the calculation of the interference (power or voltage) levels themselves at the wanted receiver's input. This is a straightforward calculation involving slant range, transmit powers, and antenna gains. In effect this has already been formulated in terms of the carrier-to-interference ratio in Section 4.1. The second aspect, which is the topic of this chapter and the next, is the determination of the signal quality Q, given specific interference levels. In these two chapters we shall present a survey of the techniques and approaches to evaluating Q as it concerns, in particular, situations of interest in the orbit utilization context, and we will present a number of curves that may be useful in developing initial estimates of the effect of interference. We shall also look into the goodness of certain approximations or simplications that are often made in practice.

6.1 The General Interference Scenario

As implied by the symbolic statement (6-1) the evaluation of interference effects depends, in general, upon the details of both the wanted signal and of the interference, including the nature of the modulating signals, the modulation methods, the frequency assignments, the definition of Q, and the transformations or processing that these signals may be subjected to. The fact that results are implementation-dependent (except for a few cases of unimportance in this context) makes it impossible to obtain truly general results, and this is especially so in networks containing nonlinear transmission elements such as power amplifiers. We can, however, define a general interference "scenario" in a topological sense, and it is instructive to do so to establish a broad perspective. Such a scenario is pictured in Figure 6-1 in which an arbitrarily chosen network is considered to be the wanted one, and the other networks as interferers. The implementation characteristics, whatever they may be, are embedded in the blocks. Solid lines represent wanted signal paths, and dashed lines represent interference paths. The frequency plan within each network can be arbitrary *vis-a-vis* that of any other network. Each earth station transmits up to its own satellite a set of carriers $\{S_{gi}\}$, $i = 1, 2, \cdots, K$ for the interfering networks and $i = w$ for the wanted network. At the input to each satellite noise is added, in general, and the satellite, which may be a simple repeater or a processing transponder, emits a set of transmissions $\{S_{ti}\}$. In general, $\{S_{ti}\}$ is not merely a translated version of $\{S_{gi}\}$ both because there is noise and interference present at the satellite input and because, even in the absence of noise and interference, "new" transmissions will appear whenever multiple carriers are amplified by the same (nonlinear) amplifier. Note that even within the wanted network itself there are three possible sources of self-interference: the intermodulation products just noted, adjacent-channel interference, and cross-polarized (generally co-channel) carrier interference. Finally, the waveform seen at the wanted receiver input is the sum of its front-end noise N_{dw} and the transmissions from the wanted and interfering satellite. Further processing, i.e., filtering, demodulation, and detection, will produce the final signal given over to the end

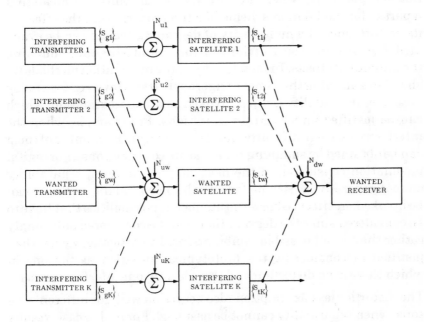

Figure 6-1 Schematic Diagram of General Interference Scenario

(human or machine) receptor. This is where the quality Q is ultimately defined meaningfully, but what is of common interest is the relation between Q and the interference power (other things being fixed) at the receiver input.

There is evidently a virtually limitless number of interesting combinations of wanted and interfering signals as well as possible implementations. Clearly, it will only be feasible to examine a limited subset of these possibilities. In choosing the members of this subset we can be guided by the following considerations. First, we shall confine our attention strictly to classes of signals and modulation schemes that are of current interest in satellite communication. For each type of wanted signal, however, we shall not necessarily investigate exhaustively the effect of all other types of wanted signal, itself included, considered as interferers: in other words, we shall consider for each wanted signal a set of interfering signal types generally smaller than the set of wanted signals itself.

Another consideration which helps to narrow our focus is the condition of separability, where it exists. By "separability" we mean that a particular modulation scheme has the property that the effect on its performance of a multiplicity of degrading factors can be evaluated by considering each of these factors one at a time and summing the individual effects. In fact, for any of the modulation methods we shall be examining this is never rigorously true. However, for analog methods, separability is often invoked as an approximation, which can be justified under certain conditions. For example, when the interferences are sufficiently small the effect of any number of them can be obtained by summing the separate contributions. (For digital modulation considered in the next chapter, however, separability cannot generally be assumed to be valid.) Separability, where it can be invoked, in effect reduces the number of possibilities that have to be considered since the degradations can be considered only singly rather than in all possible combinations. In particular, we are then justified in considering the prototypical situation as the one in which there is no distortion and there is only one interferer present.

The distortionless assumption also serves us well, for different reasons, when separability cannot be assumed. For in this case, results are not implementation-dependent. Clearly, it is not possible to consider all design variations, and in fact the distortionless assumption is the only one which does not make any specific assumptions in that regard, and hence can be regarded as an invariant reference. Note, too, that in many cases preliminary assessments of interference effects have to be made by system designers before there is a firm or even tentative design: in this case, the distortionless assumption is perhaps the most reasonable one to make initially. Finally, from an expository point of view, the distortionless assumption is satisfying because (in the non-separable situation) it is the only one in which we can uniquely assign a degradation effect to the presence of interference, which is, after all, the subject of our present inquiry.

The preceding does not imply that practical implementation effects are unimportant. Rather, that we are justified from several points of view in simplifying matters for ourselves. This serves as a convenient starting point. We shall where possible indicate the influence of non-ideal design, and the manner of extending the analyses to take that into account.

In this chapter we shall confine ourselves to the case where the wanted signal is an analog modulated carrier. Signals of this type that are of importance in satellite communication are FDM/FM telephony, FM television, and SCPC/FM. Since these are all basically angle-modulated signals, we begin by formulating in fairly general terms the effect of interference on such signals. The three subsequent sections then take up in turn the three specific types of signals mentioned just above. The next section discusses the role of energy dispersal in interference reduction, while the ensuing section addresses a number of implementation-related considerations. Finally, in the last section of this chapter we outline the idea behind as approximation that forms the basis for the initial step in coordination procedures, and discuss its implications as regards interference. In the next chapter we shall deal with the situation where the wanted signal is digital.

6.2 Angle-Modulated Analog Wanted Signal

All of the analog signals that we shall consider are angle-modulated. Hence, for the sake of generality, it is instructive to begin assuming only that property. We shall then, in subsequent sections, specialize to the cases of interest. As per our previous discussion we assume initially that the desired signal is undistorted and the detector is ideal. Thus, consider a wanted angle-modulated signal

$$w(t) = A \cos[\omega_1 t + \phi(t)] \tag{6-2}$$

$$= Re \{A \, v(t) \, exp \, (j\omega_1 t)\}$$

and an interfering signal

$$i(t) = R(t) \cos [\omega_2 t + \psi(t) + \mu] \tag{6-3}$$

$$= Re \{u(t) \, exp \, (j\omega_2 t + \mu\}$$

where $u(t) = R(t) \, exp \, [j\psi(t)]$, and the angle μ, at this stage, can be arbitrarily assigned.

Notice that we are not necessarily assuming a single interferer at this point. Any bandpass signal, which can be the sum of several signals, can be represented in the form (6-3). In addition these signals may or may not be filtered, e.g., by the front-end of the wanted receiver, for the form (6-3) still applies. In the latter case, however, it may not be simple to find an explicit expression for $R(t)$ and $\psi(t)$ in terms of the original modulating functions. We can further use the same representation for an amplitude-modulated interferer simply by letting $\psi(t) = 0$.

By straightforward trigonometric manipulations, the sum signal can be put in the form

$$s(t) = w(t) + i(t)$$

$$= A(t) \cos [\omega_1 t + \phi(t) + \varepsilon(t)] \tag{6-4}$$

where

$$A^2 (t) = \{A + R(t) \cos [\omega_d t + \psi(t) - \phi(t) + \mu]\}^2 \tag{6-5}$$
$$+ \{R(t) \sin [\omega_d t + \psi(t) - \phi(t) + \mu]\}^2$$

$$\varepsilon(t) = tan^{-1} \frac{R(t) \sin [\omega_d t + \psi(t) - \phi(t) + \mu]}{A + R(t) \cos [\omega_d t + \psi(t) - \phi(t) + \mu]} \tag{6-6}$$

$$\omega_d = \omega_2 - \omega_1$$

An ideal phase detector* operating on (6-4) produces the baseband output

$$e_o(t) = \phi(t) + \varepsilon(t) \tag{6-7}$$

whence the effect of interference is embodied in the excess angle $\varepsilon(t)$. If $w(t)$ were originally frequency-modulated, the output would be

$$e_{o,FM}(t) = \frac{d}{dt}[\phi(t) + \varepsilon(t)] \tag{6-8}$$

and indeed, in general, the output of a receiver that operates linearly on the phase is simply

$$e_o(t) = [\phi(t) + \varepsilon(t)] * h(t) \tag{6-9}$$

where $h(t)$ is the impulse response of the (cascade of) linear devices and * denotes convolution.

It should be noted that (with the assumptions made) our result so far is perfectly general. In particular, there is no requirement that the interference be smaller than the desired signal. Thus, for example, given specific values for ω_d, μ, and A, and specific functions for $\phi(t)$, $\psi(t)$, and $R(t)$, one could compute the time function $\varepsilon(t)[$ or $\varepsilon(t) * h(t)]$ and (in principle) any temporal moments of interest**. Or, given an appropriate statistical description of the quantities in (6-6) one could, again in principle, obtain the desired statistical moments of the distortion. Generally, the latter approach has enjoyed favor in

---

*We shall consider non-ideal angle demodulators in Section 6.8

**This is the "instantaneous" approach elaborated by Middleton and Spaulding [2], which we shall touch upon in Section 6.8.

the interference context. Equation (6-6), however, is not easy to manipulate as it stands. By making an assumption which is entirely reasonable in the context of this book, we can obtain more pleasing results which ultimately permit a more systematic evaluation of interference effects. This assumption is that the peak interference amplitude does not exceed the wanted signal amplitude: in practice this is more than amply met in virtually all situations.

With this assumption in mind (though not yet invoked) we return to the sum signal (6-4) which can be alternately expressed as

$$s(t) = Re\{ Aa(t) \, exp \, [\, j \, (\omega_1 t + \phi(t) + \lambda(t))] \} \tag{6-10}$$

where

$$a(t) \, exp \, [\, j\lambda(t) \,] = 1 + z(t) \, exp \, \{ \, j \, [\, 2\pi f_d t - \phi(t) + \mu] \} \tag{6-11}$$

$$z(t) = u(t)/A$$

and $f_d = \omega_d/2\pi$. As before, an ideal phase detector operating on (6-10) yields

$$e_o(t) = \phi(t) + \lambda(t)$$

where now we have labeled the excess phase angle $\lambda(t)$, which from (6-11) can be expressed as

$$\lambda(t) = Im \, ln \, \{1 + z(t) \, exp \, \{ \, j \, [\, 2\pi f_d t - \phi(t) + \mu] \} \tag{6-12}$$

when *Im* indicates imaginary part.

At this point we invoke the peak-limited interference assumption, which permits us to use the series representation

$$ln \, (l + y) = \sum_{m=1}^{\infty} \frac{(-1)^{m+1}}{m} \, y^m \tag{6-13}$$

for $|y| < 1$. Identifying

$$y = z(t) \, exp \, \{ j [2\pi f_d t - \phi(t) + \mu] \}$$

which implies $|u(t)|/A < 1$, and applying (6-13) to (6-12) yields

$$\lambda(t) = Im \sum_{m=1}^{\infty} \frac{(-1)^{m+1}}{mA^m} u^m(t) \, exp \, \{jm[2\pi f_d t - \phi(t) + \mu]\}$$

$$= \sum_{m=1}^{\infty} \frac{(-1)^{m+1}}{mA^m} R^m(t) \, sin \, [\, m(2\pi f_d t - \phi(t) + \psi(t) + \mu) \,]$$

(6-14)*

As stated, (6-14) depends on the inequality $|z(t)| = |u(t)|/A < 1$, or $R(t)_{max} < A$. In fact we usually have $R(t)_{max} \ll A$. It is occasionally possible, however, for $|z(t)| > 1$ in certain situations, namely when $w(t)$ and $i(t)$ occupy closely spaced adjacent channels. This case has been analyzed [3,4] and it has been shown that the excursions of $|z(t)|$ produce impulsive or "click" noise. We shall henceforth ignore this possibility.

In a formal sense Equation (6-14) contains all the possible information on the effect of interference. The evaluation of that effect, however, is another matter, for we are ultimately interested in characterizing the effect of interference in the same terms that the quality Q of the signal are normally specified. This, in general, may be rather difficult to do. For example, if $\phi(t)$ is a television signal, how does one relate $\lambda(t)$ to picture quality?

Thus, the problem at hand, generally stated, is what are the required manipulations on (6-14) to obtain a meaningful assessment of the interference. It is not clear whether there is a universal answer, but one approach which may have fairly general validity is the evaluation of the power spectral density (PSD) corresponding to $\lambda(t)$. This is meaningful in at least one important case (see Section 6.3) and is intuitively attractive from an engineering standpoint.

We now seek to derive this PSD, which requires us to take a statistical approach. We thus define the angle μ in (6-3) to be a random variable, uniformly distributed on $[0, 2\pi]$, and we assume $\phi(t)$ and

---

*Notice that under the condition $| z(t) | \ll 1$, the first term of (6-14) is an adequate approximation which could have been obtained immediately from (6-6) under the same condition by invoking tan x $\approx$ x. However, (6-14) gives a better appreciation of the numerical domain where these approximations can be used.

$\psi(t)$ to be wide-sense stationary random processes independent of each other and of μ.* The phase receiver's output autocorrelation function can be shown to be

$$R_o(\tau) = R_\phi(\tau) + R_\lambda(\tau) \qquad (6\text{-}15)$$

with

$$R_\lambda(\tau) = \sum_{m=1}^{\infty} c_m E\{[\bar{u}\ (t)\ u\ (t+\tau)]^m\ e^{jm[\phi(t)\cdot\phi(t+T)]} e^{jmw_d T}$$

$$+[u\ (t)\ \bar{u}\ (t+\tau)]^m\ e^{-jm[\ \Phi(t)\cdot\phi(t+\tau)]} e^{-jmw_d\tau}\}$$

$$(6\text{-}16)$$

where R_ϕ and R_λ are, respectively, the autocorrelation functions of $\phi(t)$ and $\lambda(t)$, $c_m = [4m^2A^{2m}]^{-1}$, the overbar represents complex conjugate, and E is the expectation operator. Note that the form (6-15) implies $\phi(t)$ and $\lambda(t)$ are uncorrelated, even though the latter is a function of the former. This fact evolves from the assumed properties of μ.

With the definitions

$$E[\bar{u}\ (t)\ u\ (t+\tau)] = R_{mi}(\tau) \qquad (6\text{-}17a)$$

$$E[exp\ \{jm[\phi(t)\cdot\phi(t+\tau)]\}] = R_{mw}(\tau) \qquad (6\text{-}17b)$$

$$S_m(f) = \int_{-\infty}^{\infty} R_{mi}(\tau)\ R_{mw}(\tau)\ e^{j2\pi f\tau} d\tau \qquad (6\text{-}17c)$$

the output PSD due to interference [i.e., the Fourier transform of (6-16)] can be written as

---

*As a historical note it is interesting to point out that in one of the earliest papers on interference, by Bennett et al. [5], these independence assumptions were not made. In fact, $i(t)$ was taken to an echo of $w(t)$, implying $\psi(t) = \phi(t - T)$ and $\mu = 2\pi f_1 T$, where T is the delay. As $T \to \infty$ the echo should become independent of the signal, thus yielding the situation examined here.

$$S_\lambda(f) = \sum_{m=1}^{\infty} c_m\{S_m(f\text{-}mf_d) + S_m(\text{-}f\text{-}mf_d)\} \tag{6-18}$$

From the definitions (6-17) it is easy to show that

$$S_m(f) = S_{mi}(f) * S_{mw}(f) \tag{6-19}$$

where $*$ represents convolution and $S_{mw}(f)$, which is the Fourier transform of (6-17b), represents the PSD of the mth power of the complex envelope of $w(t)$ or, equivalently, it represents the low pass equivalent spectrum of the "multiplied" signal $w_m(t) = cos\{m[\omega_l t + \phi(t)]\}$. Similarly, $S_{mi}(f)$, which is the Fourier transform of the auto-correlation function (6-17a), represents the PSD of the mth power of the complex envelope of the interference. Note that in the preceding formulation the wanted signal spectrum is normalized while that of the interference is not.

When the interference is small, the first term of (6-18) is sufficient to characterize it and the output PSD is simply

$$S_\lambda(f) = \frac{1}{4A^2}\left\{ S_i(f\text{-}f_d) * S_w(f) + S_i(\text{-}f\text{-}f_d) * S_w(\text{-}f) \right\} \tag{6-20}$$

which is the well-known formulation leading to a convolution of spectra. The "general" solution (6-18) is thus the sum of the convolution of successively higher-order spectra. Although this is a formal "solution", the computation of these spectra is not generally a simple task.* The situation is considerably simplified when the small interference condition holds for then only the "first-order" spectra are necessary and we shall generally assume in the following that this condition holds. The computation of spectra can be regarded as a separate discipline, for which a vast literature exists (see, e.g., Ref. [6-11]), and hence will not be considered here in detail.

---

*Koh and Shimbo [12] have outlined a method for obtaining the convolution without explicitly computing the *PSD* of the wanted signal when the interfering *PSD* is given in tabular form (e.g., from measurements) or can be assumed to have an otherwise specified behavior.

We have now proceeded about as far as can be fruitfully done in general terms, so we turn in the next sections to specific cases.

6.3 Interference Between FDM/FM Telephony Signals

Most of the studies published on the performance of (analog) signals in an interference environment have concerned mutually interfering FDM/FM telephony signals [Ref. 13-21]. The most extensive treatments, which are basically generalizations of earlier work, are found in [Ref. 22,23]. The main reason for this attention, aside from the practical importance of this interference scenario, is the analytic tractability of the situation which stems from the fact that (for most purposes) a multichannel FDM speech signal can be represented by a white Gaussian noise process.

We will assume in the remainder that only a single interfering carrier is present and undistorted as well. The first of these assumptions is justified by the separability with respect to multiple interferences when these are small enough: this assertion will be demonstrated in Section 6.8.4. We thus have a wanted signal

$$w(t) = A \cos [\omega_1 t + \phi(t)] \qquad (6\text{-}21)$$

and an interfering signal

$$i(t) = rA \cos [\omega_1 t + \psi(t) + \mu], \ r < 1 \qquad (6\text{-}22)$$

where now $\phi(t)$ and $\psi(t)$ are independent Gaussian processes. But, they need not be white, since the formulation applies equally for pre-emphasized basebands.

For the case in question the power in a telephone channel due to the interference has been found to be a useful measure. This power can be found directly from $S_\lambda(f)$, i.e., by integrating the latter between the channel limits. Thus, (6-18) or (6-20) apply, but their evaluation is facilitated by the Gaussian property of the modulating signals. In particular, when $\phi(t)$ is Gaussian it can be shown [5] that (6-17b) reduces to

$$R_{mw}(\tau) = exp \{-m^2[R_\phi(0)\text{-}R_\phi(\tau)]\}$$

hence the corresponding PSD is

$$S_{mw}(f) = \int_{-\infty}^{\infty} exp\,\{-m^2[R_\phi(0) - R_\phi(\tau)]\}exp\,(-j2\pi f\tau)\,d\tau \qquad (6\text{-}23)$$

where $R_\phi(0)$ is the mean-square value (average power) of $\phi(t)$. *If we write*

$$S_w(f) = \int_{-\infty}^{\infty} exp\,[R_\phi(0) - R_\phi(\tau)]\,exp\,(-j2\pi f\tau)\,d\tau$$

and notice that $R_{mw}(\tau)$ can be decomposed as the product of m^2 autocorrelation functions, it is clear that (6-23) can be expressed as

$$S_{mw}(f) = S_w(f) \overset{N}{*} S_w(f) \qquad (6\text{-}23a)$$

where $N = m^2\text{-}1$ and $\overset{N}{*}$ denotes N convolutions. Thus, the mth-order spectrum $S_{mw}(f)$ can be calculated from the first-order spectrum $S_w(f)$. The evaluation of the first-order spectrum has itself been the subject of much scrutiny [7,10]. Thus, a systematic numerical approach exists for computing either (6-18) or (6-20). In general, however, closed-form expressions are not available even for the first-order PSD. Therefore, results must ultimately be displayed in graphical form (see, e.g., [16,18,22]), and some of these will be given later for illustration. There are, however, some special cases amenable to analytic formulation, and these will be taken up presently.

It is useful to set down a general formulation for the signal quality Q in terms of the PSD $S_\lambda(f)$. It is customary in telephony to deal with the NPR (noise-power ratio), while we have found it convenient to use the receiver transfer characteristic (RTC). It will be shown in Appendix J that for FM signals using pre-emphasis one can write *

$$NPR(f_c) = \frac{(2\pi)^2 M_1^2 f_{m1} b}{2r^2(1-\varepsilon_1)} \left\{ \int_{f_c-b/2}^{f_c+b/2} (2\pi f)^2 G_d(f) S_\lambda'\,(f)df \right\}^{-1} \qquad (6\text{-}24)$$

where

    f_c = center frequency of voice channel under consideration
    b = telephone channel bandwidth (3.1 kHz)

---

*The *PSD* $S_\lambda(f)$ or $S'(f)$ is double sided; hence for any positive frequency the actual power is twice that in $S_\lambda(f)$, which accounts for the factor of 2 in the denominator.

f_{m1} = top baseband frequency of wanted signal
M_1 = rms modulation index of wanted multichannel baseband
ε_1 = ratio of lowest-to-highest frequency of multichannel baseband
$G_d(f)$ = power transfer function of de-emphasis network.

The interference-to-carrier ratio r^2 has been explicitly brought out in (6-24) so that $S'_\lambda(f)$ there corresponds to the normalized spectrum. Also note the factor $(2\pi f)^2$ which is needed to convert the output of a phase detector to that of an FM detector. It is usually safe to assume that $S_\lambda(f)$ is constant over $f_c \pm b/2$. Hence (6-24) simplifies to

$$NPR(f) = \frac{M_1^2 f_{m1}}{2(1-\varepsilon_1)\, r^2 f^2\, S'_\lambda\,(f)\;G_d(f)}$$

(6-25)

where for simplicity the subscript on f_c has been dropped. Normally, the particular channel of interest is the one for which $NPR(f)$ is minimum. The location of that channel is a function of the specifics of the case although the top channel is typically taken as a reference point.

The RTC is simply related to the NPR, as shown in Appendix J, by

$$RTC(f) = NPR(f)\,(B/b)\,L\,(n_1)\,r^2 \tag{6-26}$$

where

n_1 = number of channels on the wanted carrier
$B = f_{m1}(1-\varepsilon_1)/n_1$

and

$$L(n_1) = \begin{cases} 31.6 & ; n_1 \geq 240 \\ 1.26 n_1^{0.6} & ; 12 \geq n_1 < 240 \end{cases}$$

Thus, it will be sufficient for present purposes to restrict ourselves to the NPR.

6.3.1 High-Index FM Signals

A case particularly amenable to analysis occurs when the rms modulation index of both signals is reasonably high ($\lesssim 1$). In this case the spectrum $S_w(f)$ is given approximately by

$$S_w(f) = \frac{exp\ [-f^2/2\Delta f_1^2]}{\sqrt{2\pi}\ \Delta f_1} \tag{6-27}$$

where Δf_1 is the rms frequency deviation $(M_1 f_{m1})$ of the wanted signal; a similar expression holds for the interference spectrum. The convolution [Equation (6-20)] is straightforward, and the final result (derived in Appendix J) is

$$NPR(f) = \frac{2\sqrt{2\pi}\ M_1^2\ M\ G_p\ (f)}{(1-\varepsilon_1^2)\ r^2\ u^2\ \{exp\ [-(u+v)^2/2M^2] + exp\ [-(u-v)^2/2M^2]\}} \tag{6-28}$$

where

$$
\begin{aligned}
G_p(f) &= G_d^{-1}(f) \\
u &= f/f_{m1} \\
v &= f_d/f_{m1} \\
M^2 &= M_1^2 + M_2'^2 \\
M_2' &= M_2 f_{m2}/f_{m1} \\
f_{m2} &= \text{top baseband frequency of interfering signal} \\
M_2 &= \text{rms modulation index of interfering multichannel} \\
&\quad \text{baseband } \Delta f_2/f_{m2}.
\end{aligned}
$$

Thus, for a given channel, the NPR due to interference reduces in this case to a three-parameter problem involving M, M_1, and v. Equation (6-28) lends itself to presentation in generalized form if $NPR(r/M_1)^2$ is plotted as a function of v with M as a parameter. For the case considered the top channel is either the worst one or close to it so that $u \sim 1$. Hence universal family of curves such as that shown in Figure 6-2 (taken from Pontano, et al. [22]) can be plotted. It should be noted that a family of curves derived from (6-28) would have the same *appearance* as that in Figure 6-2 but not the same point-by-point values. point Since (6-28) is based on the high-index assumption it will coincide with the figure in the range of *(M₁, M₂)* that satisfies this assumption, but for small indices this equation will not agree with the figure. The latter is correct but derives from the basic formula (6-20); the more general conditions under which Figure 6-2 applies will be considered in Section 6.3.5.

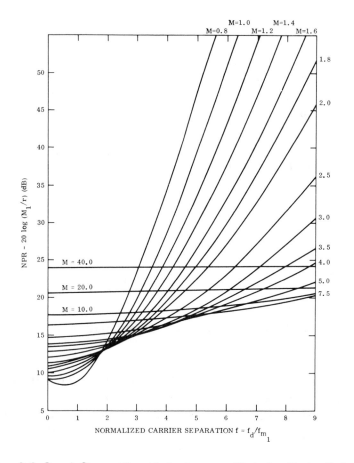

Figure 6-2 Semi-Generalized Interference Plots for Mutually Interfering FDM-FM Carriers

6.3.2 High-Index Wanted Signal: Unmodulated Interference

In the case of an arbitrary wanted signal and an unmodulated *(cw)* interferer, one obtains [as is obvious from (6-20)] in general exactly the form (6-24) with

$$25'_\lambda (f) = S_w(f\text{-}f_d) + S_w(\text{-}f\text{-}f_d) \tag{6-29}$$

irrespective of the modulation index on the wanted signal. However, in the particular case that this signal is high-index, one obtains exactly the form (6-28) with $M_2 = 0$.

6.3.3 Identical Co-channel Carriers

In many practical cases the wanted and interfering carriers have identical or very similar parameters. For high-index co-channel carriers the NPR is given by (6-28) with $v = 0$, $M_1 = M_2 = M'_2$. This yields (with $\varepsilon_1 = 0$), [see also Appendix J]

$$NPR = \frac{8.9 M_0^3}{r^2 \exp(-1/2M_0^2)} \tag{6-30}$$

where M_0 is the common multichannel rms modulation index and 4 dB pre-emphasis advantage has been assumed. For low modulation indices this expression no longer applies. However, it is still possible to obtain a single expression for all M_0. Pontano, et al. [22] have presented a curve of NPR as a function of rms modulation index for identical co-channel carriers. A single expression fits this curve fairly well for the top channel, namely

$$NPR \cong \frac{1 + 9.5 M_0^3}{r^2} \tag{6-31}$$

For $M_0 = 1$, (6-31) is 1.4 dB lower than (6-30) and for $M_0 = 2$ it is 0.2 dB lower. Equation (6-31) forms the basis of the calculations for homogeneous FDM/FM systems presented in Chapter 4.

6.3.4 Low-Index Small Separation Case

When both carriers are low index FM the dominant term (in any telephone channel bandwith) resulting from the convolution (6-20) arises from the interaction of the two residual carrier ("spike") components. When the frequency separation f_d between the two carriers is less than f_m but greater than f_l the (sinusoidal) term in question falls in a telephone channel and constitutes the limiting condition for this case. Thus, (6-20) leads straightforwardly (for this term) to *

$$S_\lambda(f) = \frac{r^2}{4} k_1 k_2 \delta(f\text{-}f_d)$$

(6-32)

where

$\delta(\cdot)$ = Dirac delta function
k_1 = fraction of power of wanted signal in carrier component
k_2 = fraction of power of interfering signal in carrier component.

Substituting (6-32) into (6-24) (recalling that in the latter the PSD is normalized to $r = 1$), one obtains

$$NPR(f_d) = \frac{2M_1^2 (f_{m1}/f_d)^2 G_p(f_d) b}{k_1 k_2 r^2 (1\text{-}\varepsilon_1) f_{m1}}$$

(6-33)

Handy curves for determining k_1 (or k_2) as a function of modulation index and ε can be found in [22, Fig. 9].

6.3.5 Semi-Generalized Interference Curves

For arbitrary wanted and interfering signal parameters it is usually necessary to solve for the NPR explicitly using (6-20) and (6-24). However, for a restricted but still useful range of parameters one can present semi-generalized results — this has been done in Figure 6-2. This figure gives exact results for the worst-channel NPR when the wanted and interfering basebands are identical, but can be extended to a larger set of cases as follows [22]:

---

*There is also an impulse at $f = -f_d$ which contributes nothing to the integral in (6-24).

(a) if the interfering signal has the smaller baseband, Fig. 6-2 provides a good (upper bound) estimate of the interference,

(b) if the interfering signal has the larger baseband, Fig. 6-2 is likely to indicate a smaller than actual interference,

(c) for small carrier separations the curves apply irrespective of baseband sizes.

Thus, Figure 6-2 can serve fairly generally, but as will be seen in Section 6.3.7 the format of this figure will not accommodate all interfering situations.

6.3.6 Carrier Interleaving Improvement

The preceding sections allow us to look into the effect of carrier frequency interleaving, i.e., $f_d \neq 0$. Consider first the high-index case, (6-28). It is clear from this equation that the improvement due to interleaving, i.e., the ratio of NPR with $f_d \neq 0$ to that with $f_d = 0$ is given simply by (ILR = interleaving ratio)

$$ILR = \frac{2 \, exp \, (-u^2/2M^2)}{\{ exp \, [-(u+v)^2/2M^2] + exp \, [-(u-v)^2/2M^2] \}} \qquad (6\text{-}34)$$

For the special case of the top channel, $u = 1$, and expressing the frequency separation as a fraction of the wanted signal bandwidth,

$$v = d(\alpha M_1 + 1)$$

one obtains after some manipulation,

$$ILR = exp \, \{d^2(\alpha M_1 + 1)^2/2M^2\} \, sech \, [2d(\alpha M_0 +)/2M^2] \qquad (6\text{-}35)$$

Assuming now identical carriers $(M_1 = M_2 = M_0)$, and for the range of interest as well as validity of M_0 $(1 \leq M_0 \leq 3)$, (6-35) can be approximated by

$$ILR \approx 2 \, exp \, \left\{ \left(\frac{d\alpha}{2} \right)^2 - \frac{d(2-d)}{4M_0^2} + \frac{2\alpha d(d-1)}{4M_0} \right\}$$

Assuming $\alpha = \sqrt{10}$ and $d = 1.2$ as in [22] (this corresponds to half the Carson rule bandwidth plus a guard band) one obtains

$$10 \ log \ ILR = 18.6 + 10 \ log \ \left(\frac{0.38}{M_0^2} - \frac{0.24}{M_0} \right) \ dB \qquad (6\text{-}36)$$

which was used in Chapter 4. Typically, of course, there are two such interference entries, in which case Eq. (6-34) to (6-36) are reduced by 3 dB. These equations were obtained initially on the basis of high-index carriers (with the additional restriction $M_0 \geq 3$ for the last of these), but actually give fairly accurate results down to $M_0 \approx 0.5$. For smaller M_0 one must use exact calculations, which have been reported in [22] and are presented here in Figure 6-3. For large M_0 one must revert to (6-35), which can be seen to converge to $exp \ (d\alpha/2)^2$, or 15.6 dB.

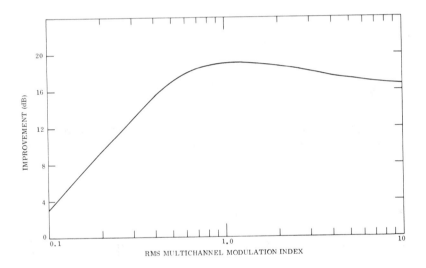

Figure 6-3 Interleaving Improvement for Single Interference

It should be noted that the interleaving improvement is an inherent property of the modulation method and does not derive from filtering (which is normally present) of the adjacent carrier since no such filtering has been assumed in the derivation.

For the larger set of conditions under which Figure 6-2 applies, one can obtain the interleaving improvement directly from that figure simply by taking the difference in the ordinate between the NPR values for $f = f_d$ and $f = 0$. For still less restricted situations one must revert to the original Equation (6-20).

6.3.7 Some Examples Requiring Exact Calculations

As implied earlier some combinations of wanted and interfering carriers cannot be simply characterized, as for example in a display such as that of Figure 6-2. For these situations one must evaluate the convolution (6-20) on a case-by-case basis. To impart some appreciation of what may be expected in this more general setting, we reproduce here in Figures 6-4 to 6-6 some results originally presented by Pontano, et al. [22]. These illustrate, respectively, several points alluded to earlier.

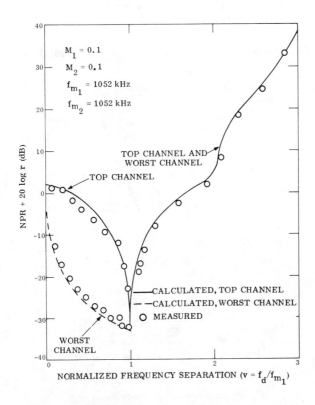

Figure 6-4 Illustration of Location of Worst Channel for Low Modulation Indices

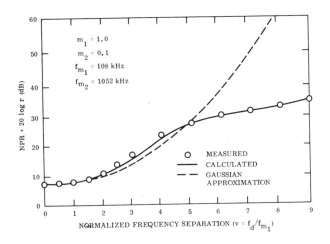

Figure 6-5 Comparison of Actual and Approximate Results for One Case

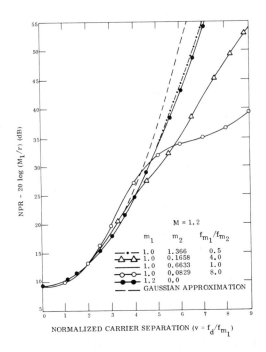

Figure 6-6 Comparison of Calculated Interference for *M* = 1.2

Source: Ref. [22], © 1973 IEEE

Figure 6-4 shows the fact pointed out in Section 6.3.4 that when both carriers have low modulation indices the worst channel is not necessarily the top channel for small $(<f_{m1})$ separations. Figure 6-5 clarifies the breakdown in the "Gaussian approximation" (i.e., Equation 6-28) as the carrier separation increases. Finally, Figure 6-6 demonstrates the effect of relative baseband sizes and the inability of the parameter M to capture the correct functional behavior for all conditions. It may be noted, however, that all the curves tend to coalesce for cochannel operation.

6.3.8 Comparison with Thermal Noise

The effect of thermal noise, because of its importance as well as intuitive value borne of long experience, provides a natural yardstick for assessing the effect of other disturbances. Therefore, a brief comparison between thermal noise and interference is useful. From Appendix A one has for the NPR due to thermal noise

$$NPR_t \approx 5 \ (C/N) \ M_0^2 \ (\alpha \ M_0 + 1) \qquad (6\text{-}37)$$

including pre-emphasis, where C/N is the carrier-to-noise ratio.

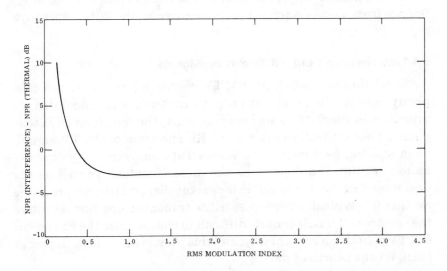

Figure 6-7 Comparison of Susceptibility to Identical Co-Channel Interference and Noise

As has been seen, a wide range of NPR behavior for interference is possible. A convenient reference case for present purposes is the identical cochannel case characterizated by (6-31). The ratio (in dB) of (6-31) to (6-37) is plotted in Figure 6-7, where we have used $\alpha = \sqrt{10}$ and assumed equal CIR and CNR, respectively. Evidently, for all but rather small indices (≈ 0.4) interference is more harmful than thermal noise for the conditions examined.

6.4 Interference Into FDM/FM Telephony from Other Types of Signal

So long as an appropriate measure of the interference is spectrally related, which generally will be the case when its effect is unintelligible, then the formulation (6-24) continues to apply irrespective of the nature of the interfering signal. The output interference PSD $S_\lambda(f)$ must then be computed from the convolution of the spectra in question. Thus, given the spectrum of an arbitrary interfer one would compute the effect on an FDM/FM telephony signal. However, no systematic investigation appears to have been carried out for classes of interfering signals other than FDM/FM signals. Some scattered results exist and we present these here. In particular, we give some limited results for interfering FM television signals and we show a general approach that can be used to compute the interference from a set of narrow-spectrum carriers, the latter representing SCPC signals.

6.4.1 Interference from FM Television Signals

As implied above, when an FM-TV carrier interferes with a telephony carrier, the resultant output interference is calculated, in principle, in much the same way as when the interference is telephony. This evidently requires the RF spectrum of the TV signal. Such spectra, as observed on a spectrum analyzer, are generally quite irregular and depend on the picture content as well as the deviation. The use of spreading (or energy dispersal) waveforms (see Section 6.7) typically used in satellite transmissions further alters the spectrum. Thus, it is quite difficult to characterize the PSD of an FM-TV signal in a general way, and this possibly accounts, in part at least, for the dearth of data.*

---

*One possible way out of this difficulty would be to define a reference FM-TV spectrum for interference calculations, based on a suitable criterion.

The only material seemingly available [24] presents graphical results for the RTC for two assumed symmetrical (TV) RF spectra, as a function of the picture deviation, the telephony rms modulation index, and the difference frequency between wanted and unwanted carriers. Examination of the data reveals that while the RTC depends to some extent on the TV spectrum, a single functional relationship can be inferred which fits all of the data with reasonably good accuracy. The result of this fitting process [25], which is in part based on the behavior of the telephony equations, is as follows:

$$NPR = \frac{0.2 + 8m^3}{r^2(1-\varepsilon)\{exp[-(1+v)^2/1.7(m^2+1.85)] + exp[-(1-v)^2/1.7(m^2+1.85)]\}}$$

(6-38)

which applies to the top channel, and where $v = f_d/f_{m1}$ as before, and m is the telephony carrier rms modulation index. The RTC is related to (6-38) through (6-26). Note that (6-38) does not contain pre-emphasis or noise weighting, which jointly would add about 6.5 dB. It should be noted that the original data [24] were based on $v \lesssim 8$ and discrete values for $m = 0.3, 0.5, 1.0, 2.0$. Use of equation (6-38) for other values of these parameters involves uncorroborated extension of these results, though for small departures there is probably little risk.

6.4.2 Interference from SCPC Signals

A situation which can be straightforwardly characterized is when the interference is composed of the sum of any number of carriers which are very narrow spectrally (ideally, impulses) in relation to the wanted signal. This situation is well approximated when the interferers are SCPC signals (either FM or digital) for these are quite narrow even in relation to the spectrally narrowest telephony carriers (12 channel carriers). We shall proceed using the impulse idealization, in which case we can write for the interference equivalent lowpass PSD [$S_i(f)$ in (6-20)]

$$S_i(f\text{-}f_d) = \sum_{k=1}^{N} r_k^2 A^2 \delta(f\text{-}f_{d_k})$$

(6-39)

where N is the number of interferers, r_k^2 is the kth interference-to-

wanted carrier power, and f_{d_k} is the difference frequency between the
kth interfering and the wanted carriers. Hence, from (6-20), one has

$$S_\lambda(f) = \sum_{k=1}^{N} (r_k/2)^2 \{ S_w(f\text{-}f_{d_k}) + S_w(\text{-}f\text{-}f_{d_k}) \} \tag{6-40}$$

which, when substituted into (6-24) yields the desired answer.* One
only needs $S_w(f)$ for the case in question. For the special cases considered earlier, i.e., high-index or low index FM, one obtains readily
computable forms. Some specific calculations have been presented
[26] for several FDM/FM signal parameters on the assumption that
the (digital) SCPC signals are equal power $(r_k = r)$ and equally spaced
in frequency across the wanted signal's bandwidth.

6.5 Interference Into FM Television Signals

Unlike the telephony situation, there are no analytic theories to
predict and explain the effect of interference into television signals.
Although equation (6-14) can be used to formally describe the output
of an FM or PM receiver with interference, the question is how to
relate it to a useful measure of performance. The basic problem lies
with the difficulty of modeling the psychophysical process of vision.
This is an extremely complex subject [27,28,29,30], even when restricted to the relatively narrow context of quality assessment.

There are two basic methods of quality assessment, implicit evaluation by subjective testing and explicit (or "objective") evaluation by
meter or calculation [31]. For establishing the quality of a transmission, the objective method is preferrable. However, there does not yet
exist an accepted procedure for establishing the quality of a picture
in the presence of interference in terms that would be most useful in
the present context, e.g., in units that are subjectively equivalent to
the SNR in the presence of thermal noise only. Ideally, such a procedure would set up a standardized calculation, e.g., the interference
power spectrum $S_\lambda(f)$, appropriately weighted, using perhaps a
"representative" spectral characteristic for the video process such as
that discussed in [32]. In point of fact, the effects of interference have

---

*Because (6-24) was based on a single interferer, a slight modification is required
here, namely replace $S_\lambda^I(f)$ by $S_\lambda(f)$ as given by (6-40) and suppress r^2.

traditionally been evaluated by using the subjective method, i.e., the determination of carrier-to-interference ratios (termed protection ratios) necessary to produce a specified subjective reaction for a given set of wanted and interfering signal parameters.* Results of specific measurements will be presented below. We shall also briefly describe a proposed objective method.

6.5.1 A Preliminary Heuristic Discussion

Before discussing the protection ratios, it may be instructive to look briefly at the information that is "objectively" available in the error signal itself, (6-14). Assuming the interference is small, as we have been doing, the error signal is given by the first term of this equation, *viz.*,

$$\lambda(t) = \frac{R(t)}{A} \sin\left[2\pi f_d t - \phi(t) + \psi(t) + \mu\right] \tag{6-41}$$

If the interfering signal is another FM TV signal, co-channel with the desired carrier, with relative amplitude r, then

$$\lambda(t) = r \sin\left[\psi(t) - \phi(t) + \mu\right]$$

Using the inequality $x \geq \sin x$, we can bound (or approximate) the above equation by

$$\lambda(t) \leq r\left[\psi(t) - \phi(t) + \mu\right] \tag{6-42}$$

Thus, to a first approximation, the wanted signal is reduced by the factor *(1-r)* and the interfering signal appears in a reduced but undistorted form at the (phase) receiver output.** This would be expected to give rise to the worst situation since the interference is "intelligible",

---

*The visibility of the interfering picture will depend on its standard relative to that of the wanted signal, and on the relative temporal synchronization of the two pictures.

**It is interesting to note, however, that CCIR *Recommendation 483-1*, which stipulates the recommended level of interference into FM-TV networks from other networks, does so in terms of a fraction (1/10) of the permissible video noise power. This certainly implies a power spectral density approach, which in practice does not appear to be followed. In subsection 6.5.5 we shall further discuss *Rec. 483-1*.

and this conclusion is confirmed by the protection ratio measurements shown in Table 6-1.

When (6-42) is approximately true, the baseband signal-to-interference ratio can be expressed as

$$\frac{S}{I} = \left(\frac{1-r}{r}\right)^2 \frac{P_\phi}{P_\psi}$$

where $P_\phi = E\,[\phi^2(t)]$, $P_\psi = E\,[\psi^2(t)]$. Since P_ϕ, P_ψ are proportional to the square of the modulation index, S/I would be expected to vary accordingly. In fact, Table 6-1 does show that the protection ratio is inversely proportional to M_w^2, the wanted modulation index squared. A similar dependence on the interfering modulation index would not be expected (except in the range where $sin\ x \approx x$ is a good approximation) since the peak limit $sin\ x \leq 1$ would become significant. Nevertheless, some decrease in S/I, or increase in protection ratio, might be expected as the interfering modulation index increases, and this is also seen to be the case in the Table.

If now $f_d \neq 0$ the baseband interference is spectrally shifted to around f_d. This will tend to reduce the annoying effect of the interference for two reasons: the picture information is "chopped" by the subcarrier $sin(2\pi f_d t)$, which should reduce its intelligibility, and if f_d is sufficiently large some of the energy in $\lambda(t)$ should be filtered out. When f_d is larger than the wanted baseband bandwidth one would expect, for given wanted M_w^2, that the protection ratio would increase with increasing modulation index of the interference. This follows from the fact that the angle-modulated interference term (6-41) would have a wider spectrum and hence extend further (to lower frequencies) into the passband of the receiver. These observations are also basically supported by the tabulated data.

When the interfering signal is FDM/FM telephony, similar comments can be made as to the trends, but overall the protection ratios would be expected to be lower since $\psi(t)$ would not represent a visually intelligible signal. (A similar statement would apply for any non-pictorial interference.) This expectation is borne out by the protection ratios shown in Table 6-2.

Evidently, the preceding considerations do not constitute a sufficient basis for a quantitative theory of assessment. However, they do provide heuristic explanations of the measured protection ratios, and to that extent shed some light on the situation.

6.5.2 Protection Ratios

It might be presumed that protection ratios *(PR)* represent definitive statements concerning tolerable interference levels, stemming as they do from direct subjective evaluation. Unfortunately this is not the case. First, the *PR* is a function of the permissible degree of impairment. Various scales have been devised [30,32] for ranking such impairments, but words must ultimately be used to define picture quality and each individual's inner interpretation of them will be different, as indeed will be a given individual's interpretation at different times. The reaction to picture impairment also depends on such variables as the viewing distance, thermal noise level (which can mask other impairments), ambient illumination, scene material and motion, and system standard [31,33]. Thus, even though subjective testing may be the ultimate measure of quality, there is by no means unanimity on the protection ratios needed to achieve that quality. Nevertheless, it is useful to illustrate the possibilities with some specific measurements.

(1) FM Television Unwanted Signal

Here we present the results of two independent series of tests.* Figure 6-8 [34] shows the protection ratio for signalling parameters representative of a broadcasting application, as evaluated by a single "expert" observer for "just perceptible" interference. Perhaps the more significant information to be drawn from this figure is the variability in PR as a function of the desired scene material.

The result of another set of tests is given in Table 6-1 [35]. Here, the pictures were rated by ten "experienced" observers using a 7-point impairment scale having Grade 1 ("not perceptible") at one end and Grade 7 ("extremely objectionable") at the other.

The thermal weighted SNR was 54 dB, and pre-emphasis/de-emphasis and energy dispersal were employed, unlike the case in the

---

*In the tests reported here system M was used. Results for other systems as well as other conditions can be found in [34].

tests leading to Figure 6-8. Here, the wanted scene was a test color slide while the interfering modulation was off-the-air programming. From these experiments the PR required to produce Grade 2 ("just perceptible") or better evaluation were determined as shown in Table 6-1. The results were fit to normal distributions whose standard deviations, σ, are shown in the Table; the mean, μ, is basically defined to be the protection ratio.

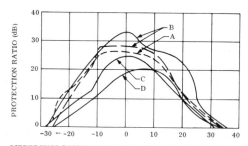

DIFFERENCE BETWEEN UNMODULATED CARRIER FREQUENCY
(UNWANTED SYSTEM) AND UNMODULATED CARRIER FREQUENCY
(WANTED SYSTEM) (MHz)

	WANTED SYSTEM	UNWANTED SYSTEM
PEAK-TO-PEAK DEVIATION	18 MHz	18 MHz
SIGNAL-TO-NOISE RATIO (WEIGHTED)	50 dB	
PRE-AND DE-EMPHASIS	NONE	NONE

CURVE	PROGRAM MATERIAL	
	WANTED SIGNAL	UNWANTED SIGNAL
A	WHITE WINDOW	OFF-THE-AIR
B	COLOR BARS	OFF-THE-AIR
C	KITCHEN SCENE	OFF-THE-AIR
D	OFF-THE-AIR	OFF-THE-AIR

Source: Ref. [34]

Figure 6-8 Protection Ratio for Just Perceptible Interference in a Frequency-Modulation Television System Subjected to Frequency-Modulation Television

There is some degree of agreement between the two sets of results. For example, at co-channel operation Figure 6-8 shows about 30 dB PR for still scenes. The closest conditions for Table 6-1 is when both signals have unity modulation index; this corresponds to about 16

MHz bandwidth, which is close to the 18 MHz used in the other tests. For these conditions the Table shows a PR of about 29 dB, which is very nearly the 30 dB figure above. On the other hand, we should not expect full agreement because the two sets of test did not use exactly the same conditions. We have already noted the difference in numbers of observers and the use of pre-emphasis and energy dispersal in one case and not the other. The viewing conditions also were not identical as were not the specific wanted and interfering picture material. The evaluation criterion, "just perceptible" interference, while ostensibly the same in both sets of tests does not necessarily imply the same subjective quality because the scales used were not the same; it is also unstated what the specific instructions were to the evaluators. This points out the need for a universal and clearly defined set of reference conditions for subjectively measured protection ratios.

Even with subjective measurements available, as just presented, there is still need for "objective" interpretation or manipulation of these results to account for situations not specifically covered. This will always be the case for one cannot perform an experiment for every conceivable measurement of interest. For example, Figure 6-8 applies only to a single common deviation, while Table 6-1 gives results only for a discrete set of deviations and frequency offsets. Working now only with the latter data, for specificity, we can extend their application most expeditiously by fitting an expression to the tabulated numbers, taking into account as much as possible the apparent trends as discussed in Sec. 6.5.1 above. The result of this process is the following equation [25]

$$\rho(M_w, M_i, f_d) = 29.5 - 20 \log M_w - f_d M_w^{-0.85} - 0.475(\mu^{-2.5})(f_d^{0.645\mu}) \log \mu$$
$$(6\text{-}43)$$

where

ρ = protection ratio (dB)
M_w = wanted signal peak modulation index
M_i = interfering signal peak modulation index
f_d = difference in carrier frequency of wanted and interfering signals
$\mu = M_w/M_i$

The values of ρ computed from (6-43) are shown in parentheses in Table 6-1 below the measured values, and are seen to agree quite well with these. We can thus conjecture that (6-43) is valid for sets of (M_w, M_i, f_d) other than those explicitly measured so long as the departure is not too great.

It might be noted that the improvement arising from frequency offsets is inherent, though it might be aided by filtering. This can be deduced, e.g., from the (3,1,10) case where even with the offset all of the interfering spectrum is essentially within the wanted signal's bandwidth.

Another situation which requires objective extrapolation from the measured data is where there are multiple interferers, noting that the results presented were obtained for a single interferer. Here, the concept of sensitivity factor [25] is useful. This is defined as

$$10 \log S(M_w, M_i, f_d) = \rho(M_w, M_w, 0) - \rho(M_w, M_i, f_d) \tag{6-44}$$

and represents the difference in protection ratio between the case when the interfering signal has the same modulation index and carrier frequency as the wanted one, and that in which it has arbitrary parameters M_i, f_d. Thus the identical interferer situation is taken as a reference. Equation (6-44) implies that an identical co-channel interferer reduced in power by $S(M_w, M_i, f_d)$ has the same subjective effect as the actual one. It is further conjectured that when the impairment is sufficiently small the aggregate interference power has the same effect as a single interferer of the same power. Let C/X represent the total carrier-to-interference ratio and $(C/X)_j$ the CIR for the jth interfer. We postulate that

$$(C/X)^{-1} = \sum_{j=1}^{N} (C/X)_j^{-1} S^{-1}(M_w, M_j, f_j)$$

where M_j, f_j, are the modulation index and difference frequency of the jth interferer. To see whether or not a protection ratio requirement is met one then compares (C/X) to $\rho(M_w, M_w, 0)$. Thus, in an actual situation, this procedure provides at least an approximate way of determining if system requirements are satisfied.

(2) FDM/FM Telephony Unwanted Signal

Measured protection ratios for this case are shown in Table 6-2 [35]. Here again some form of analytical representation is useful for interference studies. Because the interfering TV case provides somewhat greater intuition, and to relate to that case, it is convenient to define an equivalent peak modulation index for the interfering telephony signal

$$M_{eq} = \left(\frac{W_i}{2f_{mi}} -1 \right) \left(\frac{f_{mi}}{f_{mw}} \right)^{1.5} \tag{6-45}$$

where

W_i = RF bandwidth of interfering signal
f_{mi} = top baseband frequency of interfering signal
f_{mw} = baseband bandwidth of wanted television signal

The first factor in (6-45) might appear to be sufficient but it is found empirically that a baseband bandwidth factor is needed. Note that if these are the same, the equivalent modulation index is the same as the actual one. It is found [25], then, that the following equation

$$\rho(M_w, M_{eq}, f_d) = 24.1 - 20 \log M_w - f_d M_w^{-1.15} - 0.85 f_d^{0.5} \mu^{-3} \log \mu \tag{6-46}$$

gives a reasonably goof fit to the tabulated data. In (6-46) $\mu = M_w/M_{eq}$. The PRs computed from (6-46) are shown parenthetically in Table 6-2 below the corresponding measured value, and again are seen to agree very well.

As with television a sensitivity factor can be defined that permits the combination of several interferers into an equivalent one. This idea can be further extended to combine simultaneous interference from telephony and TV signals and, for that matter, any other signals for which we can define a sensitivity factor [25].

(3) Effect of Thermal SNR

It has been ascertained through subjective tests — and it makes intuitive sense — that the visibility of a given amount of interference depends on the amount of thermal noise present. As the thermal noise is reduced below the threshold of perceptibility one would expect an interference "floor," that is an amount more than which will be always perceptible. This has led to a proposal [34] to incorporate into the protection ratio a thermal noise dependence of the form

$$PR = \begin{cases} (PR)_0 - (49 - SNR), \ SNR < 49 \ dB \\ \\ (PR)_0 \ SNR \geq 49 \ dB \end{cases}$$

$$(6\text{-}47)$$

where SNR is the weighted thermal noise SNR. The first line of this equation reflects the "masking" of interference by noise. However, for the basic hypothetical reference circuit in the fixed satellite service the SNR requirements exceed 49 dB, and thus no advantage can be taken of this masking. Note that (6-47) refers to the masking of interference by noise, that is to a reduction in the conscious awareness of "foreign" picture material. This does not imply, as it might seem, that thermal and interference noise cannot be traded, if *total* satisfaction is the criterion. If fact, if such a criterion is adopted, it has been shown [34] that thermal SNR and C/I do trade in an inverse (though not one-to-one) manner as would generally be expected. The apparent anomaly is simply due to the different governing criteria.

6.5.3 An Objective Method

It would be most desirable if it were possible to express a signal-to-noise ratio due to interference that would be subjectively identical to a numerically equal thermal SNR, for then one would have a vehicle for trading thermal noise for interference. Such a formulation might take the form

$$\left(\frac{S}{N} \right)^{\mathrm{I}} = \left(\frac{C}{I} \right) R_x$$

$$(6\text{-}48a)$$

where R_x is the receiver transfer characteristic and would specifically need to reflect the characteristics of wanted and interfering signals; evidently R_x would be a complex function, most likely dependent on (C/I) as well. However, once we had such a relationship, we would have a presumably objective method of accounting for interference, that is, one devoid of explicit reference to subjective assessments.

A first step has been made [35] in giving concrete form to (6-48a). In particular, it was found that

$$R_x = 256M_w^2 \qquad (6\text{-}48b)$$

was reasonably accurate for $f_d = 0$, $1 \leq M_w$, M_I, ≤ 3, and for small (imperceptible) interference. However, much work remains to be done to generalize (6-48) to arbitrary levels of interference and arbitrary signal parameters.*

6.5.4 Concerning the Interference Specification for FSS Television Networks

In *Recommendation 483-1* the CCIR has recommended a particular way of specifying the permissible interference into a television network of the fixed satellite service (FSS), which is in principle quite different from the protection ratio method that we have already dealt with. The recommendation states that the interference noise power from all other FSS networks should not exceed 1/10 of the permissible video noise (in the hypothetical reference circuit) and that the contribution from any single interfering network should not exceed 4/10 of the total allowable interference. We shall interpret this recommendation here, and show the type of calculation that it implies.

The "video noise" in the absence of interference is taken here to mean the baseband thermal noise power in the bandwidth of the desired signal. This is given, to within a constant, by [see also Appendices A,B]

$$P_n = \int_0^B S_n(f)\, G_d(f)\, G_w(f)\, df$$

where

$S_n(f)$ = noise power spectral density at FM demodulator output
= $(N_0/C)f^2$

---

*In particular, aside from subjective effects, the pre-emphasis/weighting improvement is a function of the baseband spectral density, and this points further to the need of standardizing a video source for interference calculations.

$G_d(f)$ = power transfer function of de-emphasis network
$G_w(f)$ = power transfer function corresponding to the noise weighting characteristic for the TV standard in question
B = baseband bandwidth of television signal.

We could also write

$$P_n = P_{n,0} I(p,w)$$

where

$$P_{n,0} = N_0 B^3 / 3C$$

is the noise power at the output of the FM demodulator, and $I(p,w)$ is the improvement factor due to pre-emphasis/de-emphasis and noise weighting. As the case may be, then, we can consider the noise power at the output of the FM demodulator, at the output of the de-emphasis network, or at the output of the cascade of de-emphasis and weighting networks. The recommendation does not specifically indicate which of these possibilities should be used.

The "permissible" video noise must meaningfully be defined in relation to the signal power, P_s, which, to within the same constant that applies above, is given by $P_s = 2\Delta f^2$ where Δf^2 is the peak deviation. Thus, the permissible video noise power can be taken as

$$P_{n,p} \leqq P_s / (S/N)_{req}$$

where $(S/N)_{req}$ is the recommended baseband signal-to-noise ratio (in terms consistent with the noise power, i.e., weighted or otherwise) for the hypothetical reference circuit for the television standard under consideration.

The baseband interference power, in analogy with the foregoing, is given by

$$P_I = 2 \int_0^B S_\lambda(f) f^2 G_d(f) G_w(f) \, df$$

where $S_\lambda(f)$ is the PSD of the output phase, and is given by (6-18) or (6-20).

The permissible interference, $P_{I,p}$ is given by

$$P_{I,p} \leq \delta P_{n,p}$$

where $\delta = 0.1$ if $S_\lambda(f)$ is calculated using the PSD of the total interference, or $\delta = 0.04$ if only one interfering network is considered.

Thus, *Recommendation 483-1* implies the following steps: 1) determine $S_\lambda(f)$, which for small interference is given by the convolution of wanted and interfering RF spectra; 2) compute the integral P_I; 3) ensure that $P_I \leq P_{I,p}$, using the applicable value of δ. Clearly, this is a complicated procedure, but in principle it is identical in nature to the one that is more or less routinely made for FDM/FM telephony signals. The key distinction lies in the difficulty of defining the television RF spectrum, which in turn stems from the difficulty in defining the video process. Thus, as was alluded earlier, in order to systematically apply the Recommendation, we would need to define a "reference video spectrum."

It is possible, however, to obtain a simple bound on P_I under some circumstances. Consider for simplicity a single constant-envelope interferer. Then we have $S_\lambda(f) = (r/2)^2 S'_\lambda(f)$ where the primed quantity has unit power and r is the ratio of interfering to wanted amplitude. Let us suppose that we are interested in unweighted quantities, and further assume that the de-emphasis characteristic is idealized, i.e., that $f^2 G_d(f) = 1$. Then, one obtains

$$P_I \leq \frac{r^2}{2} = \left(\frac{C}{I}\right)^{-1}; P_n = \frac{N_0 B}{C}$$

Therefore, the Recommendation is met if

$$\left(\frac{C}{I}\right) \geq \delta^{-1}(C/N_0 B)$$

Notice that the right-most term is the carrier-to-noise ratio referred to baseband bandwidth. Defining C/N as the carrier-to-noise ratio in the RF bandwidth, one has

$$\left(\frac{C}{I}\right) \geq \delta^{-1}(C/N)\, 2\, (M+1)$$

where M is the peak modulation index. If this inequality is satisfied then the recommendation is automatically satisfied. However, the converse is not necessarily true. It is not clear how useful this bound is likely to be.

6.5.5 Summary

To summarize the situation concerning interference into FM television systems, a theory does not exist which expresses the quality Q as a continuous function of the relevant parameters. Rather, empirical results, known as protection ratios, are usually relied upon to stipulate acceptable levels of interference. As we have seen, it is possible, and frequently necessary, to extend these results to cover situations not explicitly measured. It would seem that standardization is necessary with respect to the conditions under which protection ratios are measured, perhaps most importantly concerning the evaluation scale and criteria. Clarification of the trade between thermal and interference noise needs to be made. Of course, once we have a protection ratio, interference calculations are trivial, consisting merely of the determination of carrier-to-interference ratio and comparison thereof with the PR.

6.6 Interference Into SCPC/FM Signals

Broadly speaking, interference into an SCPC/FM signal should be broachable by the methods applicable to FM systems in general. And in fact, subject to the conditions leading to its derivation, Equation (6-14) will correctly describe the (interference) output phase of the SCPC demodulator. If, further, a spectral characterization of the interference suitably captures the subjective effects, as it usually does, then (6-18) or its usual simplification (6-20) contain the essential information for quantifying the performance. In fact, there are certain considerations peculiar to the nature of SCPC signals and the way they are typically implemented that warrant additional discussion.

Let the SCPC signal be represented by its test-tone loaded form, i.e.,

$$w(t) = A \cos \left[\omega_I t + \beta \sin \omega_m t \right]$$

where

$$f_m = \omega_m / 2\pi, \text{ the test-tone frequency}$$

β = test-tone modulation index

   = $\Delta f/f_m$

Δf = peak test-tone frequency deviation.

Then, analogous to (6-24), we can write for the test-tone-to-interference (power) ratio,

$$TTI = (\Delta f)^2/2 \left\{ 2\int_{b_1}^{b_2} f^2 \, S_\lambda(f) \, G_d(f) \, W(f) \, df \right\}^{-1} \qquad (6\text{-}49)$$

where $S_\lambda(f)$ is the phase PSD at the receiver output, $b = b_2 - b_1$ is the audio channel bandwidth, $G_d(f)$ as before is the de-emphasis transfer function (different, of course, than the one implied earlier for FDM/FM), and $W(f)$ is the weighting function. Note that the latter is included here in the integrand, as opposed to the FDM case where it is legitimate to use a fixed factor because in that case the PSD can be considered constant over the bandwidth of an audio channel.

As mentioned above, there are conditions that apply specifically to SCPC which need to be addressed in order to properly evaluate the interference. First is the fact that compared to almost all other types of transmissions in satellite communication, SCPC signals are narrowband. Hence it is necessary to explicitly account for the filtering of the wider bandwidth interferers. This is frequently done, as an approximation, by treating the interference within the bandwidth of the SCPC signal as additional thermal noise. However, filtering can be straightforwardly incorporated. Note that (6-49) already includes this effect in principle since $S_\lambda(f)$ as earlier defined implicitly takes account of filtering. However, it is more useful to bring this out explicitly. In Section 6.8.2 it will be shown that we can express the phase PSD as

$$S_\lambda(f) = \frac{1}{4A^2} \left\{ S_i(f\text{-}f_d) \, | H(f) |^2 * S_w(f) + S_i(\text{-}f\text{-}f_d) | H(\text{-}f) |^2 * S_w(\text{-}f) \right\}$$

where $H(f)$ is the receiver transfer function up to the detector, and $S_i(f)$ is the interference power spectral density prior to filtering.

Another consideration relates to voice activation, a technique whereby the carrier is transmitted only when speech is present. This of course engenders more efficient loading of the satellite power amplifier. Voice activation produces random switching of the carrier, and this will have an effect on the relevant power spectral densities. In particular, the wanted power spectral density $S_w(f)$ will not be merely the PSD corresponding to modulation by a speech wave of a continuous carrier, but will reflect the activation process. If the interference also comprises other SCPC carriers and intermodulation products thereof then its PSD will be similarly affected. The influence of voice activation on the power spectral density of SCPC/FM signals has recently been examined [36].

Another commonly used technique with SCPC signals is syllabic companding. Let $F(\cdot)$ be the compressor characteristic and $F^{-1}(\cdot)$ its inverse, i.e., the expander characteristic. The output signal is $F^{-1}(s+i+n)$ where s, i, n, are, respectively, the signal, interference, and noise at the FM demodulator output. Because F is a nonlinear functional the output $F^{-1}(s+i+n)$ cannot strictly speaking be further decomposed. Even with a quasi-linear assumption, it does not necessarily follow that the response to i is the same as to n. The effect of companding on speech quality generally requires subjective assessment which does not seem to have been systematically approached with both noise and interference. However, to the extent that the interference can be treated as noise, as it often is, the compandor improvement in the presence of both can be taken as that with noise alone.

Finally, even in the absence of companding, there are interference effects which may not be best or correctly characterized in the spectral domain. In particular, periodic CW interference, as may result from a carrier modulated by an energy dispersal waveform [see next section] can produce repeated clicks which are subjectively annoying. The extent of this depends on the periodicity of the interference relative to the audio channel bandwidth if the C/I in the predetection bandwidth continues to exceed threshold. A general evaluation of this type of effect does not appear to be available although it will probably be minor for the implementations typically found in use. By contrast, periodic-type interference is much more significant for ditital SCPC because here the predetection C/I is controlling and for a given sweep rate the interference dwells longer in the predetection bandwidth than in the audio channel bandwidth. This will be discussed in greater detail in the section of Chapter 7 dealing with interference into digital SCPC signals.

6.7 On the Effect of Energy Dispersal

Energy dispersal is a technique commonly used in satellite communication for two basic purposes: reduction of peak power spectral density* of the satellite transmission, and interference reduction. The second of these, of course, is more basic and subsumes the first. However, we mention the first explicitly because from the point of view of the satellite operator, controlling the peak spectral density can be considered merely as an externally imposed requirement, devoid of interference-reduction connotation. The fact that such a regulatory imposition does exist stems in the first place from the need to protect terrestrial radio relay systems from undue interference [see Sec. 5.7]. As was seen in Section 6.3.4 the dominant interference in FDM/FM telephony, when both wanted and interfering carriers have substantial discrete components, results from the combination of these components. The normal operating parameters for terrestrial carriers does generate a significant unmodulated carrier component, while for the satellite signal such a component may result from the nature (statistics) of the modulating signal, or from partial loading of the carrier, which inevitably occurs at other than the busiest times. The partial load decreases the deviation of the carrier and consequently increases the discrete component [see Appendix H]. In the limit of no load, all of the power is concentrated in the carrier. It would be possible in principle to satisfy the spectral density limitation on a pure carrier, but then the power of that carrier would be artificially constrained to unusefully low levels: hence the need for energy dispersal.

6.7.1 Energy Dispersal Effect on Spectral Density

We first briefly consider the effect of energy dispersal on the power spectral density of a modulated carrier. Let the carrier be given by

$$w(t) = A \cos\left[\omega_1 t + \phi(t) + d_1(t)\right] \qquad (6\text{-}50)$$

---

*Typically, this "peak" spectral density is an average over some bandwidth.

where $d_1(t)$ is an energy dispersal waveform. In practice, the power in $d_1(t)$ may be related to the communications load $\phi(t)$ in different ways. It may be at a constant level, irrespective of the load; it may have two distinct levels, say one for no-load and another for whenever there is any load; or, the power in $d_1(t)$ may be inversely related to that in the information signal so as to keep the total load approximately constant.

For given average power in $\phi(t)$ and $d_1(t)$ and assuming statistical independence between these two waveforms, it is readily shown that the power spectral density of $w(t)$ is given by

$$(2/A^2)\, S_w(f) = G_\omega(f\text{-}f_1) + G_w(\text{-}f\text{-}f_1) \tag{6-51a}$$

where

$$G_w(f) = G_\phi(f) * D_1(f) \tag{6-51b}$$

$G_\phi(f)$ is the normalized lowpass equivalent spectrum of the carrier modulated by $\phi(t)$ alone, and $D_1(f)$ is the lowpass equivalent PSD of the carrier modulated by $d_1(t)$ alone.

A typical energy dispersal waveform is a triangular or sawtooth waveform of low frequency ($\sim 30\ Hz$) with peak-to-peak deviation such that the modulation index for this waveform is very high. Hence the quasi-static viewpoint is appropriate here and the PSD $D_1(f)$ approaches the limiting form for such waveforms, i.e.,

$$D_1(f) = (1/W_1),\, |f| \leq W_1/2 \tag{6-52}$$

where W_1 is the peak-to-peak deviation. Assuming the energy dispersal is properly described by (6-52), a simple bound is the obtainable for the PSD of the carrier. From (6-51) and (6-52) we can write

$$G_w(f) = \int_{-W_1/2}^{W_1/2} \frac{1}{W_1} \, G_\phi(f\text{-}f')\, df'$$

$$\leq \int_{-\infty}^{\infty} \frac{1}{W_1} \, G_\phi(f\text{-}f')\, df' = \frac{1}{W_1} \tag{6-53}$$

Hence the application of energy dispersal having a flat spectrum guarantees that the normalized peak power spectral density of the modulated carrier will be no greater than the reciprocal bandwidth of the spreading signal, irrespective of the communications load.

6.7.2 Energy Dispersal Effect on Interference

Let the wanted signal be given by (6-50) and the interfering signal by

$$i(t) = rA \cos [\omega_2 t + \psi(t) + d_2(t) + \mu]$$

where $d_2(t)$ is a possible energy dispersal waveform. Retracing the steps which led to (6-14) readily yields for the excess phase due to the presence of interference,

$$\lambda(t) = \sum_{m=1}^{\infty} \frac{(-1)^{m+1}}{m} r^m \sin [m(2\pi f_d t - \phi(t) - d_1(t) + d_2(t) + \mu]$$

which for small interference reduces to

$$\lambda(t) = r \sin [2\pi f_d t - \phi(t) - d_1(t) + d_2(t) + \mu] \tag{6-54}$$

If the energy dispersal waveforms are independent of each other as well as of the information signals, the PSD of $\lambda(t)$ is obtained as a straightforward extension of (6-51), namely

$$S_\lambda(f) = G_v(f\text{-}f_d) + G_v(\text{-}f\text{-}f_d) \tag{6-55a}$$

where

$$G_v(f) = G_\phi(f) * D_1(f) * G_\psi(f) * D_2(f) \tag{6-55b}$$

and $G_\psi(f)$ and $D_2(f)$ are the lowpass equivalent spectra of $\psi(t)$ and $d_2(t)$, respectively. It is instructive to look at the combined effect of the energy dispersal waveforms, namely

$$D(f) = D_1(f) * D_2(f)$$

which is sketched in Figure 6-9 under the conditions of (6-52) and

$$D_2(f) = (1/W_2), |f| \le W_2/2$$

and the assumption $W_1 \le W_2$. From this figure we can deduce that whatever the beneficial effect from energy dispersal, it is not mate-

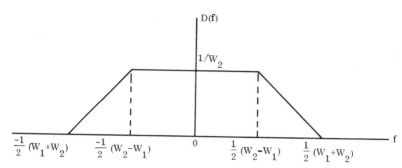

Figure 6-9 Illustrative Combined Spectrum of Energy Dispersal
         Waveforms

rially aided by the joint action of the two dispersal waveforms compared to what would be the case if either of them were singly present.

To see this, suppose $\phi(t)$ is a multichannel telephony signal, and let k_1 and k_2 represent, respectively, the fractional power in the residual carrier components of $w(t)$ and $i(t)$. Then, from (6-32) and (6-55) the baseband term due to the interaction of the spectral spikes and the energy dispersal waveforms is given by

$$\frac{r^2}{4} k_1 k_2 [D(f\text{-}f_d) + D(\text{-}f\text{-}f_d)]$$

(6-56)

From this we see that the interference in the worst channel (assuming this was due to the beating of the spectral spikes) is reduced by a factor proportional to $(1/W_2)$. If $W_1 \approx W_2$, which would be typical, the interference reduction would be about the same if only one of the carriers had energy dispersal or if both did. Note that if only one of the carriers had energy dispersal, the effect of it would be felt in the receivers of both signals. Thus, if the worst channel interference is due to the interaction of discrete carrier components, that channel's performance as indicated by (6-56) can be substantially improved. In other situations, where the worst channel performance is not necessarily proscribed by the carrier components, the effect of energy dispersal can only be ascertained by evaluating (6-55) for the case at hand. Generally, if $w(t)$ is wide-deviation, little or no improvement will be gotten from energy dispersal.

Equation (6-56) describes the effect of energy dispersal in the spectral domain. Subjectively, this description may be inadequate, depending upon the specifics of the energy dispersal waveform and the transfer function of the final receptor. To illustrate this possibility, consider a wanted multichannel telephony baseband and for the moment set $d_2(t) = 0$. Assume $d_1(t)$ is a triangular or sawtooth waveform. Looking at (6-54) one can see that $2\pi f_d t - d_1(t)$ can be considered to be a linearly time-varying periodic "carrier" with average frequency f_d. If the period of the energy dispersal waveform were sufficiently long, one could view a baseband discrete component [i.e., $k_1 k_2 \delta(f - f_d)$] as being swept over $f_d \pm W_1/2$. In any voice channel ($\sim 4kHz$) within this band, this would induce a periodic disturbance, and the question is whether the subjective effect would be the same as for an equal average interference power constantly present. The answer is in the affirmative as the sweep rate becomes large relative to the channel bandwidth; in the limit of high sweep rate the channel will respond only to the dc value of the "pulsed" interference. [These considerations are especially important in digital SCPC, as will be seen in Section 7.6]. The preceding considerations apply generally when both energy dispersal waveforms are present, except that here the relative synchronization and sweep rates of these waveforms are important. The sum $d_2(t) - d_1(t)$ will generally consist of linear segments, but in the extreme case where $d_1(t)$ and $d_2(t)$ are identical and synchronized the effect of energy dispersal will be eradicated.

If now we take the wanted signal to be a television signal, the time-domain viewpoint* predicts that the effect of energy dispersal should be minor if the peak-to-peak deviation of the latter is small (as it typically is) compared to the bandwidth of the modulated FM-TV carrier. This conclusion can be deduced from (6-43), (6-46), and Tables 6-1 and 6-2.

6.8 Some Implementation-Related Considerations

Here we discuss briefly several topics broadly related to the effect of implementation on interference, our main intention being to point out these inter-relationships. First we consider a non-ideal frequency detector and its possible effects. Then we look into the consequences

---

*In effect, treating the effect of energy dispersal as a time-varying carrier offset.

of (RF or IF) filtering. Next we examine the interference generation produced by nonlinear satellite amplifiers. And finally the question of separability of interference and impairment effects is taken up.

6.8.1 Non-Ideal Frequency Detection

A more realistic model of frequency detection than the (idealized) one we have been using is illustrated in Figure 6-10 [2]. While this model does not necessarily account for all the details of the behavior of any frequency detection method, it is sufficiently representative.

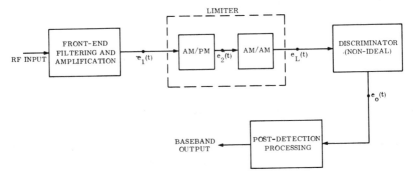

Figure 6-10 Representation of FM Demodulation Process

We do not explicitly consider here the front-end filter's effect. Our main purpose here is not to offer a design guide but rather show approach as to the general relationship between interference output and receiver characteristics, without explicit numerical evaluation.

The radio-frequency (RF) input is considered to be the desired FM signal*

$$w(t) = A \cos [\omega_0 t + \phi(t)]$$

and some number of like interferers

$$i(t) = \Sigma r_j A \cos [\omega_j t + \psi_j(t) + \mu_j], i = 1,2, \cdots N.$$

---

*For simplicity we assume here that the receiver's center frequency is tuned to the desired signal carrier frequency.

As a result of the front-end filtering, these signals can be respectively represented in the form

$$w_1(t) = A(t) \cos [\omega_0 t + \phi(t) + \delta(t)]$$

$$i_1(t) = \Sigma A_j(t) \cos [\omega_j t + \psi_j(t) + \delta_j(t) + \mu_j]$$

The sum signal (input to the limiter) can therefore be written as

$$\begin{aligned} e_1(t) &= w_1(t) + i_1(t) \\ &= E(t) \cos [\omega_0 t + \phi(t) + \delta(t) + \lambda(t)] \end{aligned} \qquad (6\text{-}57)$$

where

$$E(t) = \{ c^2(t) + s^2(t) \} \qquad (6\text{-}58\text{a})$$
$$\lambda(t) = tan^{-1} s(t) / c(t) \qquad (6\text{-}58\text{b})$$
$$c(t) = A(t) + \Sigma A_j(t) \cos \gamma_j(t) \qquad (6\text{-}58\text{c})$$
$$s(t) = \Sigma A_j(t) \sin \gamma_j(t) \qquad (6\text{-}58\text{d})$$
$$\gamma_j(t) = \Delta \omega_j t + \psi_j(t) + \delta_j(t) - \phi(t) - \delta(t) + \mu_j \qquad (6\text{-}58\text{e})$$
$$\Delta \omega_j = \omega_j - \omega_0 \qquad (6\text{-}58\text{f})$$

The limiter, as shown, is modeled as two-part zero-memory nonlinearity, after Shimbo and Berman and Mable [37,38].* Thus, the limiter output can be expressed as

$$e_L(t) = E_L(t) \cos [\omega_0 t + \phi(t) + \delta(t) + \lambda(t) + \alpha(t)] \qquad (6\text{-}59)$$

where

$$E_L(t) = g[E(t)]$$
$$\alpha(t) = f[E(t)]$$
$$g(\cdot) = \text{single-carrier amplitude transfer characteristic}$$
$$f(\cdot) = \text{single-carrier phase transfer (AM/PM) characteristic.}$$

---

*While this model was initially developed for traveling-wave tubes, it appears applicable in form to a general bandpass nonlinearity.

Note that if we have a description of the limiter's instantaneous transfer characteristic $G(\cdot)$, i.e.,

$$e_{out,\,L}(t) = G[e_{in,\,L}(t)]$$

then one can obtain $g(\cdot)$ from the first-order Chebyshev transform [39]

$$g(u) = \frac{2}{\pi} \int_0^{\pi} G(u \cos v) \cos v \, dv$$

though the single-carrier transfer functions are often directly available from measurements. Ideally, of course, we wish to have $g(E) = constant \times E$ and $f(E) = 0$ (or a constant).

The waveform (6-59) is input to the discriminator, for which different functional models are possible. One such model that has been proposed [2] expresses the discriminator output as (ignoring scale)

$$e_0(t) = \frac{\dot{\theta}\,E_L(t)}{1 + b^2|\dot{\theta}|^{\nu}},\ \nu > 1 \tag{6-60}$$

where

$$\dot{\theta} \triangleq \frac{d}{dt}[\phi(t) + \delta(t) + \lambda(t) + \alpha(t)] \tag{6-61}$$

and b, ν embody the non-idealness of the discriminator action. Further baseband processing, e.g., through a de-emphasis circuit, can be straightforwardly modeled but for our purposes we need not proceed beyond (6-60). Further, from (6-58a) - (6-58d) Equation (6-61) can be given the more specific form

$$\dot{\theta} = \dot{\phi}(t) + \dot{\delta}(t) + \dot{\alpha}(t) + \frac{c(t)\,\dot{s}(t) - s(t)\dot{c}(t)}{E^2(t)} \tag{6-62}$$

An extensive study of the behavior of $\dot{\theta}$ under various conditions, including some numerical results, has been performed by Middleton and Spaulding [2]. However, as stated above, our objective here is to show in a general way the extent to which interference is affected by non-ideal receiver behavior. Toward this end, we note that in a well-designed receiver the denominator of (6-60) must satisfy $b^2|\dot{\theta}|^{\nu}\ll 1$, whence the denominator can be expressed as a series expansion so that (6-60) becomes

$$e_o(t) = \dot{\theta}E_L(t) \sum_{n=0}^{\infty} (-1)^n b^{2n}|\dot{\theta}|^{\nu n}$$

$$= \dot{\theta}E_L(t) + E_L(t)\, sgn\,(\dot{\theta}) \sum_{n=1}^{\infty} (-1)^n b^{2n}|\dot{\theta}|^{\nu n + 1} \tag{6-63}$$

Again, in well-designed receivers we can expect the limiting to be nearly ideal so that one can write as a reasonable approximation

$$E_L(t) = L + d\,[E(t)]$$

where the function $d(\cdot)$ depends upon the design details, but whatever its form, $d(E)/L \ll 1$, where L is the average limiter output. Hence, we get

$$e_0(t) = L\dot{\theta} + d(E)\dot{\theta} + E_L(t)\, sgn\,(\dot{\theta}) \sum_{n=1}^{\infty} (-1)^n b^{2n}|\dot{\theta}|^{\nu n + 1}$$

$$\tag{6-64}$$

and we have succeeded in isolating to a first-order the contributions due to the receiver imperfections. Hence, to the extent that $\dot{\theta}$ contains terms due to interference, we can see that additional interference is added (with amplitude modulation) by non-ideal limiting alone, and a complex form of interference is added through the combined action of non-ideal limiting and imperfect discrimination. The importance of these additions relative to the interference which exists in their absence can only be determined on a case-by-case basis. We shall return to (6-64) in Section 6.8.4 where we examine it from the point of view of separability.

6.8.2 On the Effect of Predetection Filtering

In principle, the effect of filtering is incorporated in the development of the preceding subsection; however, this is only in a symbolic sense. Here we provide a more explicit treatment specifically reflecting the filter's properties. To focus on the effect of filtering, *per se,* we now assume the receiver is otherwise ideal and, further, we shall concentrate on the case where the interference effect is adequately characterized by its output PSD.

With $h(t)$ the filter's impulse response, the input to the detector is simply

$$e_1(t) = [w(t) + i(t)] * h(t) \qquad (6\text{-}65)$$
$$= w_1(t) + i_1(t)$$

It is reasonable in this context to assume that the filter has been designed compatibly with the desired signal so that distortion on the latter is minimal. Making this assumption, we have

$$e_1(t) = w(t) + i_1(t) \qquad (6\text{-}66)$$

Now, $i_1(t)$ is still of a bandpass nature, hence the development leading to (6-18) is completely applicable. However, in order to explicitly isolate the filter's effect we need to express the lowpass equivalent PSD of $i_1(t)$, $S_{mi1}(f)$, in terms of the unfiltered PSD and the filter properties. This can be done as follows [40]:

$$S_{mi1}(f) = E\left\{\int_{-\infty}^{\infty} d\tau e^{-j2\pi f\tau} \int_{-\infty}^{\infty} dx_1 \int_{-\infty}^{\infty} dy_1 \cdots \int_{-\infty}^{\infty} dx_m \int_{-\infty}^{\infty} dy_m \right.$$

$$\overline{h}(x_1)\,\overline{h}(x_2)\cdots\overline{h}(x_m)\,h(y_1)\,h(y_2)\cdots h(y_m)$$

$$\cdot exp\left[j2\pi f_d(x_1 + x_2 + \cdots + x_m - y_1 - y_2 - \cdots - y_m)\right]$$

$$\left. \cdot \overline{u}(t\text{-}x_1)\,\overline{u}(t\text{-}x_2)\cdots\overline{u}(t\text{-}x_m)\,u(t + \tau\text{-}y_1)\cdots u(t + \tau\text{-}y_m)\right\}$$

where E is the expectation operator, the overbar indicates complex conjugate, and $u(t)$ is the unfiltered interference complex envelope.

For arbitrary *u(t)* and *h(t)* it does not appear possible to usefully reduce the above expression. For the case *m = 1*, however, this can be done, and if the interference is small then this is the only case necessary, i.e., we need only the first term of (6-18) or (6-20).

Then we can show that

$$S_{i_1}(f) = S_i(f) | H(f + f_d) |^2$$

where *H(f)* is the filter transfer function. Defining $G(f) = | H(f) |^2$, we then obtain the extension of (6-20), *viz.*,

$$S_\lambda(f) = \frac{1}{4A^2} \left\{ S_i(f\text{-}f_d) G(f) * S_w(f) + S_i(\text{-}f\text{-}f_d) G(\text{-}f) * S_w(\text{-}f) \right\} \qquad (6\text{-}67)$$

Notice that in (6-67) the interference can be arbitrary. At this point a simple example is instructive. We take the case of a constant-envelope single interferer with relative amplitude *r* (prior to filtering) and assume both wanted and interfering spectra are Gaussian-shaped, i.e., wide-deviation multichannel telephony carriers:

$$S_w(f) = \frac{exp\,[f^2/2\sigma_w^2]}{\sqrt{2\pi}\,\sigma_w} \,;\, S_i(f) = \frac{exp\,[f^2/2\sigma_i^2]}{\sqrt{2\pi}\,\sigma_i}$$

where σ_w, σ_i are the corresponding ms frequency deviations. We further assume an ideal lowpass filter

$$G(f) = \begin{cases} 1, \, |f| \leq B \\ \\ 0, \, elsewhere. \end{cases}$$

Hence, after a little manipulation

$$S_\lambda(f) = \frac{r^2}{4} \int_{-B}^{B} \{ S_w(f + p) + S_w(f\text{-}p) \} \, G(p) \, S_i(p\text{-}f_d) \, dp$$

and the FM output, $S_0(f)$, is $f^2 S_\lambda(f)$, which after integration can be shown to equal

$$S_0(f) = \frac{r^2 f^2}{\sqrt{32\pi}\, M f_{m1}} \left\{ C_1 \exp\left[-(x+y)^2/2M^2\right] + C_2 \exp\left[-(x-y)^2/2M^2\right] \right\}$$

where f_{m1} is the upper baseband frequency of the desired signal;
$x = f/f_{m1}$; $y = f_d/f_{m1}$; $M^2 = M_w^2 + M_e^2$; $M_w = \sigma_w/f_{m1}$; $M_e = M_i f_{m2}/f_{m1} M_e = \sigma_i/f_{m2}$; and f_{m2} is the upper baseband frequency of the interfering signal.

The terms C_1 and C_2 account for the filter's presence and are given by

$$2C_1 = erf\left[(A + x - \rho y)/\sqrt{2 + 2\rho}\, M_w\right] - erf\left[(-A + x - \rho y)/\sqrt{2 + 2\rho}\, M_w\right]$$

$$2C_2 = erf\left[(A - x - \rho y)/\sqrt{2 + 2\rho}\, M_w\right] - erf\left[(-A - x - \rho y)/\sqrt{2 + 2\rho}\, M_w\right]$$

where $\rho = (M_w/M_e)^2$ and $A = (1 + \rho)B/f_{m1}$. Note that when $B \to \infty$, $C_1 = C_2 = 1$.

To be specific consider now the top channel, i.e., $x = 1$. Further, assume 2B is given by Carson's rule, i.e., $2B = 2f_{m1}(\alpha M_w + 1)$, where we take $\alpha = \sqrt{10}$. With these stipulations, $S_0(f_{m1})$ has been computed with and without filtering and the ratio R is shown in Figure 6-11 as a function of the fractional frequency offset $Y = f_d/2B$ for various conditions. It can be seen that the effect of the filter is negligible for $Y < 0.5$ for any reasonable set of conditions.

No general conclusions as to the effect of filtering can be drawn. The improvement to be gained in any particular case will depend on the filter in question and the signals considered. A number of examples of this effect have been published [41] for realistic (Chebyshev) filters and non-Gaussian spectra. Recently, the effect of filtering for different types of filters has been computed by Stojanović et al. [42] in conjunction with the performance of a proposed improved FM demodulator which in a sense also acts as an interference canceller [see also Section 8.4].

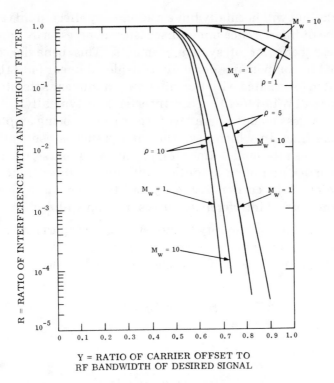

Figure 6-11   Reduction of Output Noise Due to Pre-Detection Filtering
            for Example Case

6.8.3 Interference Generated by Equipment Nonlinearities

The basic limitations on orbit utilization stem from interference
directly radiated from one network into another. However, there is
another source of potential interference that is equipment-induced
which must be taken account of in the design of a network. While this
interference is primarily intra-network in nature, and thus does not
directly affect satellite spacing, it does affect orbit utilization as a
whole since it acts as another source of noise and ultimately limits
the inter-network interference which might otherwise be permitted
to exist. Equipment-generated interference can also affect other

networks, but will normally have a secondary effect in this respect. The type of interference under discussion is necessarily caused by nonlinear behavior of system elements. This behavior can be observed in components normally considered linear [43,44], but is more often associated with the inherently nonlinear satellite power amplifier (TWT) [45-49]. This interference typically arises in multiple-access modes where several carriers are being amplified at the same time, in which case the interference is conventionally referred to as intermodulation (IM) products. Of course, even with a single carrier, nonlinear amplifiers will produce interference at the harmonics of the carrier frequency, but these should be kept well below the normal interference sources by proper filtering.

Let the satellite nonlinearity be modeled as a two-part zero-memory nonlinearity (AM/PM conversion and gain saturation) as in (6-59), and let the input consist of N angle-modulated carriers,

$$V_I(t) = \sum_{m=1}^{M} A_m \cos \theta_m(t) \tag{6-68a}$$

where

$$\theta_m = (\omega_0 + \omega_m)t + \psi_m(t) + \mu_m \tag{6-68b}$$

Extensive investigations of the properties of the RF and baseband properties of the output of the nonlinearity with (6-68) and noise as input have been made [45-49]. For present purposes it is sufficient to note that the bandpass output can be represented as [47]

$$V_0(t) = Re\left\{ \sum_{\{k_m \varepsilon s\}} M(k_1, k_2, \cdots k_M) \cdot exp\left[j \sum_{m=1}^{M} k_m \theta_m(t) \right] \right\}$$

where $\{k_m \varepsilon S\}$ denotes the integer set

$$\{ \forall k_m : -\infty < k_1, k_2, \cdots, k_m < \infty \ U \sum_{m=1}^{M} k_m = 1 \} \tag{6-70}$$

and $M(k_1, k_2, \cdots, k_M)$ is a complex amplitude that depends on the input amplitudes $\{A_m\}$ and the nonlinearity transfer characteristics. Subject to the combinations in (6-70), it can be seen that the output consists of the sum of angle-modulated carriers which, except for the original input terms, represent additional interference to a receiver whose passband intercepts these products. Note in particular that IM terms may fall spectrally where there might otherwise be no energy.

The quantity $\Sigma\,|\,k_m\,|$ is called the "order" of any term in (6-69) while the set $\{k_1,\,k_2,\cdots,\,k_m\}$ is called the "type." Evidently, different types can result in the same order. It will also be noted that, depending on the type, a given IM term may or may not be independent of the input carrier near which it falls. The situation that has been of typical interest is where (6-68) represents the desired multiple-access input to the satellite, in which case the third-order products predominate. These products represent self-interference in the network where they are generated. But they also represent additional interference to carriers of another network. Admittedly the nonlinearity will be operated so that the IM terms are small compared to the input terms, but they may not be negligible with respect to another type of carrier in another network that is expressly positioned so as to avoid the main carrier terms.

More generally, we may take (6-68) to represent not only the desired inputs to the nonlinearity but also interfering carriers from other networks. In this case the interference produces still more interference through the action of the nonlinearity. Again, this additional (IM) interference will normally be small with respect to the inputs that produced it, but depending on its actual level and spectral location, need not be negligible with respect to the wanted signals whose passband it falls into.

In any case, since the IM-produced interference consists of angle-modulated carriers, its effect on a wanted signal can be found using the methods studied in the previous sections, taking care in the present situation to account for the possible dependence of wanted and IM terms.

6.8.4 Concerning the Separability of Interference and Other Effects

The notion of separability is important not only for purposes of analysis, but has obvious repercussions on the meaning and uniqueness of an interference allowance. If a particular effect is separable from all others, then it can be unambiguously specified: if it is not separable, then a specification on it is unique only under a fixed set of conditions, and any departure from these conditions will to some degree change the magnitude of that effect. Thus, when we speak of an interference allowance for a wanted analog signal, it is of some interest to see whether such a specification can be made without any reference to other conditions. Strictly speaking, of course, this cannot be true for any but a linear system. But it will be seen that for "small" deviations from ideal (a quantification which is vague but cannot generally be made more precise) the effects of interference can generally be separated from other degradations, and further the effects of multiple interferers can be evaluated one at a time.

Basically, separability requires that even in an otherwise ideal system, the effect of interference is small relative to the desired signal. In such an ideal system, we saw earlier that the system output would be

$$e_0(t) = \dot{\phi}(t) + \dot{\lambda}(t)$$

where $\lambda(t)$ is the excess phase due to the interferers and is given by Eq. (6-58b) to (6-58f) except that in the ideal case we sould set $\delta(t) = \delta_j(t) = 0, \forall j$, $A_j(t) = A_j$. We shall henceforth assume $\dot{\lambda}(t) \ll \dot{\phi}(t)$.

If now we add non-ideal front-end filtering, the output would be modified to

$$e_0(t) = \dot{\phi}(t) + \dot{\delta}(t) + \dot{\lambda}(t)$$

The effect of filtering on the signal itself, $\delta(t)$, is thus separable from its effect on the interference. Now, however, the interference term $\lambda(t)$ does contain the effect of filtering since both $\delta(t)$ and $\delta_j(t)$ appear in the arguments γ_j [see (6-58e)], and the amplitudes $A_j(t)$ fluctuate because of the filtering. However, we also saw that the effect of filtering on (the PSD of) $\lambda(t)$ is separable in the sense that it can be accounted for by a weighting function on the interfering PSD. The numerical effect, of course, cannot be independent of the filter characteristics nor of the nature of the interference.

The additional excess angle, $\alpha(t)$, is due strictly to the presence of AM/PM conversion. It is a function of both the wanted and interfering signals and thus cannot be strictly expressed as separate functions of the wanted signal and interference. However, the effect should be rather small in any case because there should be minimal AM/PM conversion in the receiver.

The effect of imperfect discriminator action can only be separated from all others if the limiter action is ideal [see the right-most term of (6-64)]. However, as just above, the effect itself cannot be separated into contributions strictly depending on the wanted and interfering signals, respectively. We should expect this effect as a whole to be small, and the proportion of it due to the presence of interference as smaller still. The effect of imperfect limiting is likewise separable if the discrimination is ideal, and if so, that effect is contained in the second term of (6-64) which can be seen to act separately on the terms of (6-62)

When the implementation is ideal, the effect of interference is contained in $\lambda(t)$ in the manner [see (6-58)]

$$\lambda(t) = tan^{-1} \frac{\Sigma A_j sin \, \gamma_j}{A + \Sigma A_j cos \, \gamma_j} \tag{6-71}$$

which, for small interference, reduces to

$$\lambda(t) \approx A^{-1} \Sigma A_j sin \, \gamma_j \tag{6-72}$$

and therefore the effect of multiple interferences is separable into the sum of the individual effects. Note that the approximation involved in going from (6-71) to (6-72) requires that the total interference be small, not merely that each interferer be small.

Although we have used generally qualitative arguments, we can conclude that the effects of individual impairments can generally be separated, except that imperfect limiting and discrimination, when they both exist, are intertwined. These impairments cause a degradation that is generally a function of both the signal and interference. In a well designed system, however, we can expect that the impairments will be kept small, and in normal operating conditions the interference is small in relation to the desired signal. Thus, the additional "noise" caused by the presence of interference in a nonideal system should be small with respect to the primary contribution that it would induce in an ideal system. Typically, the orbit spacings are based on interference calculated under idealized implementations. The actual interference observed will thus generally be greater than the "nominal," but should not be greatly so.

6.9 Concerning An Approximation Used in Coordination: The ΔT/T Method

When a new satellite network is being planned, the *Radio Regulations* specify certain procedures to be followed [see Part I] in order to ensure interference compatibility with the existing networks. As has been seen — and this applies in the main to telephony signals* —interference calculations are not necessarily simple and may become time-consuming when repeated for many carriers. In order to minimize this burden, the procedures mentioned above contain a preliminary approximate (hence simplified) calculation, the purpose of which is to perform an initial screening of the cases to be considered into two categories: one in which more detailed interference calculations are required, and the other in which no further calculations are deemed necessary, i.e., the conditions for acceptable interference are well met. The calculation in question is contained in *Appendix 29* of the *Radio Regulations,* and is based on the so-called "$\Delta T/T$ method" which will be described shortly. In order to confidently screen out those cases to which further consideration need not be given, the method is necessarily and intentionally conservative.** On the other hand, it should not be overly so, otherwise it will initiate detailed interference calculations in an unnecessarily large number of cases.

In this section we shall outline the $\Delta T/T$ method and compare its predictions to more exact interference calculations so as to gain an appreciation of the degree of its approximativeness. It will be convenient to treat FDM/FM telephony and FM-TV separately.

---

*Within the class of analog signals; exact interference calculations are generally complex for any digital signal, but this will be considered in Chapter 7.

**There is one notable exception to this statement, namely when digital SCPC circuits are interfered with by TV carriers modulated by frame-rate energy dispersal waveforms. This situation is discussed in the next chapter.

6.9.1 The ΔT/T Method*

The $\Delta T/T$ method is based upon a simplified treatment of the interference in which both an overestimation and an approximation are involved. Let $G_u(f)$ be the PSD of an interfering signal at the antenna terminals of a wanted satellite. Let $G_{u,m}(f)$ be the PSD averaged over the worst B_a Hz, as illustrated below. The averaging bandwidth, B_a, is 4 kHz for carriers less than 15 GHz and 1 MHz for carrier frequencies greater than 15 GHz. Generally, B_a is comparatively small next to the spectral extent of the signal, and for continuous spectra, there will be little difference between the maximum value of $G(f)$ and that averaged over B_a. The main reason for averaging is to account for spectral spikes.

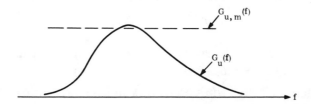

Let C_u denote the wanted signal power at the satellite antenna input. Define an "equivalent" noise temperature through the relation

$$k\Delta T_s = G_{u,m}(f) \ \ (Watts/Hz)$$

where k is Boltzmann's constant. Then the (uplink) carrier-to-equivalent interference noise temperature ratio, denoted $(C/\Delta T)_u$, is given by

$$(C/\Delta T)_u = C_u/\Delta T_s$$

Now let $G_d(f)$ and $G_{d,m}(f)$ be the PSD and its value averaged over the worst B_a Hz, respectively, at the antenna input of the wanted earth terminal, and let C_d be the wanted signal power at the same point. As in the foregoing, we define an equivalent temperature as

$$k\,\Delta T_e = G_{d,m}(f)$$

---

*Here we shall confine ourselves primarily to examining the theoretical connection between this method and more rigorous interference calculations. In Chapter 3 we have already outlined this method from the application point of view.

and a consequent carrier-to-equivalent interference noise temperature ratio for the downlink as

$$(C/\Delta T)_e = C_d/\Delta T_e$$

The total carrier-to-equivalent interference noise temperature ratio, $(C/\Delta T)$, is then

$$(C/\Delta T)^{-1} = (C/\Delta T)^{-1}_u + (C/\Delta T)^{-1}_d$$

or,

$$\left(\frac{C}{\Delta T}\right) = \frac{C_d}{\Delta T_e}\left(\frac{1}{1 + \gamma\,\Delta T_s\,/\,\Delta T_e}\right)$$

where $\gamma = C_d/C_u$. Thus, an equivalent noise temperature increase*

$$\Delta T = \Delta T_e + \gamma\Delta T_s \qquad (6\text{-}73)$$

can be defined straightforwardly from knowledge of the link parameters and the peak values of the interfering spectra.

We next define T, the "equivalent link noise temperature" (ELNT). Basically, the ELNT is the noise temperature that the earth station would have in order to account for the observed performance, if the network were otherwise ideal. Let T_s and T_e be the actual receiver noise temperature of the satellite and earth station receivers, respectively. Then, the carrier-to-noise temperature ratio at the demodulator input is

$$\left(\frac{C}{T}\right) = \frac{C_d}{T_e}\left(\frac{1}{1 + \gamma T_s\,/\,T_e}\right)$$

$$(6\text{-}74)$$

or

$$T = T_e + \gamma T_s$$

Equation (6-74) gives the ELNT due to thermal noise alone. More generally T can also include contributions from other sources of noise, not necessarily Gaussian, so long as these can be equated to some amount of thermal noise.

---

*This assumes a translating repeater and interference from the same network on both uplink and downlink. Equation (6-73) is straightforwardly modified if, e.g., interference exists only on one path, or if remodulation is used on the satellite.

The $\Delta T/T$ method stipulates that the ratio $\Delta T/T$ be computed and compared to a threshold value τ. If $\Delta T/T \leq \tau$, then the interference is considered to be acceptable and no further calculations are necessary. If $\Delta T/T > \tau$, this does not mean that interference is unacceptable, but only that more detailed calculations must be undertaken. These need not be confined to calculations of the type described earlier in this chapter [e.g., evaluation of spectral convolutions] but may simply repeat the $\Delta T/T$ computation with, say, more precise antenna patterns. Clearly, the treatment of interference in the manner described involves an overestimation of interference power by (in effect) assuming that its spectral density is constant at its peak-averaged value, and it involves an approximation by implicitly equating it to thermal noise. The latter approximation can actually be optimistic or pessimistic depending upon the specifics of the case [see, e.g., Figure 6-7]. The key to the process, however, lies in the choice of τ. If τ is made too small, detailed calculations will be triggered unnecessarily; and if too large, the opposite will be true. Thus, the ideal value of τ will sift only the borderline cases. Currently the *Radio Regulations* specify $\tau = 0.04\,(4\%)$, which corresponds to the ratio of maximum single entry interference to total performance. In the next subsections we compare numerically the $\Delta T/T$ method to more exact ones, from which we can gain a better perspective of the adequacy of certain values of τ.

6.9.2 Interference Between FDM/FM Telephony Signals

As was seen, the $\Delta T/T$ method is tantamount to treating the interference as thermal noise with temperature ΔT. Thus, the NPR in a channel can be obtained from Appendix A. In the current instance, we can state the result as *

$$\widetilde{NPR} = \frac{1}{r^2 \overline{G}} \frac{M_1^2\, G_p(f)}{(r-\epsilon_1) f_{m1}\, u^2} \tag{6-75}$$

where

$\overline{G}$ = maximum PSD of the interfering signal, normalized to unit power and averaged over B_a Hz;

---

*We shall use the tilde ($\sim$) here to indicate results associated with the $(\Delta T/T)$ method.

and the other terms have the same meaning as before. Equation (6-75) assumes a translating repeater. Note that ΔT enters through the fact that

$$(r^2 \overline{G})^{-1} = (C/\Delta T)/k$$

We now compare (6-75) to several special cases for telephony.

A. *High-Index Carriers*

This is the case studied in Section 6.3.1. The result [Eq. (6-28)] is slightly rephrased here in the form

$$NPR = \frac{2\sqrt{2\pi} M_1^2 M G_p(f)}{r^2 (1-\varepsilon_1) u^2 D(u,v)}$$ (6-76)

where

$$D(u,v) = exp\left[-(u+v)^2/2M^2\right] + exp\left[-(u-v)^2/2M^2\right]$$

Since the spectrum in this case is given by the Gaussian shape (6-27) one directly obtains $\overline{G} = 1/(\sqrt{2\pi}\Delta f_2)$, whence (6-75) becomes

$$\widetilde{NPR} = \frac{\sqrt{2\pi}\,\Delta f_2 M_1^2 G_p(f)}{r^2 (1-\varepsilon_1) f_{m_1} u^2}$$ (6-77)

To compare the previous two equations we assume the top channel $(u=1)$ which is the worst for (6-77) and either the worst or nearly so in (6-76) when M_1, M_2 are such as to satisfy the Gaussian spectrum assumption. The ratio of exact to approximate NPR is therefore given by

$$R = NPR/\widetilde{NPR} = 2M/[M_2' R(1,v)]$$

$$= \frac{2\sqrt{1+(\Delta f_1/\Delta f_2)^2}}{D(1,v)}$$

Now, the function $D(1,v)$ can take on different values depending on v and M, i.e., $0 \le D(1,v) \le 2$. The $\Delta T/T$ method is (intentionally) incapable of accounting for interleaved operation, so that a "fair" comparison requires $v=0$. Since high modulation index is assumed in this discussion, $M \gtrsim 1$, hence $D(1,0) \gtrsim 1.2$. As M increases $D(1,0)$

approaches 2, which is the most favorable comparison for the $\Delta T/T$ method. Hence,

$$R \approx \sqrt{1 + (\Delta f_1 / \Delta f_2)^2}$$

measures the accuracy of the latter method in the current context. It can be seen that it is always necessarily pessimistic, by a large amount if $\Delta f_1 \gg \Delta f_2$, or only slightly in the reverse situation.

B. Low-Index Carriers

This is the case considered in Section 6.3.4 and capsuled in (6-33). This attains its worst value when $f_d = f_{m_1}$, namely

$$NPR = \frac{2bM_1^2 G_p(f_{m1})}{r^2 k_1 k_2 (1 - \varepsilon_1) f_{m1}}$$

In the case at hand $\overline{G} = k_2 / B_a$, so that (6-75) becomes

$$\widetilde{NPR} = \frac{B_a M_1^2 G_p(f_{m1})}{r^2 k_2 f_{m1} (1 - \varepsilon_1)}$$

Hence

$$R = NPR / \widetilde{NPR} = (2/k_1)(b/B_a)$$

If $b = 3100$, $B_a = 4000$, and in the limit $k_1 \to 1$, $R = 1.55 \,(\approx 2dB)$.

This is the smallest value of R, which increases as k_1 decreases, i.e., the $\Delta T/T$ method becomes more pessimistic. Of course we cannot let k_1, get too small without violating the assumption under which we are working, but in any event we see that 2dB is as close a match as is possible here.

C. Low-Index Carriers with Energy Dispersal

The case discussed just above might not actually occur in practice since it is likely that energy dispersal would be used, at least as k_1, $k_2 \to 1$. Assuming this is so, and for simplicity taking the spreading bandwidth, B_s, to be the same for both carriers, we have $\overline{G} = 1/B_s$; hence

$$\widetilde{NPR} = \frac{B_s M_1^2 G_p(f_{m1})}{r^2 f_{m1} (1 - \varepsilon_1)}$$

For the exact case, we need to revert to (6-25), wherein $S'_\lambda (f)$, the baseband phase PSD is given in the present situation in the sketch below.

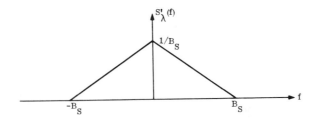

The worst channel occurs at the maximum value of

$$S'_\lambda (f) (f/f_{m1})^2 = M(f).$$

From the sketch,

$$S_\lambda (f) = -f / B_s^2 + 1 / B_s , f \geq 0$$

hence

$$M(f) = -f^3 / B_s^2 f_{m1} + f^2 / B_s f_{m1}$$

Differentiating $M(f)$ and setting to zero yields $f_w = 2B_s /3$ as the worst location. Hence the worst NPR is given by

$$NPR = \frac{(27/2) M_1^2 f_{m1} G_p (f_w)}{r^2 (1-\varepsilon_1) B_s}$$

for $B_s \leq 1.5 f_{m1}$. If $B_s > 1.5 f_{m1}$ the worst channel cannot be higher than f_{m1}. In this case $S'_\lambda (f_{m1}) = (1-f_{m1} /B_s)(1 /B_s)$ and the corresponding NPR is

$$NPR = \frac{2M_1^2 (B_s / f_{m1}) G_p (f_{m1})}{r^2 (1-\varepsilon_1) (1-f_{m1} /B_s)}$$

Combining the previous results yields

$$R = NPR / \widetilde{NPR}$$

$$= (27/2) \left(\frac{f_{m1}}{B_s}\right)^2 \frac{G_p(f_w)}{G_p(f_{m1})}, \quad B_s \leq 1.5 f_{m1}$$

$$= \frac{2}{(1 - f_{m1} / B_s)}, \quad B_s > 1.5 f_{m1}$$

$$(6\text{-}78)$$

Equation (6-78) is plotted in Figure 6-12.

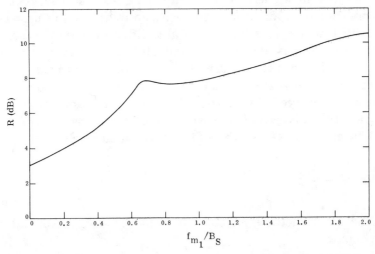

Figure 6-12 Comparison of NPR Results Using Exact and ΔT Methods for Low-Index Carriers with Energy Dispersal

D. *Some Cases Requiring Numerical Evaluation*

Here we consider a number of cases representing examples of actual operational situations. In general, these cases are not representable by simple analytic forms, as discussed just above, but must be numerically evaluated case by case. This has been done, using the data given in [22], and the results are shown in Table 6-3. The "peaking factor" shown in one of the columns is the amount by which the actual peak PSD of the interference exceeds that which would have been computed using the Gaussian-spectrum assumption.

6.9.3 Interference Between FM-TV Signals

Unfortunately, for the case of television there are no rigorous analytical models in the sense that they exist for telephony. In fact, here we are in somewhat of a dilemma because the generally accepted measure of interference susceptibility is the protection ratio (PR), which is a fundamentally different way of assessing interference than is the $\Delta T/T$ method. The latter compares the interference level to the noise level while the former method compares the interference level to the signal level. However, for present purposes a meaningful bridge between these two viewpoints appears obtainable from the "objective" assessment method described in Section 6.5.3, from which we can write for the weighted SNR

$$(S/N) = \frac{256\,M_w^2}{r^2}$$

(6-79)

For the $\Delta T/T$ method, we can use Equation (A-18), Appendix A, which gives

$$(S/N) = 6\,M_w^2\,(C/N_o\,f_v)\,I\,(p,w)$$

for the thermal SNR, where N_o is the noise PSD. This can be expressed for the $\Delta T/T$ method as

$$(\widetilde{S/N}) = \frac{6M_w^2\,I}{r^2\,\overline{G}\,f_v}$$

(6-80)

A probable worst-case interference is an unmodulated carrier with energy dispersal alone. In that case, let B_s be the associated bandwidth, whence $\overline{G} = (1/B_s)$. Therefore,

$$R = (S/N) / (\widetilde{S/N}) = 42.7 \, f_v / (B_s \, I) \tag{6-81}$$

Equation (6-81) is plotted in Figure 6-13 for $f_v = 4 \, MHz$ and $I = 12.5 \, dB$, as a function of B_s. For typical values of B_s in current use, i.e., 1-2 MHz, it can be seen that the $\Delta T/T$ method grossly overestimates the interference.

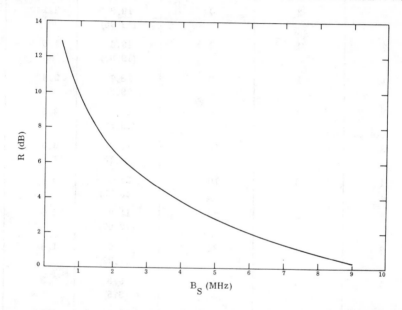

Figure 6-13 Ratio of SNR: Exact Method Relative to ΔT Method for Television Interfered with by Carrier Modulated by Energy Dispersal Waveform

As a last point of passing interest, one should note that if indeed the PR is adopted as the measure of acceptable interference, the $\Delta T/T$ method provides no particular computational advantage. This is in contrast to telephony where calculations can become rather complex and where a simplified procedure can be given some justification even if it is approximate. The calculation involving C/I, on the other hand, is certainly as simple as that used to determine $\Delta T/T$.

TABLE 6-1

Measured Protection Ratios and Fit from Equation in Parentheses

Modulation Index		Carrier Separation MHz	Protection Ratio	
Wanted Signal	Unwanted Signal		μ (dB)	σ (dB)
3	3	0	19.9 (19.95)	3.1
3	1	0	19.6 (19.95)	1.5
1	1	0	28.7 (29.5)	2.3
1	3	0	29.2 (29.5)	2.5
3	1	10	14.7 (14.75)	3.0
1	3	10	25.5 (25.25)	3.7
3	3	20	11.9 (12.07)	1.6
3	1	20	7.2 (7.22)	1.6
1	1	20	9.8 (9.5)	2.3
1	3	20	16.5 (16.2)	1.7
3	1	3.24	20.6 (18.53)	3.0

Source: Ref. [35]

*The Original data were given in terms of the peak-to-peak deviation. However, for convenience we have assumed a 4 MHz baseband and expressed the results in terms of the modulation index.

TABLE 6-2

Measured Protection Ratios and Fit from Equation

Wanted Signal (Television)		Interfering Signal (Telephony)			Carrier Offset, MHz	Protection Ratio (ρ) dB	
Δf_{pp}(MHz)	M_w	W(MHz)	Δf_{pp}(MHz)	$M_1 = M_{equiv.}$		μ	σ
24	3	14.45	6.39	0.81	0	14.9 (14.55)	2.5
24	3	14.45	6.39	0.81	10	11.1 (11.045)	1.5
8	1	27.2	19.14	2.4	10	21.4 (21.3)	3.2
24	3	33.3	18.2	3.16	0	14.2 (14.55)	1.9
24	3	33.3	18.2	3.16	20	9.7 (8.98)	1.5

Source: Ref. [35]

TABLE 6-3

Numerical Comparison Between Exact and ΔT Methods for Some Selected Cases

Wanted Carrier		Interfering Carrier			Composite Mod Index (M)	Ratio of NPR -dB (Exact to ΔT)
No. Channels	RMS Mod Index	No. Channels	RMS Mod Index	Parking Factor (dB)		
72	0.87	72	0.87	2	1.23	4.1
132	0.96	132	0.96	1	1.35	3.2
192	0.94	192	0.94	1	1.33	3.2
192	0.57	192	0.57	4	0.81	7.4
252	0.96	252	0.96	1	1.36	4.1
252	0.70	252	0.70	3	0.99	5.6
312	0.77	312	0.77	2	1.09	4.4
432	0.82	432	0.82	2	1.16	4.4
792	0.54	792	0.54	4	0.76	7.8
252	1.55	252	0.7	3	1.7	7.0
252	0.96	432	0.82	2	1.7	3.1
252	0.7	132	1.61	0	1.1	2.0
432	1.27	432	0.82	2	1.5	5.2
432	0.82	312	1.32	0	1.26	1.8
432	0.82	192	0.94	1	0.93	5.7
792	0.54	312	1.32	0	0.75	3.9
792	0.54	432	1.5	0	0.98	1.9
792	0.54	612	0.79	2	0.82	5.2

Chapter Seven
THE PERFORMANCE OF DIGITAL SIGNALS IN AN INTERFERENCE ENVIRONMENT

7.0 Introduction

The behavior and characterization of the performance of digital systems in an interference environment differs markedly from that of analog systems. The dominant fact is that, irrespective of the channel characteristics, a digital system is fundamentally and inherently nonlinear. By this we mean that the performance measure, almost universally taken to be the bit error probability* (or, synonymously the bit error rate or BER), is not linearly related to the analog waveform present at the detector input. This, of course, implies that the BER depends specifically and inextricably on the detailed joint description of the system's design and environment, interference being only one aspect. In our previous terms, this means that separability does not hold. Thus, the effect of interference cannot be explicitly nor uniquely defined as can be done for analog signals. The only situation where degradation can be solely attributed to interference is when all of the implementations are idealized (distortionless). For an existing system design which exhibits a certain degradation D_1 (at a given BER) without interference and a degradation D_2 with interference, it is fair to say that $D_2 - D_1$ is degradation that can be attributed to the presence of interference with the given set of conditions. However, a problem arises when we try to conceptually isolate this effect — in fact, as we have said, this cannot be done. This problem is of more than academic interest because in preliminary system planning one wishes to do just that. That is, one would like to make allocations for various sources of impairment. If many sources of impairment are included it becomes more difficult to extract one effect from another. Thus, it is problem-

---

*Sometimes the symbol (or word or character) error probability is important, but this is easily related to BER. In some applications, burstiness is also of interest, but we shall not consider this situation.

atical how to set aside a precise margin for interference without a specific design, which in turn may not yet be available. An associated problem in the regulatory sphere is how to meaningfully specify permissible levels of interference.

While it is possible to set down a fairly general (symbolic) formulation, numerical results are intrinsically conditioned on the assumptions made. Thus, no general or nearly general results exist, in the sense that they do for FDM/FM signals, as captured in Figure 6-2. We will therefore present a number of examples culled from the literature, which together should begin to provide a certain insight into the nature of digital system performance subjected to interference. Obtaining the numerical results themselves is really the heart of the problem and is significantly complex. We will generally confine ourselves to pointing out the salient features of the different computational approaches that have been taken. These latter approaches attempt to compute the BER exactly for whatever model is being considered and a fairly elaborate computer program is generally needed. Because of the complexity of the exact approach, a variety of computationally simpler bounds have been developed, and we shall provide an overview of these bounds. It should be noted that even the most sophisticated of the exact approaches have considered only relatively simplified models of the way actual systems operate. It would seem that a full evaluation of the effects of interference can practically be done only by computer simulation, and in fact the application of this technique, which has been used for some time in the general performance evaluation of digital systems [50], has more recently begun to incorporate interference as well [51,52].

The preceding discussion forms more or less an outline of this chapter. First we formulate the problem in fairly general terms to indicate its scope and nature. We then specialize somewhat more explicitly to the linear and the undistorted channels. By and large we restrict ourselves to coherent phase-shift-keyed (CPSK) signals, departing from this only to show examples of computed results for minimum-shift-keying (MSK). (These signaling schemes are currently the ones of greatest interest in satellite communications because they are constant envelope, and implementation is well advanced.) We out-line the various computational and bounding approaches that have been used in this context. We next consider the case of (digital) SCPC signals interfered-with by carriers modulated by energy dispersal waveforms; this important special case requires separate considera-tion. We then present a collection of computed results in an attempt to paint a quantitative perspective, and draw such general conclu-sions as may be had in this context. Finally, brief comments are made concerning the specification of acceptable interference into digital systems.

7.1 General Problem Formulation

It is convenient for present purposes to expand somewhat the block diagram of Figure 6-1, as shown in Figure 7-1. This will permit us to formulate a reasonably general problem, which will have its prim-ary usefulness in helping us to appreciate how specific cases that have been studied relate to the general context. (Unfortunately the notation is unavoidably cumbersome; the reader wishing to omit the mathematical details may proceed to the discussion following (7-23).) Figure 7-1 shows a general but typical network configuration.

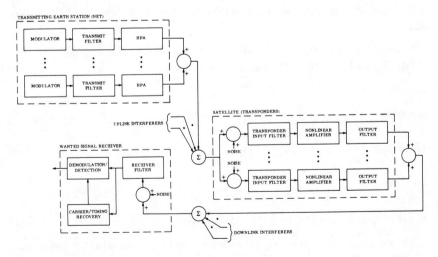

Figure 7-1 Detail of Network Configuration and Interference Scenario

At the (transmit) earth station, a number of independent transmitting chains are summed at the antenna;* spectral overlap between adjacent carriers produces adjacent-channel interference.

For the kth network let the lth modulator output (lth "channel") be written as

$$m_{lk}(t) = a_{lk}(t) \, cos \, [\omega_{lk} t + \theta_{lk}(t) + \mu_{lk}] \tag{7-1}$$

which represents a single "wideband" carrier ($a_{lk}(t) = a_{lk}$ if it is constant envelope). It may be that the lth channel is itself composed of a number of constituent "narrowband" carriers, such as SCPC signals, in which case we have

$$m_{lk}(t) = \sum_j a_{jlk}(t) \, cos \, [\omega_{jlk} t + \theta_{jlk}(t) + \mu_{jlk}] \tag{7-2}$$

The output of the transmit filter is then

$$m^{(1)}_{lk}(t) = a^{(1)}_{lk}(t) \, cos \, [\omega_{lk} t + \theta^{(1)}_{lk}(t) + \mu_{lk}] \tag{7-3}$$

where

$$a^{(1)}_{lk}(t) \, exp[\, j\theta_{lk}(t) \,] = a_{lk}(t) \, exp[\, j\theta_{lk}(t) \,] * h^{(t)}_{lk} \tag{7-4}$$

$h^{(t)}_{lk}$ is the transmit filter's impulse response and $*$ represents convolution.

The element following is a high power amplifier (HPA) with complex gain $G^{(t)}_{lk}$ so that its output is expressible as **

$$m^{(2)}_{lk}(t) = a^{(2)}_{lk}(t) \, cos \, [\omega_{lk} t + \theta^{(2)}_{lk}(t) + \mu_{lk}] \tag{7-5}$$

where

$$G^{(t)}_{lk} = g[a^{(1)}_{lk}(t)] \, exp \, \{jf[a^{(1)}_{lk}(t)]\} \, / \, a^{(1)}_{lk}(t) \tag{7-6}$$

so that

$$a^{(2)}_{lk}(t) \, exp[\, j\theta^{(2)}_{lk}(t) \,] = g[a^{(1)}_{lk}(t)] \, exp \, \{j\theta^{(1)}_{lk}(t) + jf[a^{(1)}_{lk}(t)]\} \tag{7-7}$$

---

*The model also applies if we wish to consider a multi-user network in which a net of earth stations exists so that different transmitting chains may be associated with different earth stations. The formulation is unaffected but the actual carrier amplitudes $a_{lk}(t)$ may then be independently affected by propagation factors.

**Conventionally, the HPA is modeled as a zero-memory device.

and $g(\cdot)$ is variously referred to as the gain or saturation or AM/AM function and $f(\cdot)$ is the AM/PM function: these functions should be denoted $g_{lk}^{(t)}(\cdot)$ and $f_{lk}^{(t)}(\cdot)$, in general, though we have omitted the sub-and superscripts to avoid overburdening the notation.

We omit here for simplicity the possible inclusion of an output filter on the transmit chain. If there is such an element its transfer function can be considered included in the transponder front-end filter, which should produce negligible error. (There is no error, of course, as far as the filter's action on the signal is concerned, but only with respect to the transponder input-noise.) In any case the effect of such a filter can be taken into account in exactly the manner indicated by Eq. (7-4).

The signal emanating from the kth earth station directed towards its satellite can be put in the form

$$S_k^{(u)}(t) = s_k^{(u)}(t) \cos [\omega_u t + \theta_k^{(u)}(t)] \tag{7-8}$$

where $s_k^{(u)}(t)$, $\theta_k^{(u)}(t)$ are obtained from an obvious manipulation of the identity

$$S_k^{(u)}(t) \triangleq \sum_j m_{lk}^{(2)}(t) \tag{7-9}$$

Note that the frequency $f_u = \omega_u / 2\pi$ can be chosen arbitrarily, and in particular it can be considered to be centered on the center frequency of the "wanted" carrier. The envelope and phase terms are, of course, dependent on this choice.

At the input to the lth transponder of the kth satellite we have the signal intended for that transponder, the other signals intended for that satellite (the "adjacent" channel interference), the signals intended for the other satellites (the "adjacent" satellite interference) and the transponder input noise. Thus,

$$M_{lk}(t) = m_{lk}^{(2)}(t) + S_{lk}^{(u)}(t) + I_k^{(u)}(t) + n_{lk}^{(u)}(t) \tag{7-10}$$

where

$$S_{lk}^{(u)} \triangleq \sum_j{}' m_{ik}^{(2)}(t)$$

describes the adjacent channel interference, the prime on the summation indicating that the term in the summand corresponding to the (lth) transponder in question is omitted. In actual calculation one would normally take only the one or two terms that correspond to the transponders flanking the one under scrutiny. Further,

$$I_k^{(u)}(t) = \sum_i{}' r_i^{(u)} s_i^{(u)}(t)$$

where, as above, the prime on the summation indicates sum over i excepting $i = k$ the satellite under consideration, and $r_i^{(u)}$ is the relative level of the ith uplink interference. Finally,

$$n_{lk}^{(u)}(t) = n_{c,lk}^{(u)}(t) \cos \omega_{lk} t + n_{s,lk}^{(u)}(t) \sin \omega_{lk} t \tag{7-11}$$

is the input noise to the lkth transponder expressed in its usual quadrature decomposition.

As usual, Eq. (7-10), being a bandpass signal, is representable as

$$M_{lk}(t) = b_{lk}(t) \cos[\omega_{lk} t + \alpha_{lk}(t)] \tag{7-12}$$

This signal encounters the transponder input filter with lowpass equivalent impulse response $h_{lk}^{(s)}(t)$, whose output is

$$M_{lk}^{(1)}(t) = b_{lk}^{(1)}(t) \cos[\omega_{lk} t + \alpha_{lk}^{(1)}(t)] \tag{7-13}$$

where

$$b_{lk}^{(1)}(t) \exp[j\alpha_{lk}^{(1)}(t)] = b_{lk}(t) \exp[j\alpha_{lk}(t)] * h_{lk}^{(s)}(t) \tag{7-14}$$

Next, the signal is amplified, typically by a TWTA, so that the signal now becomes*

$$M_{lk}^{(2)}(t) = b_{lk}^{(2)}(t) \cos[\omega_{lk} t + \alpha_{lk}^{(2)}(t)] \tag{7-15}$$

where

$$b_{lk}^{(2)}(t) \exp[j\alpha_{lk}^{(2)}(t)] = g[b_{lk}^{(1)}(t)] \exp\{j\alpha_{lk}^{(1)}(t) + jf[\alpha_{lk}^{(1)}(t)]\} \tag{7-16}$$

and as before, it is understood that $g(\cdot)$ and $f(\cdot)$ may be specific to the transponder in question.

Finally the transponder output is **

$$M_{lk}^{(3)}(t) = b_{lk}^{(3)}(t) \cos[\omega_{lk} t + \alpha_{lk}^{(3)}(t)] \tag{7-17}$$

---

*The nonlinear amplifier representation is still valid under some circumstances when there is a cascaded nonlinearity such an an (ideal) limiter—TWT combination.

**In this development we have assumed for simplicity that the uplink and downlink frequencies are identified by the same symbol w_{lk} since ideal frequency conversion will have no effect on the results. We do not consider here non-ideal frequency conversion, which would add yet another nonlinear element to the model.

where

$$b^{(3)}_{lk}(t) \, exp\,[\, j\alpha^{(3)}_{lk}(t)\,] = b^{(3)}_{lk}(t) \, exp\,[\, j\alpha^{(2)}_{lk}(t)\,] * h^{(s_2)}_{lk}(t) \qquad (7\text{-}18)$$

and $h^{(s_2)}_{lk}$ is the lowpass equivalent impulse response of the output filter.

The signal emanating from the kth satellite intended for its cooperating receiving earth station is thus

$$S^{(d)}_k(t) = \sum_l M^{(3)}_{lk}(t)$$

which is also expressible as

$$S^{(d)}_k(t) = S^{(d)}_k(t) \, cos\,[\omega_d t + \alpha^{(d)}_k(t)\,] \qquad (7\text{-}19)$$

Hence the total waveform at the input to the kth receiver intended to receive the lth transponder's signal, is

$$V(t) = M^{(3)}_{lk}(t) + S^{(d)}_{lk}(t) + I^{(d)}_k(t) + n^{(d)}_{lk}(t) \qquad (7\text{-}20)$$

$$= v_{lk}(t) \, cos\,[\omega_{lk} t + \phi_{lk}(t)\,]$$

where

$$S^{(d)}_{lk}(t) = \sum_i{}' M^{(3)}_{ik}(t)$$

$$I^{(d)}_k(t) = \sum_i{}' r^{(d)}_i S^{(d)}_i(t)$$

and

$$n^{(d)}_{lk}(t) = n^{(d)}_{c,lk}(t) \, cos \, \omega_{lk} t + n^{(d)}_{s,lk}(t) \, sin \, \omega_{lk} t$$

represent, respectively, the adjacent-channel interference, the adjacent-network interference (with relative levels $r^{(d)}_i$), and the receiver noise. The receiver input filter then acts upon the signal to yield at its output

$$V^{(1)}(t) = v^{(1)}_{lk}(t) \, cos\,[\omega_{lk} t + \phi^{(1)}_{lk}(t)\,] \qquad (7\text{-}21)$$

where

$$v^{(1)}_{lk}(t) \, exp\,[\, j\phi^{(1)}_{lk}(t)\,] = v_{lk}(t) \, exp\,[\, j\phi_{lk}(t)\,] * h^{(r)}_{lk}(t)$$

and $h^{(r)}_{lk}(t)$ is the filter's lowpass equivalent impulse response.

The receiver synchronization circuitry generates a carrier reference

$$u(t) = 2 \, cos\,[\omega_{lk} t + \zeta_{lk}(t)\,] \qquad (7\text{-}22)$$

so that the baseband output is

$$V_o(t) = v^{(i)}_{lk}(t) \, cos\,[\phi^{(1)}_{lk}(t) - \zeta_{lk}(t)\,] \qquad (7\text{-}23)$$

The detector then operates on $V_o(t)$ to produce a decision

$$D[V_o(t)]$$

as to the value of the transmitted symbol. Evidently, the reliability of that decision will depend on the statistics of $V_o(t)$ and the sampling process.

At this stage no further progress can be made without specific assumptions as to the nature of the wanted signal and of the other variables concerned. The purpose of the largely symbolic preceding development has been to provide a general (though not completely general) setting, both to impart an appreciation of the complexities involved and to place in context the work that has been done. No single study has broached a problem of the generality outlined here, while many studies have attacked various parts of the problem individually. Perhaps the greatest attention has been paid to the effect of interference in an otherwise idealized environment [53-58]. A number of authors have extended these results to the linear channel, i.e., one in which the signal and interference may be subjected to linear filtering [59-62]. In the nonlinear channel, the problem has been considered without filtering [63-65] for multiple interferers, and with filtering for a single uplink interferer [66].

The most extensive studies, in the context of system optimization, have been carried out through simulation; Fang [52] considered the effect of adjacent-channel interference and fading in the filtered nonlinear channel, and this was extended in [67] to include co-channel interferences as well. In the analytical studies no account has generally been taken of the behavior of real synchronization circuits, especially as they may be affected by the presence of interference. Only one attempt [68] has been made to incorporate the effect of imperfect phase tracking but without explicitly reflecting the role of interference, and in an otherwise idealized system. The effect of interference on synchronization has, however, been observed experimentally [69,70] and some authors [71-73] have studied the influence of interference on phase-locked loop behavior without further relating these results to BER performance. Practically all of the preceding work has been done for *M-ary* coherent phase-shift-keying (CPSK) modulation, with some recent attention being paid to MSK [52,61], and some earlier and more recent work considered

DPSK [53,74,75]. Other modulation types have been addressed in the literature [76-81] but these are not within the intended purview of this book. Also, the work reported has considered only conventional single-bit observation detectors; the effect of more complicated receiver structures, such as maximum-likelihood sequence detectors, on interference does not appear to have been addressed. Similarly the possible use of adaptive equalization to mitigate the effects of interference does not seem to have been investigated. (An exception to this, in the context of cross-polarization interference in dual-polarized systems, is briefly addressed in Chapter 8.) Of course, an extensive literature exists on the performance of digital systems subjected to a variety of impairments in the absence of interference [82] and this literature is useful both as illustrative of analytical techniques and to obtain approximate results in the presence of interference by treating the latter as thermal noise of equal power.

Even in the simplest of the situations outlined above it has not been found possible to frame the error probability in a simple closed form expression. It has therefore been necessary to resort to numerical evaluation of the relevant expressions. Generally, this requires somewhat sophisticated numerical techniques and significant computing power. (This fact is one of the major motivations for developing "simple" bounds, and we shall spend some time examining such bounds.) The approaches used are almost as varied as the number of contributions, though they can be grouped into a few broad categories. It is beyond our objective here to give a detailed description of these numerical techniques, though we shall briefly outline the main ideas involved. Rather, we shall concentrate on presenting and interpreting some of the actual results that have been obtained. Since most of these results have been obtained for CPSK and for the linear or distortionless channel it will first be useful to reduce the general scenario to these cases.

7.2 The General Linear Scenario for CPSK

The general scenario of Figure 7-1 can be reduced to the block diagram of Figure 7-2 when it is assumed that no nonlinear elements are present and ideal synchronization is provided. The noise source in the wanted earth station receiver is then a composite of that receiver's front-end noise and re-transmitted noise from the uplinks, and

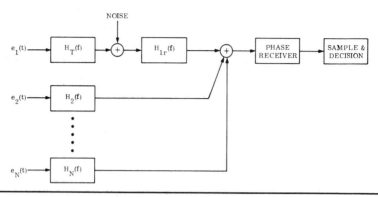

Figure 7-2 Block Diagram of General Linear Interference Problem

may be "colored." The transfer function seen by any one signal is, of course, simply the product of the transfer functions in its path. Note that the (wanted) receiver transfer function $H_{1r}(f)$ is common to all signals, and is understood to be incorporated into $H_i(f)$, which is the end-to-end transfer function for the ith interferer.

Rather than the general form of Eq. (7-1) for the wanted signal, we now represent it as*

$$v_1(t) = \sum_{k=-\infty}^{\infty} p(t-kT) \cos\left[\omega_1 t + a_k \theta(t-kT)\right] \tag{7-24}$$

where

$p(t)$: possible amplitude shaping function,
$\theta(t)$: possible phase pulse shaping function,
T : symbol interval,
a_k : kth phase symbol.

For M-*ary* CPSK one normally chooses a_k in the set

$$a_k \varepsilon \quad \frac{2\pi n}{M} \quad , n = 1,2,\ldots, M \tag{7-25}$$

The ith interfering signal can be represented as

$$v_i(t) = R_i(t) \cos\left[\omega_i t + \psi_i(t) + \mu_i\right], \, i = 2,3,\ldots, N \tag{7-26}$$

*For the purpose of this section we no longer need the double subscript notation.

where

$R_i(t)$: envelope of ith interfering signal

f_i : ith interfering carrier frequency $= \omega_i / 2\pi$

$\psi_i(t)$: phase of the ith interfering signal

μ_i : phase angle of the ith interfering signal, assumed to be independent of one another and of other functions, and uniformly distributed on $(0, 2\pi)$.

Note that here we may take $v_i(t)$ to be a composite of several carriers from one earth station, e.g., one co-channel and two adjacent-channel carriers; or, we can equally assign a separate identity (index i) to each individual carrier. The latter course will normally be simpler. When the interfering signals are themselves of the same nature as the wanted signal, they take on the form

$$v_i(t) = \sum_{k_i = \infty}^{\infty} r_i p_i(t - k_i T_i - \tau_i) \cos[\omega_i t + a_{ik_i} \psi_i(t - k_i T_i - \tau_i) + \mu_i]$$

$$(7\text{-}27)$$

where T_i is the symbol duration, τ_i is the relative origin or symbol slip, r_i is the relative interference level, and a_{ik_i} is in a set similar, but not necessarily identical, to that in (7-25).

The receiver is assumed to be an ideal phase receiver that, once per symbol, samples the instantaneous phase β and decides that $a_k = (2\pi n / M)$ was sent if

$$(2n - 1) \pi / M \leq \beta \leq (2n + 1) \pi / M$$

Defining the complex envelopes

$$e_1(t) = \sum_{k = -\infty}^{\infty} p(t - k T) \exp[j a_k \theta(t - kT)]$$

$$(7\text{-}28\text{a})$$

$$e_i(t) = R_i(t) \exp\{j[\omega'_i t + \psi_i(t) + \mu_i]\}, i = 2, 3, \ldots, N \qquad (7\text{-}28\text{b})$$

where $\omega'_i = \omega_i - \omega_1$, the complex envelope at the input to the phase detector is

$$e(t) = n_c(t) + j n_s(t) + \sum_{i=1}^{N} e_i(t) * h_i(t)$$

$$(7\text{-}29)$$

where $n_c(t)$ and $n_s(t)$ are the in-phase and quadrature components of the noise at the output of the receiver filter, and $h_i(t)$ is the equivalent lowpass impulse response of the ith signal. Suppose, without loss of generality, that the *zero*th symbol a_o for which the sampling occurs at $t = t_o$, is to be detected. The decision variable is

$$\beta_o \triangleq \beta(t_o) = \tan^{-1} \frac{e_s(t_o)}{e_c(t_o)} \qquad (7\text{-}30)$$

where $e_c(t)$ and $e_s(t)$ are the real and imaginary parts, respectively, of $e(t)$:

$$e_c(t_o) = s_{co} + n_{co} + x_{co} + y_{co} \qquad (7\text{-}31a)$$

$$e_s(t_o) = s_{so} + n_{so} + x_{so} + y_{so} \qquad (7\text{-}31b)$$

and the subscript o implies t_o, s is the "signal" component, x is the intersymbol interference (ISI) component and y is the interference component. These functions are obtained from Eq. (7-28) to (7-29) as follows:

Let

$$C_k(t) = p(t - kT) \cos[a_k \theta(t - kT)] \qquad (7\text{-}32a)$$

$$S_k(t) = p(t - kT) \sin[a_k \theta(t - kT)] \qquad (7\text{-}32b)$$

$$A_i(t) = R_i(t) \cos[\omega_i^1 t + \psi_i(t) + \mu_i] \qquad (7\text{-}32c)$$

$$B_i(t) = R_i(t) \sin[\omega_i^1 t + \psi_i(t) + \mu_i] \qquad (7\text{-}32d)$$

$$h_i(t) = h_{ic}(t) + j h_{is}(t) \qquad (7\text{-}32e)$$

Then,

$$s_c(t) = C_o(t) * h_{1c}(t) - S_o(t) * h_{1s}(t) \qquad (7\text{-}33a)$$

$$s_s(t) = C_o(t) * h_{1s}(t) - S_o(t) * h_{1c}(t) \qquad (7\text{-}33b)$$

$$x_s(t) = \sum_{k \neq o} C_k(t) * h_{1c}(t) + S_k(t) * h_{1s}(t) \qquad (7\text{-}33c)$$

$$y_c(t) = \sum_{i=2}^{N} A_i(t) * h_{ic}(t) - B_i(t) * h_{is}(t) \qquad (7\text{-}33d)$$

$$y_s(t) = \sum_{i=2}^{N} A_i(t) * h_{is}(t) + B_i(t) * h_{ic}(t) \qquad (7\text{-}33e)$$

It can be seen that if the interferers are digital, as in (7-27), Equations (7-33e) and (7-33f) consist of *(N-1)* summations each of which is identical in form to (7-33c) or (7-33d). Thus the external interference problem becomes formally identical with the intersymbol interference (ISI) problem, which is why many of the techniques recently developed for the latter problem [83,84] have been brought to bear on the former.

In order to minimize the cumbersome notation, we shall now primarily deal with binary wanted systems. More importantly, as shown in Appendix K, the character error probability of an M-ary system is bounded to within at most a factor of 2 by an expression identical in form to that of the binary system. For a binary system the decision is incorrect if $e_c(t_o)$ has the wrong sign. Thus the probability of error is

$$P_2 = \tfrac{1}{2} Pr \left[e_c(t_o) < 0 / a_o = 0 \right] + \tfrac{1}{2} Pr \left[e_c(t_o) > 0 / a_o = \pi \right]$$

In particular, when $n_c(t)$ is a Gaussian process with variance σ^2, we can write

$$P_2 = \tfrac{1}{4} E \left\{ erfc \left(\frac{s_1 + x_{co} + y_{co}}{\sqrt{2}\,\sigma} \right) + erfc \left(\frac{-s_2 - x_{co} - y_{co}}{\sqrt{2}\,\sigma} \right) \right\} \tag{7-34}$$

where

$s_1 = s_{co}$ *given* $a_o = 0$,

$s_2 = s_{co}$ *given* $a_o = \pi$,

erfc = complementary error function evaluated for given values of x_{co}, y_{co},

E = expectation operator with respect to x_{co}, y_{co}.

Implicit in the above is the assumption that s_1 and s_2 are equiprobable; only a trivial change is required if this is not the case. Note that P_2 represents the average probability of error if the samples $\beta(kT + t_o)$ can be considered independent. Otherwise a more elaborate formulation, which we shall not pursue, is necessary.

Thus the computational problem reduces to a conditional expectation of *erfc*. The ways in which this problem has been attacked constitute the essential difference between the various approaches that exist.

7.3 The Distortionless Scenario for CPSK

We now idealize the preceding scenario one step further by assuming that all the filters are distortionless, so that all signals appear unmodified at the detector input. As mentioned earlier this assumption, while not realistic, does at least allow us to uniquely attribute the degradation to interference. This assumption is also not totally compatible with the fact that some filtering at the front-end $[H_{1r}(f)]$ is necessary to permit us to assume the noise power is finite. In practice this simply means that the filter bandwidth is large enough to accommodate all signals essentially without distortion. In any case, the detection process always effectively band-limits the noise. Thus, for our present purposes, we may write $h_i(t) = \delta(t)$. Assuming that all the transmitted signals are constant-envelope signals then it is consistent that they are still constant-envelope at the receiver input. We further assume here for simplicity that $p(t) = \sqrt{2S}\, q(t)$ and $\theta(t) = q(t)$ in (7-24), where $q(t)$ is a unit pulse. Thus, we may write for the wanted signal

$$w(t) = \sqrt{2S}\, \cos[\,\omega_1 t + \alpha\,]$$

where in any symbol interval α is chosen in the set (7-25), and the interference is given by

$$i(t) = \sum_{j=1}^{K} i_j(t) = \sum_{j=1}^{K} \sqrt{2I_j}\cos[\,\omega_j t + \psi_j(t) + \mu_j\,]$$

Thus, from (7-33), for binary CPSK

$$s_1 = -s_2 = \sqrt{2}\,S$$

and

$$y_{co} = \sum_{j=1}^{K} \sqrt{2I_j}\,\cos\lambda_j$$

where

$$\lambda_j = (\omega_j - \omega_1)\,t + \psi_j(t) + \mu_j \tag{7-35}$$

It can be shown [25] that, with the μ_j independent and uniformly distributed, the angles λ_j are also independent and uniformly distributed. It is useful now to define the normalized quantities

$$\rho = \sqrt{S} / \sigma \qquad (7\text{-}36a)$$

$$r_j = \sqrt{I_j / S} \qquad (7\text{-}36b)$$

$$R = \sum_{j=1}^{K} r_j \cos \lambda_j \qquad (7\text{-}36c)$$

It can be shown [85] that the probability density function *(pdf)* of R is

$$p(R) = \frac{1}{2\pi} \int_{-\infty}^{\infty} \cos Rt \, \prod_{j=1}^{K} J_o(tr_j) \, dt \qquad (7\text{-}37)$$

where $J_o(\cdot)$ is the Bessel function of the first kind and order zero. Since $p(R)$ is an even function, we can therefore reduce (7-34) in the present case to [recalling that $x_{co} = 0$ because the filters are distortionless]

$$P_2 = \frac{1}{2} E\{erfc(\rho + \rho R)\} \qquad (7\text{-}38a)$$

$$= \frac{1}{2} \int_{-\Omega}^{\Omega} erfc(\rho + \rho R) \, p(R) \, dR \qquad (7\text{-}38b)$$

where

$$\Omega = \sum_{j=1}^{K} r_j \qquad (7\text{-}38c)$$

More generally, for *M-ary* PSK, one can show [55] that the symbol error probability P_M is bounded by

$$\overset{\vee}{P}_M \le P_M \le \hat{P}_M$$

where

$$\overset{\vee}{P}_M = E\{erfc(\rho \sin \frac{\pi}{M} + \rho R)\} \qquad (7\text{-}39)$$

and $\overset{\vee}{P}_M$ can be given different forms depending upon the assumptions made. Without any assumptions it can be shown that

$\overset{\vee}{P}_M = \frac{1}{2} \hat{P}_M$, and with slightly more restrictive but very reasonable assumptions one can show that

$$\overset{\vee}{P}_M = \left(\frac{M\text{-}1}{M} \right) \hat{P}_M$$

These bounds will be demonstrated in Appendix K. Thus, for most purposes the upper bound is adequately tight. Note that (7-39) can be written as

$$P_M = E\{ \, erfc \, (\rho' + \rho' R') \}$$

where

$$\rho' = \rho \, sin \, (\pi \, / \, M) \, ; \, R' = R \, / \, sin \, (\pi \, / \, M)$$

Thus, the BER for the *M-ary* case is obtainable from the binary case by decreasing the signal-to-noise and carrier-to-interference (power) ratios by $sin^2 (\pi / M)$, and noting the factor of *(½)* in going from (7-38) to (7-39).

It can be observed from the foregoing that the random variables λ_j are statistically independent of the particulars of the modulation on the interfering carriers and of the frequency offset. Therefore, no advantage can be expected from carrier interleaving, in direct contrast to the situation with analog signals.

7.3.1 An Orbit Utilization Example

Although we shall present a collection of numerical results in a single section, 7.7, it is worthwhile considering here separately an example of the application of (7-38) and (7-39) to the orbit utilization problem [86]. Specifically we consider the homogeneous model and adopt the simplifying assumptions leading to Equation (4-30). We further take here $\eta = 0.5$ and the CCIR sidelobe reference pattern. Recognizing that r_j is the square-root of $(C/I)_j^{-1}$, we thus have

$$r_j \approx \frac{18}{|j \, \Delta \, \theta|^{1.25} (D / \lambda)} \tag{7-40}$$

for the *j*th satellite east or west of the desired one. Equation (7-40) can apply to either uplink or downlink using the appropriate value of D/ λ. For simplicity we shall consider in this example only either direction. Thus, if (7-40) is substituted into (7-36c) and thence to (7-38) or (7-39), P_M can be explicitly related to $\Delta\theta$. This has been done for

$M = 2$, using (7-38b), and for $M = 4, 8, 16$ using (7-39), and the results are shown in Figures 7-3 to 7-6. For these figures $K = 10$ (5 satellites on either side of the wanted one) and $D/\lambda = 100$ were used.

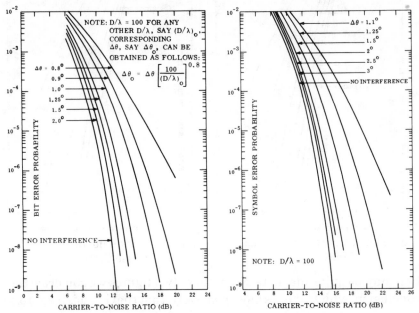

Figure 7-3 Error Rate for Binary CPSK

Figure 7-4 Error Rate for 4-Phase CPSK

Some additional results of interest are shown in Figures 7-7 and 7-8, which show for $M = 2$ the effect of the number of interfering satellites for two different given spacings. As can be seen there is relatively little difference between 2 and 40 satellites and this is somewhat to be expected in light of (7-40) since the two nearest satellites are the major contributors. The progression of the curves, from $K = 2$ to $K = 40$, also indicates that invoking a "large" number of interferers, at least in the present situation, may not justify the Gaussian approximation. This is made evident by the Gaussian approximation curve (for $K = 10$), which is simply obtained by adding the interference power to the thermal noise power, i.e., the BER is given by *(½) erfc (ρ_e)* where $\rho_e = \sqrt{S/\sigma_e}$ *and* $\sigma_e^2 = \sigma^2 + \Sigma I_j$. Thus, as shown in Figure 7-7 the Gaussian approximation may lead to pessimistic results. On the other hand, Figure 7-8 indicates that this approxima-

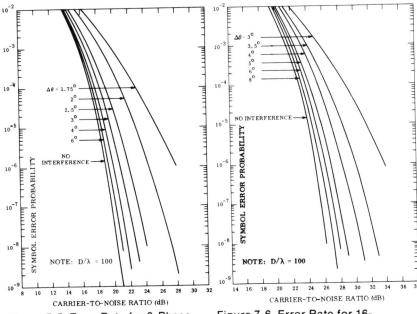

Figure 7-5 Error Rate for 8-Phase CPSK

Figure 7-6 Error Rate for 16-Phase CPSK

tion may be quite adequate. From these two figures we can draw the conclusion, which applies more generally, that the Gaussian approximation can yield fairly accurate results for relatively low carrier-to-noise ratios and improves as the carrier-to-interference ratio increases (or, equivalently, as the ratio of interference power to thermal noise power decreases). However, it is clear that as ρ increases the Gaussian approximation asymptotes to a minimum error probability while the actual error rate can be reduced indefinitely. This will be true, however, only below a certain level of interference; above this level, regardless of ρ, the error probability will be interference-limited. There is thus "threshold" spacing $\Delta\theta_t$ which separates the noise-limited from the interference-limited region. This spacing can be obtained from the following considerations. First we note that

$$\lim_{\rho \to \infty} erfc\,[\,p\,(sin\,\frac{\pi}{M}+R)\,] = \begin{cases} 0 & ,\,|R|<sin\,\pi/M \\ 1.0, R=-sin\,\pi/M \\ 2.0, R<-sin\,\pi/M. \end{cases}$$

Denoting the limiting value by $\hat{P}_M(\infty)$ and using (7-39) we find

$$\hat{P}_M(\infty) = 2 \, Pr\, [\,\text{-}\,\Omega \leq R < \text{-}sin\, \pi\, /\, M\,] \qquad (7\text{-}41)$$

(For $M = 2$, the exact BER is given by (7-41) by omitting the factor of 2.) The value of $\hat{P}_M(\infty)$ is obtained by integrating $p(R)$ over the appropriate region, and in particular we observe that if $\Omega \leq sin\, \pi\, /\, M$ the error probability can be made as small as desired. When r_j is given by (7-40) we see that

$$\Omega = \Sigma\, r_j = \Sigma\, \frac{18}{|\,j\Delta\theta|^{\,1.25}\,(D/\lambda)} = \frac{36\,(\lambda\,/\,D)\,S(K)}{\Delta\theta^{1.25}} \qquad (7\text{-}42)$$

where

$$S(k) = \sum_{j=1}^{K/2} j^{\text{-}1.25} \text{ , whence the threshold spacing is}$$

$$\Delta\theta_t = \left\{ \frac{36\,(\lambda\,/\,D)\,S(K)}{sin\,(\,\pi\,/\,M)} \right\}^{0.8}$$

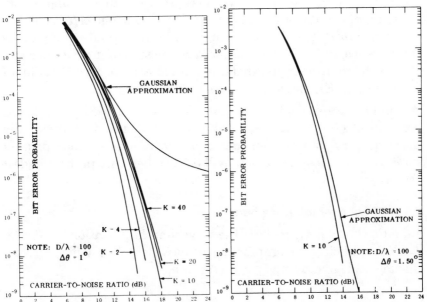

Figure 7-7 Comparison of Actual Binary Error Rates for Different Numbers of Interfering Satellites and Gaussian Approximation

Figure 7-8 Comparison of Actual Binary Error Rates and Gaussian Approximation

7.4 An Outline of Numerical Methods

It has been seen that the error probability cannot be expressed in terms of a simple closed-form function. Consequently, numerical evaluation of the relevant expression is inevitably required, and the associated mathematical considerations then constitute in effect the heart of the problem. While these considerations must ultimately be grasped in detail in order to construct a computer program, a description of these details is beyond our intended scope. Rather, our objective here is simply to outline the various numerical approaches which have been taken so as to impart to the reader a broad appreciation of the nature of the problem. This problem, basically, is the evaluation of an expression of the form *

$$I = E \left\{ erfc \left(\frac{S+u}{\sqrt{2}\,\sigma} \right) \right\} \tag{7-43}$$

where E denotes expectation with respect to u or (equivalently) its constituents, and u is the random variable on which $erfc$ is conditioned: in the distortionless environment, u is the sampled value of the interference; in the linear channel u is the sum of interference and ISI; in the nonlinear channel u is a more complicated function of interference, ISI, and uplink noise.

A number of approaches are possible and have been taken to the solution of (7-43). While the published accounts are individually distinct, they can be considered as variations of four distinct methods that, as stated above, will be sketched only in the broadest terms.

7.4.1 Direct Averaging Method

Equation (7-43) is symbolic in the sense that the operation implied by E is not explicitly brought forth. An equivalent form which directly displays this averaging is, by definition,

$$I = \int_{-\infty}^{\infty} erfc\,(s+u)\,p(u)\,du \tag{7-44}$$

as was also given (to within a constant) in (7-38b) for a specific case.

---

* While Eq. (7-43) is fairly general, it does not explicitly reflect synchronization errors, nor does it necessarily apply to all modulation/detection schemes. It does apply to M-ary PSK, and by extension to QAM.

The greatest obstacle to solution through (7-44) is the difficulty of finding the *pdf p(u)*. If that were known, there would likely be no insuperable difficulties to the numerical evaluation itself. If fact, it has not been possible to find an explicit expression for *p(u)* for any case but the distortionless one. In that case the pdf for the interference has the form already given in (7-37). With the latter substituted into (7-37) one can then directly compute *I;* the details of the numerical procedure are described in [25,86].

As a variant of this approach it can be seen that if the *pdf* of the detected angle β [see Eq. (7-30)] were known, one could integrate over the appropriate region to obtain the BER. This *pdf* is also difficult to obtain generally, but in the case of a single interferer this has been so done [56,87] and the error probability evaluated in this fashion.

7.4.2 Series Expansion Method

A class of approaches is based on the series expansion of *erfc*. Specifically, since

$$erfc\ (s+u) = erfc\ (s) + \frac{2}{\sqrt{\pi}}\ e^{-s^2} \sum_{n=1}^{\infty} (-1)^n\ H_{n-1}(s)\ \frac{u^n}{n\ !}$$

where $H_n(\cdot)$ is the Hermite polynomial of order n, following the prescription of (7-43) we then take the expectation of each term on the RHS so that

$$I = erfc\ (s) + \frac{2}{\sqrt{\pi}}\ e^{-s^2} \sum_{n=1}^{\infty} (-1)^n\ H_{n-1}(s)\ \frac{E(u^n)}{n\ !} \tag{7-45}$$

where for notational convenience we have redefined $s = s\ /\ \sqrt{2}\ \sigma$ and $u = u\ /\ \sqrt{2}\ \sigma$ in (7-45). Hence the computational problem in this instance reduces to finding an efficient way to compute the moments of u and of bounding the error which results from (inevitably) truncating the summation. The fullest discussions of this appraoch have been articulated in [55,57] which consider only the distortionless case, while an interesting formulation within a more general framework appears in [87a].

On the surface it may appear that (7-45) is skirting the issue since the moments $E(u^n)$, which are required, are defined by

$$E(u^n) = \int_{-\infty}^{\infty} u^n p(u) \, du \qquad (7\text{-}46)$$

so that it appears $p(u)$ is needed after all. The answer, of course, is that u is the sum of several other random variables, $u = \Sigma u_i$, and that

$$\int du_1 \int du_2 \cdots \int du_K (\Sigma u_i)^n p(u_1, u_2, \cdots u_K)$$

is equivalent to (7-46), where $p(u_1, u_2, \cdots u_K)$ is the joint pdf of the constituents $\{u_i\}$, which is often much more easily determinable than $p(u)$ itself. In particular, if the u_i are independent, then the previous equation can be evaluated as the sum of products of terms like

$$E(u_i^k) = \int_{-\infty}^{\infty} u_i^k p(u_i) \, du_i$$

which involves only the moments of the constituents of u. It is on such devices that the application of the series method rests.

7.4.3 The Characteristic Function Method

This method, which has had its chief exposition in [58,62] is based on formulating I in terms of the characteristic function rather than the probability density function; this of course can be done because these two functions are Fourier inverses. However, whereas the *pdf* is usually extremely difficult to express in any form, the characteristic function for sums of independent random variables can usually be set down in useful form.

We first note that

$$erfc \left(\frac{s+u}{\sqrt{2}\,\sigma} \right) = \int_{\frac{s+u}{\sqrt{2}\,\sigma}}^{\infty} \frac{2}{\sqrt{\pi}} e^{-y^2} dy$$

$$= \frac{2}{\sqrt{2\pi}\,\sigma} \int_{-\infty}^{0} e^{-(y-s-u)^2/2\sigma^2} dy$$

so that

$$I = \frac{2}{\sqrt{2\pi}\,\sigma} \int_{-\infty}^{\infty} du\, p(u) \int_{-\infty}^{0} e^{-(y-s-u)^2/2\sigma^2}\, dy$$

By definition, the characteristic function $\phi(v)$ and $p(u)$ are related by

$$p(u) = \frac{1}{2\pi} \int_{-\infty}^{\infty} \phi(v)\, e^{jvu}\, dv$$

so that

$$I = \frac{2}{(2\pi)^{3/2}\,\sigma} \int_{-\infty}^{\infty} du \int_{-\infty}^{0} e^{-(y-s-u)^2/2\sigma^2}\, dy \int_{-\infty}^{\infty} \phi(v)\, e^{jvu}\, dv$$

Since

$$\frac{-1}{2\sigma^2}\left(y-s-u-j\sigma^2 v\right)^2 = \frac{-1}{2\sigma^2}\left(y-s-u\right)^2 + jvy - jvs - jvu + \frac{v^2\sigma^2}{2}$$

the composite exponent in the last integral is given by

$$\frac{-1}{2\sigma^2}\left(y-s-u\right)^2 - jvu = \frac{-1}{2\sigma^2}\left(y-s-u-j\sigma^2 v\right)^2 - jvy + jvs - \frac{v^2\sigma^2}{2}$$

Therefore, we have

$$I = \frac{2}{2\pi} \int_{-\infty}^{\infty} \frac{du}{\sqrt{2\pi}\,\sigma} e^{-(y-s-u-j\sigma^2 v)^2/2\sigma^2} \int_{-\infty}^{\infty} dv\, \phi(v)\, e^{jvs}\, e^{v^2\sigma^2/2} \int_{-\infty}^{0} e^{jvy}\, dy$$

$$= \frac{2}{2\pi} \int_{-\infty}^{\infty}\int_{-\infty}^{0} \phi(v)\, e^{-\sigma^2 v^2/2}\, e^{-jv(y-s)}\, dy\, dv \tag{7-47}$$

since the first integral is unity.

Equation (7-47) is the point of departure for the characteristic function method. Typically, the method continues by further expanding $\phi(v)$ or $\phi(v)$ times an exponential factor into a series, following which the actual numerical integration takes place.

As a sidelight, it is interesting to illustrate the earlier remark concerning the relative ease of finding an expression for the characteristic function. Consider, as an example, that u is the ISI alone. The ISI itself is composed of many individual contributors, typically having the form

$$u = \sum_{k \neq 0} a_k p(t - kT) * h(t)$$

where $p(t)$ is the signaling pulse and $h(t)$ the system impulse response. With $x(t) = p(t) * h(t)$, we have

$$u = \sum_{k \neq 0} a_k x(t - kT) = \Sigma u_k$$

If the a_k are independent and, say ± 1 with equal probability, then it is easy to show that

$$\phi_k(v) = \int_{-\infty}^{\infty} p(u_k) e^{ju_k v} du_k = \cos v x_k$$

where $x_k \triangleq x(t - kT)$. Hence

$$\phi(v) = E(exp \, juv) = \prod_{k=1}^{\infty} \cos v x_k$$

which demonstrates what was to be shown, namely the utility of the characteristic function in this context.

7.4.4 Gaussian Quadrature Method

This method [87b] treats the problem of solving (7-43) from the point of view of numerical mathematics rather than from the more probabilistic one of the previous approaches. That is, one can approximate the integral

$$I = \int_{-\infty}^{\infty} p(u) \, erfc(s + u) \, du$$

by the finite sum

$$I \approx \sum_{j=1}^{L} w_j \, erfc\left(\frac{s + u_j}{\sqrt{2}\,\sigma}\right)$$

$$(7\text{-}48)$$

and the approximation can be made as close as desired by increasing L; in the limit $L \to \infty$, the RHS of (7-48) converges to I. The set of pairs (w_j, u_j) is called a quadrature rule. The crucial fact in this context is that the rule can be obtained without explicit knowledge of the function $p(u)$; in fact one needs only the first $2L + 1$ moments of u.

One of the few direct comparisons of different approaches is given in [83], which indicates an apparent advantage for the quadrature method over the series method, at least for the examples chosen. However, this reference treats ISI only. When there is both ISI and adjacent channel interference, [59] and [88] treat virtually the same case, the former using the quadrature method and the latter the series method. The numerical results are quite consistent, but no information is available to assess the relative computational cost.

7.5 Bounding Approaches

Because of the substantial numerical complexity of the "exact" approaches outlined in the preceding section, it is natural to seek bounds on the error probability which are perhaps less accurate but easier to compute. Consequently, a number of authors [89-100] have investigated upper and/or lower bounds on the BER in the presence of interference, usually in the distortionless environment but also in the linear channel. No bounds appear to have been studied for the general (nonlinear) case. The utility of a bound can be gauged by its tightness, simplicity, and generality. The tightness, of course, cannot be ascertained before the fact, although comparison of the bound to known results can provide a measure of its goodness. Where both an upper and a lower bound exist they of course jointly quantify the degree of uncertainty. While simplicity in the computational sense is a desirable goal, it cannot be considered a compelling requirement without tightness, particularly since the ever-increasing power and speed of computing machines has blunted to a large extent the need for "simplicity." On the other hand, the generality of a bound is most important, since to be useful it must subsume many cases of interest. As mentioned, since the known bounds do not apply generally, and mostly consider only the distortionless case, their usefulness for design purposes in the nonlinear satellite channel is limited. However, they may be useful in more restricted situations and are theoretically interesting in their own right. In this section we briefly describe several of the bounds that have been published, the derivations for which can be found in the cited references.

7.5.1 Chernoff Upper Bounds

  a) Approach I

The point of departure for this approach is the inequality

$$P_e \leq e^{g(\lambda)} \qquad (7\text{-}49)$$

where $g(\lambda) = ln\, E\,(e^{\lambda V})$, λ is any positive integer, and V is the decision variable. The tightness of the bound cannot be determined without studying specific cases. Equation (7-49) has been applied to the distortionless case in [89,90,97]. For this situation we can write [89]

$$P_e = Pr\,[\,V > 0\,|\text{-}1\,]$$

where

$$V = \text{-}1 + \eta + \sum_{i=1}^{N} r_i \cos \theta_i$$

and we have assumed a data zero *(-1)* is sent and the normalization $|s| = 1$ has been made. The noise η has variance σ^2 and $\tfrac{1}{2} r_i^2$ is the ith interference-to-carrier ratio. The angles θ_i are independent and uniformly distributed. Straightforward manipulation leads to

$$g(\lambda) = ln\, e^{-\lambda} + ln\, E\, e^{\lambda \eta} + \sum ln\, E\, e^{\lambda r_i \cos \theta_i}$$

$$= \lambda\,(\,\lambda \sigma^2 /2 \text{-} 1\,) + \sum_{i=1}^{N} ln\, I_o(r_i \lambda) \qquad (7\text{-}50)$$

where I_o is the zero-order modified Bessel function. The bound is tightest for the value of λ, $\hat{\lambda}$, which minimizes (7-50). When N is small $\hat{\lambda}$ might perhaps be found numerically or graphically. In general this is tedious and prompts one to seek alternate means. Using inequalities for $I_o(\cdot)$ one can show that

$$g(\lambda) \leq \lambda \left\{ \frac{\lambda}{2} \left(\sigma^2 + \tfrac{1}{2}\sigma_I^2 \right) - \left(1 \text{-} v_I \right) \right\} \qquad (7\text{-}51)$$

where I is any set of integers drawn from the set *(0, 1, $\cdots$, N)* and

$$\sigma_I^2 = \sum_{i \in I} r_i^2\,; \quad v_I = \sum_{i \in I} r_i$$

Equation (7-51) involves a partition of the interference into a portion that effectively increases the noise power, and another that directly reduces (coherently cancels) the signal amplitude. The numerical problem is now reduced to finding the set I that minimizes (7-51). This is discussed in detail in [89,90]. The conditions can be arranged into two categories according as to whether or not the peak interference envelope exceeds the desired envelope (i.e., $v_N \triangleq \sum_N r_i \gtrless 1$). We cite here only a few interesting and succintly stated results:

(1) when $v_N < 1$, there exists a signal-to-noise ratio p^* such that for $p \geq p^*$ the bound is optimized when $v_I = v_N, \sigma_I^2 = 0$;

(2) when $v_N < 1$ and the noise spectral density N_o is such that

$$N_o > (\sqrt{P_c / P_{i,\,max}} - 1) \sum_{i=1}^{N} P_i T$$

where P_c is the desired carrier power, $P_i = (r_i^2/2)$, we can set $v_I = 0$, $\sigma_I^2 = \sigma_N^2 = \sum_{i=1}^{N} P_i$, i.e., all the interference can be allocated as thermal noise;

(3) when $v_N > 1$ and the condition

$$\sqrt{P_c P_{i,\,max}} < \sum_{i=1}^{N} P_i$$

holds, we can once again add all the interference power to the noise power.

Numerically the bound is not very tight for relatively large interference but improves as the C/I ratio increases. Perhaps the greatest virtue of this approach is that it gives some insight into the behavior of the interference with respect to thermal noise, i.e., when the former can be treated as the latter. Incidentally, this approach also demonstrates [Equation (7-51) with $v_I = 0$] that treating all of the interference as thermal noise gives a worst case for the conditions assumed.

b) Approach II

The Chernoff bounding technique has been extended [91] to include ISI as well as interference. In particular it is shown that for 2- or 4-phase PSK interfered with by like carriers,

$$P_e \leq exp \left\{ \frac{-s^2}{2 \, (\sigma^2 + \sigma_w^2 + \sigma_I^2)} \right\} \tag{7-52}$$

where

> s^2 = wanted signal power excluding intersymbol interference
> σ^2 = noise power
> σ_w^2 = intersymbol interference power
> σ_I^2 = total interference power in the wanted receiver passband.

If $H(f)$ is the transfer function encountered by the interferers, which have PSD $P_i(f)$, then

$$\sigma_I^2 = \sum_i \int_{-\infty}^{\infty} |H(f)|^2 P_i(f)\, df$$

If $x(t)$ is the typical wanted signalling pulse and $\{a_j\}$ the symbol sequence, assumed mutually independent, then

$$\sigma_w^2 = E\,(a_j^2)\sum_j x^2\,(t\text{-}jT)$$

A bound more elaborate than (7-52) is also given in [91]. However, (7-52) is particularly pleasing in form and of all the bounds probably most qualifies as "simple." It consists essentially of an exponential bound to *erfc* combined with treatment of the interference (ISI and external) as thermal noise of equal power. While this replacement has often been made as an approximation, the present context actually shows it to be a bound, which proves that equating total interference to thermal noise gives a worst-case result. However, this does not imply that treating external interference as thermal noise will give an upper bound on P_e in the filtered channel. That is, if one had a function $P_e(\sigma^2)$ giving the BER in a particular filtered channel as a function of thermal noise power, the previous bound does not state that $P_e(\sigma^2 + \sigma_I^2)$ is also a bound. In fact, it is not, as will be further discussed in Section 7.7.

7.5.2 Maximizing-Distribution Upper Bound

This approach is interesting because it requires only the barest statistical information about the interference, namely the peak value V_i and the variance (power) σ_i^2 of the interference envelope, which often may be the only information readily available. The idea of the bound is then to find the distribution, among all distributions with parameters V_i and σ_i^2, that maximizes the probability of error. The approach is best elucidated by considering an undistorted binary CPSK wanted signal and a general interference $r(t) \cos [\omega_o t + \psi(t)]$ where $r(t)$ and $\psi(t)$ are independent and the latter is uniformly distributed on $(0, 2\pi)$. The interference is general in the sense that it can represent a filtered (adjacent channel) interferer or the sum of several interferers. Now, from (7-34) and (7-38a) we can write

$$P_2 = \frac{1}{2} E \left\{ erfc \left(\frac{s_0 + r_0 \cos \psi_0}{\sqrt{2}\,\sigma} \right) \right\} \tag{7-53}$$

where subscript 0 implies, as before, sampled value at time t_0. From (7-45) we can write (7-53) as

$$P_2 = \frac{1}{2} erfc(s) + \frac{e^{-s^2}}{\sqrt{\pi}} \sum_{n=1}^{\infty} \frac{(-1)^n}{n!} H_{n-1}(s)\, E(r^n)\, E(\cos^n \psi_0)$$

where $s = s_0/\sqrt{2}\,\sigma$, $r = r_0/\sqrt{2}\,\sigma$. For uniformly distributed ψ_0, the preceding equation reduces to

$$P_2 = \frac{1}{2} erfc(s) + \frac{e^{-s^2}}{\sqrt{\pi}} \sum_{n=1}^{\infty} H_{2n-1}(s) \frac{E(r^n)}{2^{2n}(n!)^2} \tag{7-54}$$

The main result from [98] is that for all distributions $F(r_o)$ with constraints $r_o \le V_i$ and $E(r_o^2) \le \sigma_i^2$, the maximizing distribution is

$$F_m(r_o) = (1-p)\,u(r_o) + p\,u(r_o - V_i)$$

where $p = \sigma_i^2 / V_i^2$ and $u(\cdot)$ is the unit step function. The *pdf* corresponding to F_m is a pair of delta functions which, when applied to (7-54) yields

$$P_2 \le \frac{1}{2} erfc(s) + p \frac{e^{-s^2}}{\sqrt{\pi}} \sum_{n=1}^{\infty} H_{2n-1}(s) \frac{(V_i/\sqrt{2}\,\sigma)^n}{2^{2n}(n!)^2} \tag{7-55}$$

Some relatively mild restrictions on the maximum value of V_i are described in [98]. Equation (7-55) is not computationally trivial and might fail the test for simplicity which could justify its use. However, that justification comes about because a single expression, (7-55), now applies to any interference situation. Moreover families of curves can be developed with the peak factor (PF), $-10 \log p$, and the C/I ratio, $20 \log s_o/\sigma_i$ as parameters. These calculations need to be done only once if a sufficiently large range of parameters is used, and values not specifically computed can be obtained by interpolation. In this sense the present approach is a simplification. The accuracy of the bound is good when the peak factor is relatively small but deteriorates as the peak factor increases. Equation (7-55) has been plotted in [98] and with different coordinates in [101]. From the latter we reproduce here in Figure 7-9 a typical result which exemplifies the nature of the solution. While this figure applies specifically to binary PSK it can be converted to *M-ary* PSK by using the observation following (7-39). That is, $P_m = 2P_2$ (as a bound here) if the CNR and CIR are both increased by *20 log | sin (π / M) |*. The bound turns out to be reasonably tight for relatively small PF, in fact exact for *PF = 0 dB*, but as mentioned above deteriorates for large peak factor. The method described has been expanded in [92] to permit tighter bounds to be obtained and also to incorporate ISI. However, with the latter it is no longer possible to present generalized results, *a priori*, as in Figure 7-9.

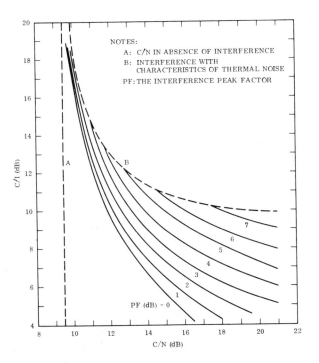

Figure 7-9 C/I *vs* C/N for *10⁻⁵* BER

7.5.3 Fourth Moment Lower Bound

A lower bound to the error probability, which can be used in conjunction with the upper bounds, has been derived [93-96] and applies to external as well as intersymbol interference. Its name arises because of the use of the fourth moment of the interference components.

Let

$$d_1 = \sum_{i \neq o} a_i x(t_o - iT)$$

$$d_2 = \sum A_j \cos \theta_j$$

be the ISI and interference, respectively. The A_j are constant and, as before, the θ_j are independent and uniformly distributed on $(0, 2\pi)$. The signal s is $x(t_o)$. It is then shown that

$$P_e \geq \frac{1}{2} \left\{ Q\left(\frac{s + \sigma_2}{\sigma}\right) + Q\left(\frac{s - b}{\sigma}\right) \right\} \tag{7-56}$$

where the Q function is the cumulative normal probability integral, $Q(u) = (½) \, erfc \, (u / \sqrt{2})$, σ^2 is the noise variance, and

$$\sigma_1^2 = E(d_1^2); \; \sigma_2^2 = E(d_2^2); \; \sigma_z^2 = \sigma_1^2 + \sigma_2^2; \; b = \sigma_z^3 / \sqrt{m_4}$$

$$m_4 = E(z^4) = E(d_1^4) + 6\sigma_1^2 \sigma_2^2 + E(d_2^4)$$

Equation (7-56) is straightforward to evaluate and gives a reasonably tight bound for relatively small interference and ISI. A more elaborate but still computationally not too complex extension of (7-56) is given in [94]. This was further generalized in [96], and the result is as follows:

$$2^N P_e \geq Q\left(\frac{s - \sum\limits_{i}^{N} b_i}{\sigma}\right) \tag{7-57}$$

where N is the number of interferences and

$$b_i = [E(x_i^2)]^{3/2} / \sqrt{E(x_i^4)}$$

where x_i is the ith interferer. In this as well as the preceding bound it is assumed that the signal exceeds the peak interference, i.e., $s \geq \sum max \, x_i$. To illustrate the simplicity of Eq. (7-57) consider the case where only external interference exists, i.e., $x_i = A_i \cos \theta_i$. We can show readily that

$$E(x_i^2) = \int_{-\pi}^{\pi} A_i^2 \cos^2(\theta_i) \, d\theta_i / 2\pi = A_i^2 / 2$$

$$E(x_i^4) = \frac{A_i^4}{4} \int_{-\pi}^{\pi} (1 + \cos 2\theta_i)(1 + \cos 2\theta_i) \, d\theta_i / 2\pi = 3 A_i^4 / 8$$

whence $b_i = A_i / \sqrt{3}$, which when substituted into (7-57) yields a very simple result.

7.5.4 Moment Space Upper and Lower Bounds

In an elegant application of finite-dimensional geometry, Tobin and Yao [100] have derived both upper and lower bounds on error probability in the presence of interference in the distortionless environment. An allied bounding theory has recently been described in [102]. Without going into the derivation, we state here one of several results from [100]. For the lower bound (for binary PSK)

$$P_2 \geq \frac{erfc[\rho\sqrt{2}(h-\Omega) - erfc(\rho\sqrt{2}(h+\Omega)}{exp[c_o(h-\Omega) - exp[c_o(h+\Omega)} \quad \frac{m_1(c_o) - exp[c_o(h+\Omega)}{+ erfc[\rho\sqrt{2}(h+\Omega)]} \tag{7-58a}$$

where $u = h - \Omega$, $U = h + \Omega$; and for the upper bound

$$P_2 \leq erfc[(\rho\sqrt{2}/c_o) \ln m_1(c_o)] \tag{7-58b}$$

In (7-56), ρ is given by 7-36a, Ω is given by 7-37c, $h = 1$ for the binary case, $c_o = -2\rho^2(h + \Omega)$, and

$$m_1(c_o) = exp(c_o h) \prod_{j=1}^{K} I_o(c_o r_j)$$

where r_j is given by (7-36b). The bounds tend to be tighter for smaller values of K and larger values of C/I. For the situation to which these bounds apply, they appear to be quite good and generally superior (tighter) than the other bounds considered.

7.6 Interference into Digital SCPC

The effect of interference into digital single-channel per carrier (SCPC) transmissions can in principle be evaluated by the same methods as have been discussed earlier. However, because the bandwidth of SCPC carriers is relatively quite small with respect to most other types of transmission in satellite communication, certain problems arise which are better treated on an ad hoc basis. When the interfering carrier is randomly modulated the approaches outlined earlier, as just mentioned, can be put to use. If the bandwidth of the interferer is much larger than that of the SCPC signal, then a reasonable approximation would be to add the interfering power intercepted by the SCPC receiver to the receiver thermal noise. The one special case which needs separate discussion is when the interference consists of a swept unmodulated carrier, a situation which arises in television transmission when there is no program information,* during which time the carrier is swept by the energy dispersal waveform that is customarily applied. This situation has been studied by several authors [103-105], most comprehensively in [105], and in the following we give a brief summary of the salient results.

7.6.1 Interference Into SCPC from a Swept CW Carrier

Consider Figure 7-10 which illustrates the situation at hand.

The SCPC channel bandwidth is nominally B Hz and the peak-to-peak energy dispersal deviation is Δf. Two types of dispersal waveform are typically considered, triangular and sawtooth. It will be useful to define the following quantities:

T = sweep period
τ = time interval during which the swept component lies within the SCPC channel bandwidth (per traversal)
δ = percentage of time during which the swept component lies within the SCPC channel bandwidth.

It is clear that

$\delta = B / \Delta f$

Defining $T = 1/f_s$, we have

$\tau = T\delta/2 = B/(2f_s\Delta f)$

for triangular sweeping, and

$\tau = T\delta = B/(f_s\Delta f)$

for sawtooth sweeping.

---

*Even in this case, there are normally sync pulses so that the carrier is not truly unmodulated.

The situation depicted requires special care because errors will not be uniformly distributed in time. In general, error "bursts" will occur whenever the swept carrier traverses the SCPC signal bandwidth. This is best appreciated by considering the limiting (quasi-stationary) case $f_s \rightarrow 0$. Here, the full interference power is intercepted a fraction δ of the time; during that time denote the error probability P_{e1}. In the remainder of the sweep cycle, the interference is outside the receiver bandwidth and nominally does not degrade performance; in this interval denote the error probability by P_{eo}. The average bit error rate is then

$$P_e = \delta P_{e1} + (1 - \delta) P_{eo}.$$

In the other extreme, $f_s \rightarrow \infty$, the SCPC filter will not be able to respond to the instantaneous interfering waveform and the interfering power at the filter output will be constant, corresponding to the average power sweeping across the bandwidth, i.e., equal to a fraction δ of the unmodulated interfering carrier power. Only in this case is the resulting error probability equal to that predicted from spectral considerations.

Between these extremes the interfering power at the filter output will depend specifically on f_s and of course on the value of δ. This power can be determined from the simplified model shown in Figure 7-11. There, the pulse train represents the interfering carrier envelope that would be seen at the input to the filter. Roughly speaking we can expect the power intercepted by the filter to depend on the separation between the spectral components of the periodic interferer, that is on f_s, for given δ, and on δ for given f_s. This power has been obtained by Yam [105] and the results reproduced here in Figure 7-12. Two cases were considered for f_s, one which corresponds to the standardly used

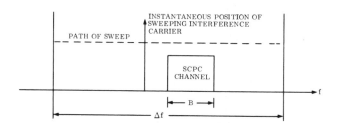

Figure 7-10 Illustration of Swept Interference Scenario

frame-rate triangular dispersal, $f_s = 30\ Hz$; and one which corresponds to half-line rate triangular dispersal, $f_s = 7.85\ kHz$ (the latter case is also equivalent to line-rate sawtooth dispersal). The filter was a 7-pole Butterworth with 3 dB bandwidth of 36.48 kHz. For the usual $\Delta f = 10^6$, $\delta = 36.48 \times 10^{-3}$ and we could expect the power reduction to approach about 14 dB as $\tau \to o$.

As can be seen, considerable improvement is to be expected from increasing the sweep rate. The ultimate test in the digital domain is the bit error rate. This was obtained experimentally by Yam [105]; representative results from this work are reproduced here in Figures 7-13 and 7-14. The general observations made above are confirmed in these figures. From these (and others in [105]) one is provided with data that help to form a fairly comprehensive picture of the effect of swept CW interference into digital SCPC.

Source: Ref. [105]

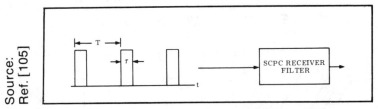

Figure 7-11 Simplified Representation of Filtering Effect

Source: Ref. [105]

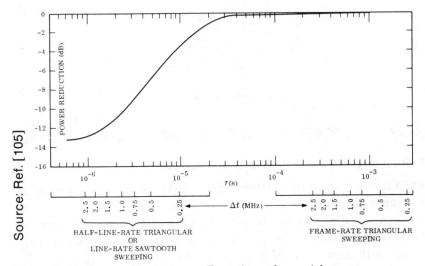

Figure 7-12 Power Change as a Function of τ or Δf

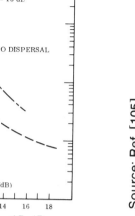

Figure 7-13 SCPC Bit-Error Rate *vs* C/I, *C/N = 15 dB*

Source: Ref. [105]

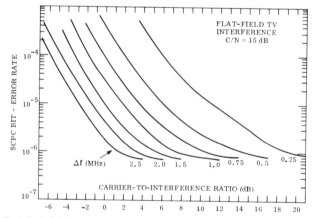

Figure 7-14 SCPC BER *vs* C/I with Line-Rate Dispersal for C/N μ 15 dB

Source: Ref. [105]

7.7 Summary and Collection of Numerical Results

As mentioned earlier the nature of digital systems makes it impossible to generate a generalized set of curves from which one could ascertain the degradation in performance due to the presence of interference. However, some generally applicable observations can be made and these do lend some insight into the problem. In addition, many independent studies have been made and the published results (curves) taken collectively add a further substantial measure of understanding. We present here a small collection of these results which should aid in forming such an understanding.

First in Figure 7-15, we show a QPSK example (augmented from [62]) that illustrates three main points. One, which will repeatedly stand out, is the fact that for the same total C/I ratio, a single interferer has a more benign effect than several interferers. We mention that, in this example, the interferences are co-channel and CW, with random phases independent of one another and of the wanted carrier; in the 4-interferer case, the interferers are taken to be of equal power. The second point shows the non-additivity of degradation (i.e., non-separability in the sense discussed earlier), i.e., the degradations with both ISI and interference is not the sum of the degradations individually induced, but may be more or less than that sum

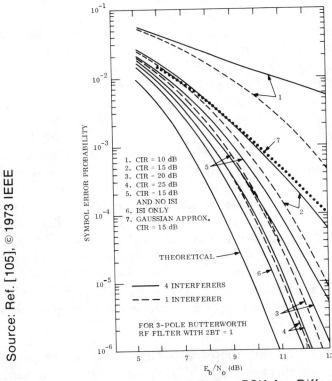

Figure 7-15 Performance of 4-Phase PSK for Different Number of Interferers and C/I ratios in a Linear Channel

depending on the BER. For small BER the sum is generally optimistic. The final point, and a significant one, raised by Figure 7-15 concerns the Gaussian approximation to interference. We have pointed out more than once that treating interference as Gaussian (thermal) noise gives an upper bound on BER in the distortionless channel. In the section on bounds we have also indicated that, in the filtered channel, an upper bound on BER results if *total* interference is treated as an equal amount of Gaussian noise, where total interference is external interference plus ISI. Now, if the ISI itself is treated rigorously, the question arises whether or not equating interference to thermal noise still results in a bound. This question does not appear to have been theoretically treated in the literature. But the results earlier derived and plotted in Figure 7-15 permit us to make some observations on the subject. In particular, curve 6 of the figure gives us the function $P_e(C/N)$ for ISI only, where C/N is 3 dB higher than E_b/N_o since we are dealing with 2 bits / symbol. Hence, treating the interference as thermal noise would be equivalent to stating that the performance, at a given value of C/N, would be given by $P_e(\rho)$ where $\rho^{-1} = (C/N)^{-1} + (C/I)^{-1}$. This has been done for $C/I = 15\ dB$ and the result is shown as the dotted curve 7 on Figure 7-15. The behavior of this curve is interesting; it can be seen that there is a crossover C/N below which it predicts better performance than actual, and above which it overbounds the actual BER. The specific behavior of this curve is of course directly connected to the particular function $P_e(C/N)$ in the linear channel that we started with, i.e., the behavior is channel-dependent. Thus, in some channels the crossover may be lower or higher than in others, and in yet other channels it may be that none exists. But while the above example cannot be considered a general result, it does tend to suggest that above a certain C/N, which corresponds to BERs of general interest, treating interference as Gaussian noise will still yield an upper bound on the BER in the linear channel. The crossover phenomenon has been observed experimentally in a test set-up which is equivalent to what we refer to as a linear channel [70, Fig. 3].

Figure 7-16, taken from [61], shows results in a format similar to the previous figure, but this time for MSK. Here only a single interferer is used, but in this case it is digitally modulated at the same rate as is the wanted carrier; both the carrier phase and bit timing of the

interferer are assumed random and independent of the corresponding quantities of the wanted signal. (We may mention in passing
that the results are somewhat dependent on the latter assumptions.
Benedetto et al. [87b] have presented some results both for random
bit timing and for synchronous data streams on the interfering and
wanted carriers.) The (linear) channel, as in Figure 7-15, is a 3-pole
Butterworth filter. The filter bandwidth is almost the same, though
not quite, but the detection filtering* details are different in the two
cases. Nevertheless, the two sets of results can be seen to be very
consistent. For example, for a BER of *10^-5* and *C/I = 20 dB,* Figure
7-15 shows a required E_b / N_o of about 12 dB while Figure 7-16 indicates about 11 dB.** This observation may seem to imply a lesser

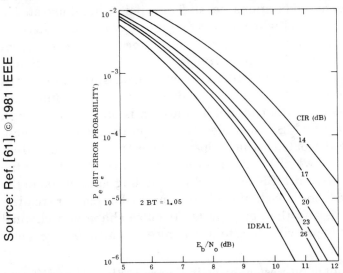

Figure 7-16 Probability of Error of an MSK System with a Triple-Pole
Butterworth Receiving Filter and a Single MSK Interferer
of the Same Rate

*By detection filtering we mean that which follows demodulation and precedes
decision. This filtering is sometimes called the data filter and is perhaps most widely
identified as matched filtering.

**The two figures must be reconciled for error probability definition. The first shows
symbol error rate, P_4, which we can take $\approx 2P_2$, while the second shows P_2. The two
figures have also been reconciled with respect to SNR definition. In both cases we
have labeled the abscissa E_b / N_o, which is actually the SNR at the sampling instant.
Because of the somewhat different definitions the predetection C/N may not be quite
the same in both cases. Also note that different computational methods were used,
which may not have quite the same accuracy.

sensitivity of MSK to interference than has QPSK. In fact, such an inference may not be made unqualifiedly. The fact is that the performance of a given modulation scheme is dependent upon the details of the implementation of the end-to-end link for the wanted carrier as well as upon the system context (multiple access scheme; number, type, and spectral location of interferers; fading conditions, etc.).

For a fixed and specified environment and set of constraints, which may not be the same for two given modulation methods, it is fair to make statements of superiority of one over the other so long as it is understood that these statements may apply only under the conditions assumed. Thus, generally, one cannot attribute relative goodness to different modulations, *per se*. These points are illustrated in Figure 7-17, from [61], which shows for given E_b / N_o, the difference in performance with respect to interference for two different "matched" filters. The upper curve is for a theoretically optimum filter (also used in Figure 7-16) in the white Gaussian channel; the lowest curve is for a "suboptimum" filter in the latter channel, but the performance in the channel in question is obviously superior. Both curves are for MSK, so evidently a unique interference immunity cannot be attributed to that property alone. In fact, as pointed out by Fang [52], if we allow optimization of the system filtering the distinction between MSK and QPSK (including as a variable the timing between the two channels*) disappears. For both are realizations of QAM, and the optimum QAM pulse shape and channel timing would in principle be attained using either MSK or QPSK as a starting point.

A further result of interest is shown in Figure 7-18 [61]. Here different specific realizations of quadrature signaling are shown to possess significantly different immunity to interference in the channel in question (still the third-order Butterworth filter); in each case a single interferer identical to the wanted signal was used. The previous comments still apply here. That is, the relative goodness of these schemes reflects the specific context. By a relatively trivial change such as the nature of the detection filter the comparative evaluation could be drastically altered. Since in principle one pulse

---

* Strictly speaking, of course, the two channels in MSK are offset by one-half bit, while in QPSK the offset can be arbitrary.

shape (modulation method) can be changed to another by a linear transformation, the sequence of curves can be thought of as successive attempts at optimization.

Now we show some results [63] obtained for the nonlinear channel. Figures 7-19 and 7-20 show the BER for a channel with no filtering but a satellite TWT operated at 2 dB input backoff. The first of these figures applies for downlink interference only, which means they are unaffected by the nonlinearity, while the second figure used only uplink interferers. In both cases the interferers are incoherent with the wanted carrier, and the nature of the modulation on the interfering carriers plays no role. The latter fact, at least concerning Figure 7-19, is consistent with the observations made earlier for the distortionless channel. It is seen that a small-signal suppression effect takes place so that, for the assumptions made, uplink interference of a given power has a weaker effect (degradation) than downlink interference of the same power. The figures also show the BER corresponding to the Gaussian approximation. The preceding results were extended by simulation [63] as shown in Figure 7-21 to include

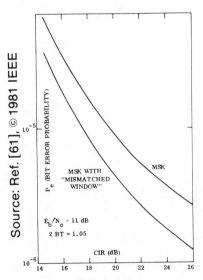

Figure 7-17 Probability of Error *vs* CIR for Optimum and Sub-Optimum MSK Detection

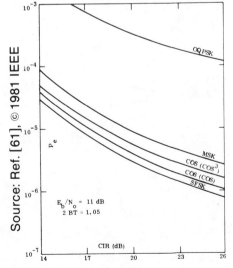

Figure 7-18 Probability of Error *vs* CIR for SQPSK, MSK, COS(*cos*³), COS(*cos*), and SFSK

the effect of filtering. Except as indicated the satellite nonlinearity was a hard-limiter, and idealized "brick-wall" filters were used, with a bandwidth/data rate ratio of 1.15. For this case only a single interferer modulated at the same data rate as the wanted signal was used. The results shown represent an average over modulating sequences and bit timing, corresponding to the "random" conditions usually assumed in the analytical studies. As can be seen, the uplink suppression effect is smaller here than in the no-filter case. The authors further report [63] that with both a transmit and receive filter, the uplink interference was more suppressed when it was modulated than when unmodulated.

It is important to recall that performance results for a digital system are intimately tied to the assumptions made in deriving them, and this is all the more so if the channel is nonlinear. Thus, the results from [63] outlined above must be associated with the conditions stated, and do not necessarily represent generalizable findings. In

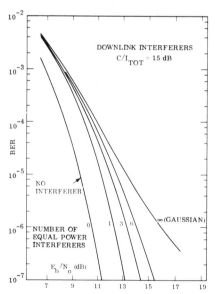

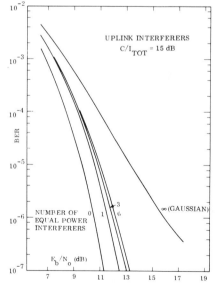

Figure 7-19 BER as a Function of *C/N* for Downlink Interferers with *C/I* + *15 dB*

Source: Ref. [63], © 1981 IEEE

Figure 7-20 BER as a Function of *C/N* for Uplink Interferers with *C/I* + *15 dB*

Source: Ref. [63], © 1981 IEEE

fact, as will be seen below, the results of another study, which was experimental, appear to be in conflict with those of [63] regarding the relative effects of uplink and downlink interference in the nonlinear channel. As will be further discussed, however, the "conflict" seems to be merely a manifestation of the fact that the channels evaluated in the two cases were not identical.

The performance characterization and behavior of a digital system in the presence of interference have been further elucidated by Wachs and Weinreich [69,70] in an extensive series of measurements, the results of which we shall summarize.* The tests were performed using hardware that may be considered representative of links to be found in practice (at least at the time the measurements were made). In particular, a commercial 4-phase PSK modem and the

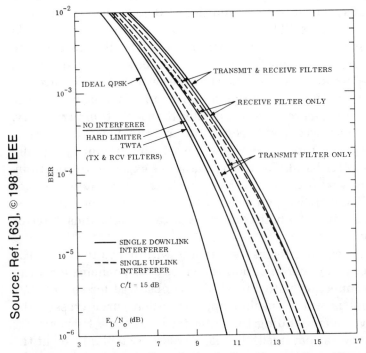

Figure 7-21 Simulation Results Including Transmit and Receive Filters

---

*Only parts of the results of the experiment have been published in [69,70]. Results reported here which cannot directly be obtained from these papers have been kindly supplied to us by M.R. Wachs.

INTELSAT IV engineering model transponder were used. The satellite transponder was operated at three levels of input backoff, thus creating three versions of a nonlinear channel. In addition, tests were made on a linear channel, which was essentially the nonlinear channel with the transponder TWT bypassed. A most instructive feature of the experiment is that measurements were made both with recovered carrier and clock synchronization and with the synchronization circuits hardwired, thus permitting an evaluation of the interference-induced degradation on the carrier and timing recovery. Except in one instance, as noted, all the results reported below were obtained with unmodulated (CW) interferers.

For the linear channel, the following results were observed:

A typical result is illustrated in Figure 7-22 [69], which shows the error probability for a single co-channel interferer for various combinations of synchronization conditions (RCRC = recovered carrier, recovered clock; RCHC = recovered carrier, hardwired clock; HCHC = hardwired carrier, hardwired clock). The dB figure next to each curve represents the C/I ratio. As can be seen, and expected, additional degradation is produced by the perturbations induced by the interference on the synchronization circuits. It appears that the carrier recovery is more seriously affected than is the timing recovery.

Tests were carried out with from one to six equal-power interferers. In one case, the interferers were spaced approximately equally across a 24 MHz band. For this situation as expected the degradation increases with the number of interferers at constant C/I for both RCRC and HCHC synchronization conditions, and the Gaussian approximation for both conditions always overbounds the error probability.

In one case, six interferers were spaced within a band of 300 kHz. For the HCHC synchronization the cross-over phenomenon mentioned in connection with Fig. 7-15 was observed for several C/I values. That is, the Gaussian approximation curve crossed the actual curve, overbounding the latter for large C/N. However, for RCRC the reverse occurred; that is, a crossover took place but the Gaussian approximation was more optimistic than the actual curve at high C/N. Evidently, in addition to the other relevant variables, the temporal variation of the interference complex envelope has a bearing on the degradation due to interference.

For the nonlinear channel (which means interference was introduced on the uplink only) the following observations apply:

Another representative result is given in Figure 7-23 which shows the BER when a nonlinear transponder (0 dB backoff) is used, with the same set of C/I ratios and combinations of synchronization recovery as were used in Figure 7-22. It is quite obvious that the performance in general is considerably poorer in the second of these figures. However, since the basic (interference-free) link degradation is higher for the nonlinear than for the linear channel, it is not immediately obvious whether the effect of interference is more damaging in the nonlinear channel. We shall return to this point shortly.

Some tests were made of the effect of modulation on a single interfering carrier. In addition to the CW interferer, an FM carrier

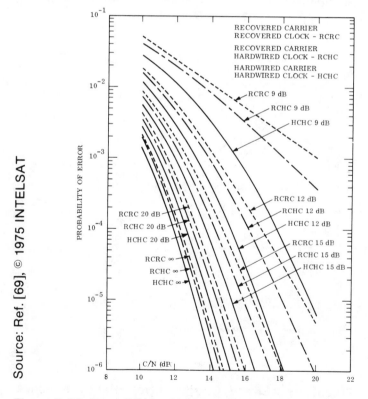

Figure 7-22 P_E *vs* C/N for Unmodulated Co-Channel Interference (linear channel)

and a PSK carrier were used as interferers. The synchronization for these tests was RCRC, and three backoffs (0 dB, 6 dB, 14 dB) were examined. Overall, there was no consistent ordering of the severity of the effects as a function of the modulation, although in most of the measurements (combination of backoff and C/I) the PSK interferer had the worst effect. For reasonably large C/I ($\geq 15\,dB$) however, all three interferers produced reasonably similar performance curves (within about 0.5 dB at $10^{-4}\,BER$) at a given backoff.

The effect of multiple interferers was also studied. Up to 4 interferers spaced at 50 kHz were generated. For HCHC synchronization and for all backoffs the degradation increased with the number of interferers and in all cases the Gaussian approximation yielded an upper bound. However, with RCRC synchronization the Gaussian approximation did not produce such a bound. A result [70] showing this phenomenon is given in Figure 7-24.

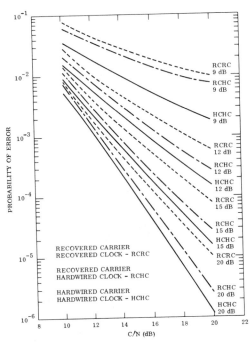

Source: Ref. [69], © 1985 INTELSAT

Figure 7-23 Probability of Error *vs* Carrier-to-Noise Ratio, Non-Linear Transponder, Un-Modulated Co-Channel Interference, Zero dB Input Backoff

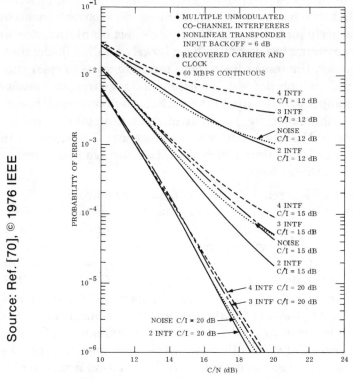

Source: Ref. [70], © 1976 IEEE

Figure 7-24 Probability of Error *vs* Carrier-to-Noise Ratio, Multiple Un-modulated Co-Channel Interferers, Non-Linear Transponders, six dB Input Backoff

As mentioned above, curves such as those in the three preceding figures do not directly yield the degradation due to interference (although, of course, they contain such information) and the figures cannot be directly compared since the performance in the absence of interference is a function of the channel and the synchronization conditions. We define the degradation due to interference as the difference in C/N, for constant BER, between that required under the given interference condition and that required without interference. Table 7-1 lists the degradation at 10^{-4} *BER* for a number of cases.

One readily available conclusion from the table is the fact that the operation of the recovery circuits can be substantially impaired, especially as the C/I decreases and the number of interferers increases. In fact it was found [69] that the carrier recovery ceased to function properly for $C/I < 9\ dB$. Since the effect of interference on the carrier or timing recovery has been largely ignored in the theoretical analyses, the lesson from these measurements is clear that this is an area which needs to be explored. Of course, the results shown here apply specifically to the particular circuits which were employed in the experiment [69]. It may be that other recovery techniques would be less sensitive to the presence of interference. In fact, this is one of the points which should be clarified in the further explorations of this subject.

A second set of inferences which can be drawn from the Table concerns the influence on the degradation of the nature of the channel. At high C/I there is little difference whether the channel is linear or nonlinear. If anything, there does seem to be somewhat of a "suppression" effect in the nonlinear channel for the RCRC condition. These particular results, however,(that is, for $C/I = 20\ dB$ in general, but specifically for the measured data point at 4 interferers [RCRC] 0 dB B.O.) should probably be viewed a bit cautiously as the closeness of the numbers may be within the measurement tolerance. But for the lower C/I values there seems to be no question that the nonlinear channel is more susceptible to degradation from interference than is the linear channel both for RCRC and HCHC. This is seen to be true, specifically, for $C/I = 15\ dB,$ the same value of C/I used in Figures 7-19 to 7-21. The conclusion from these latter figures, however, was just the reverse. This is the "conflict" referred to earlier. Evidently the two sets of results are in apparent contradiction. We should stress "apparent," however, because these results were not obtained under identical conditions. For example the filtering in Figure 7-21 was idealized while in the experiment real filters were used, and the bandwidth-to-bit rate ratio probably was not the same. We have already seen in connection with Figure 7-17 that apparently small changes in the system can produce significant changes in performance. Thus, the seeming inconsistency noted above appears to be merely a reflection of the sensitivity of performance to the channel characteristics. It has recently been observed [106] in a very related context that quite opposite conclusions have been drawn by different authors concerning the suppression or enhancement of uplink noise by a nonlinearity, and these differences must be attributed to the specific assumptions made in the analyses. The traditional difficulty in dealing with nonlinear systems still persists when we introduce interference. Thus, further detailed studies are required to systematically characterize the behavior of a "nonlinear" channel (which can mean many things) in an interference environment.

From a design point of view the effect of interference manifests itself as a necessary increase in E_b / N_o, for given BER requirement. It is therefore useful to interpret the known results in that light because it is likely that specifications on acceptable interference will reflect this aspect of system design since the BER is presumably fixed for a given application. Of course, universal results cannot be given, but the following examples should prove useful in developing some feeling for what is to be expected. First, from [58] we show in Figure 7-25 the degradation induced by the presence of interference in the distortionless channel for different MPSK systems and symbol error probability of 10^{-4}. In Figure 7-26 the effect of the number of interferers is shown by replotting the results of Rosenbaum [57], which apply here only for binary PSK. For the values of CIR shown the

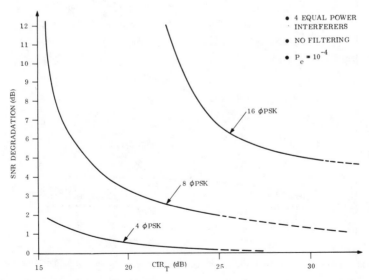

Figure 7-25 Degradation as a Function of Interference Level for MPSK

degradation is upper bounded for a fixed number of interferers when these are equal in power. It is interesting to have some idea about the tightness of this bound in relation to unequal power interferers, which is certainly likely to be the case in practice. This we can partially answer from the orbit utilization example of Section 7.3.1. From this example we see that $CIR_T = (\Sigma\, r_j^2\,)^{-1}$, which equals 18.3 dB for $\Delta\theta = 2^o$ and *15.2 dB* for $\Delta\theta = 1.5^o$. From Figure 7-3 we can get the corresponding degradation for binary PSK. These two points are plotted in Figure 7-26 at the applicable number of interferers (i.e., 10). Since in this case the interferers are far from equal power, this example suggests that the equal-power case may be a reasonably tight upper bound and is therefore useful in system studies.

Figures 7-27 and 7-28 deal with the combined filtering-interference problem. The first of these clearly quantifies (for the channel in question) the relative influence of ISI and interference. It can be seen that the desired value of P_e has not an overly strong bearing. However, the nature of the curves for large CIR indicates that if too small a degradation is required, this could lead to excessively high demands on CIR. The second figure merely isolates the differential degradation induced by the introduction of interference into the

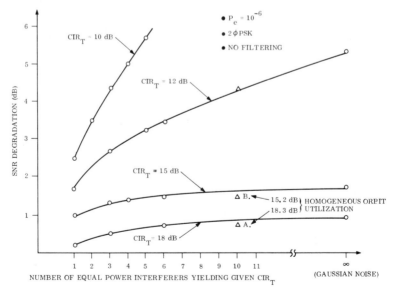

Figure 7-26 Degradation as a Function of Number of Interferers

linear channel. This curve is obtained from the preceding one by subtracting the appropriate curves from one another.

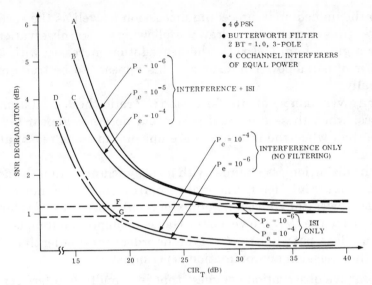

Figure 7-27 Comparison of Degradation Due to Interference and ISI

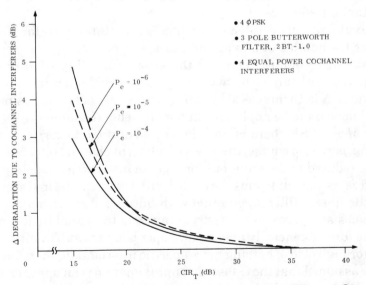

Figure 7-28 Degradation Due to Interference in a Filtered Channel

Finally, Figure 7-29, using results from Table 7-1, gives an indication of the degradation induced in the synchronization circuitry due to the presence of a single interferer.

From the immediately preceding discussion as well as the earlier results of this chapter we can draw the following general observations:

For a given total C/I ratio the degradation increases with the number of equal-power interferers; this appears to be true quite generally;

For a given number of interferers and C/I ratio; the degradation is greatest when these are equal power, but does not appear much worse than when the interferers are unequal in some reasonable fashion;

In the distortionless channel with ideal synchronization the degradation due to interference is smaller than that due to thermal noise of equal power (the Gaussian approximation); when the C/I is large compared to the C/N, however, the Gaussian approximation is reasonably close, but when the C/I is of the order of or smaller than the C/N, the Gaussian approximation is pessimistic;

The above observations suggest that in a multiple interference environment, a reasonable approach may be to add the small interferers on a power basis to the thermal noise and treat the larger ones as actual interference;

The vulnerability to interference increases substantially for *M-ary* PSK as the number of phases increases;

In the distortionless channel the error probability of M-phase systems can be bounded by twice the binary P_e evaluated at values of C/I and C/N both increased by $-20 \log \sin (\pi / M)$, but no such simple bound appears to be applicable in a more general channel;

For *M-ary* PSK there is no inherent advantage to carrier interleaving, *per se* ; of course, the effect of adjacent channel interference can be reduced by filtering but for a given intercarrier spacing the interference reduction must be traded with ISI on the desired signal;

In the linear (filtered) channel with hardwired synchronization the Gaussian approximation curve may cross the actual BER curve, but the former generally acts an an upper bound at sufficiently low BER (or high C/N); barring more specific information, therefore, it can be assumed that the Gaussian approximation is an upper bound in the BER range of interest;

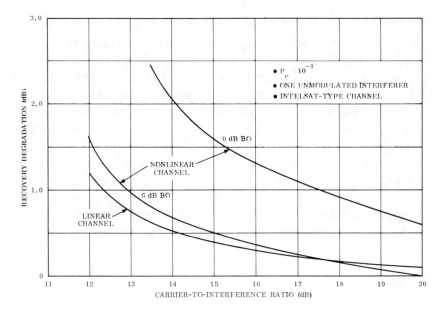

Figure 7-29 Degradation Due to Carrier and Timing Recovery with Respect to Interference-Free Channel

In the nonlinear channel with hardwired synchronization the Gaussian approximation seems to provide an upper bound;

In both the linear and filtered-nonlinear channel with recovered synchronization, the Gaussian approximation will generally no longer bound the actual performance, except perhaps for one or two interferers;

The presence of interference can have a substantially degrading effect upon synchronization performance;

The susceptibility to interference can differ markedly for different signalling schemes in specified system contexts;

The susceptibility to interference of a given signalling scheme can be significantly improved by optimization;

In a general (nonlinear) channel, the effect of interference is imperfectly understood and further study is required to characterize its effect.

7.8 Concerning the Specification of Interference

Since the additional degradation due to interference can vary significantly depending on the particulars, it could be problematical to establish regulatory limits on the acceptable degree of interference. Furthermore, such limits cannot depend upon any specific assumptions concerning the implementation, while the latter has a definite bearing on the effect of a given level of interference—thus a circular situation exists. Furthermore, the specific interference environment, i.e., the number and relative magnitudes of interferers, may not be known *a priori*. It is clear, therefore, that to make progress some simplified assumptions must be made. Although the end-to-end performance of a link is normally specified in terms of the distribution of the relevant performance parameter, the BER in this case (and a specific recommendation exists for PCM telephony [107]), the contribution of interference to this performance objective cannot be isolated.

From a practical standpoint it is more sensible to control the interference power rather than the effect on performance it may have. This is the approach taken thus far [108]. That is, the interference can be specified as, say, some fraction of the receiver equivalent noise temperature. Once this is known, the designer can optimize his system so that under any interference environment meeting the mandated limit, the total BER performance requirement is satisfied. This approach does not necessarily imply an equivalence between interference and thermal noise. However, if this equivalence is made it will tend to bound the effect, and then as a simplification one can think of this effect as requiring an additional $-10 \log (I/N)$ decibels of signal power to maintain a fixed BER.

TABLE 7-1

Measured Degradations (in dB) for Different Combinations of C/I,
Channel Conditions, and Number of Interferers for $BER = 10^{-4}$ (1)

CASE DESCRIPTION	C/I = 20 dB		C/I = 15 dB		C/I = 12 dB	
	HCHC	RCRC	HCHC	RCRC	HCHC	RCRC
1 Interferer-Linear Channel	0.8	0.9	1.9	2.3	3.3	4.5
4 Interferers-Linear Channel	1.3	2.0	2.9	4.5	6.8	(3)
1 Interferer-Nonlinear (6 dB BO)	0.8	0.8	2.0	2.5	4.1	5.7
4 Interferers-Nonlinear (6 dB BO)	1.0	1.1	3.1	6.2	11.6(2)	(3)
1 Interferer-Nonlinear (6 dB BO)	1.0	1.6	3.0	4.7	7.0(2)	11.0(2)
4 Interferers-Nonlinear (6 dB BO)	1.0	0.7	3.7	8.5(2)	(3)	(3)

Notes: (1) In every case the degradation is measured with respect to the required C/N in the interference-free situation for the case in question; thus the degradations reflect exclusively the effect of interference.

(2) Represents extrapolation; actual value may be worse

(3) Represents cases for which extrapolation would be unreliable; the BER curves appear to "bottom-out" and may never each 10^{-4}.

Chapter Eight

INTERFERENCE CANCELLATION/REDUCTION TECHNIQUES

8.0 Introduction

We have seen that the susceptibility to interference, or the ability to cause interference, varies widely as a function of the modulation method, the type of signal, and various other design choices. However, except where mandated or generally accepted (e.g., station-keeping tolerance, minimum earth station antenna sidelobe decay, etc.), it has not generally been the practice to design links with particular regard to inter-network or inter-system interference.* Rather, the design decisions are typically forced by traffic, cost, connectivity, and operational requirements. Since the permissible level of interference between two networks is delimited, this practice has not tended to impose penalties on individual networks. Instead, the "burden" has fallen on the orbit itself; that is, the intersatellite spacing has generally been adjusted to be compatible with the interference requirement. However, there are limitations with this approach. Service arc constraints may conflict with the necessary spacing to the nearest satellite, and in a dense satellite environment the flexibility of positioning to avoid interference may simply be unavailable. Already, in the 4/6 GHz bands, satellite separation alone has been insufficient to ensure compatibility between networks and some degree of traffic coordination has had to be resorted to. It thus becomes important, at least for possible future application, to consider more positive measures for interference reduction (without unduly constraining system design or operational flexibility) than has previously been deemed necessary or cost-effective.

---

*A somewhat different situation applies to intra-network interference where more conscious steps have been taken to minimize interference, both because this tends to optimize the capacity/cost ratio and because the interference is under the system designer's control. Even though reducing intra-network interference does not directly affect satellite spacing it can be so used by making a greater interference budget available for inter-network interference.

In this section we give a brief overview of one class of techniques that may be applicable in this context.* Except for polarization compensation, these techniques would not normally be contemplated in the design of a conventional satellite system, but as mentioned the onset of congestion in certain bands and certain arcs, if it occurs, may well necessitate their use. Collectively, these are broadly referred to as cancellation techniques since they typically (but not exclusively) rely on generating an independent replica of the interference which, properly scaled and signed, can be used to "cancel" the interference imbedded in the wanted signal. Of course, this can only be done approximately so perhaps a different term, like reduction, may be somewhat more accurate though less connotative. In any case, since the interference may be unknown or time-varying simultaneously in amplitude, delay, or polarization, these techniques will generally have to employ adaptive or adjustable mechanisms. An additional complication arises from the fact that a network is generally subjected to multiple independent and physically separated interference sources. Some of the techniques under discussion have shown promise in experiments or under certain operational situations though their applicability to geostationary satellite networks must still be considered in the developmental stages. However, the degree of interest recently evidenced [110-115] suggests that practical solutions may not be long in coming.

8.1 RF (IF) Cancellation With Auxiliary Source

Probably the most commonly studied group of methods of interference reduction revolves around an attempt to cancel (subtract) the interference accompanying the wanted signal by appropriate manipulation of an independently available (and relatively uncorrupted) replica of the interference at the receiver input. The subtraction is

---

*We confine our attention here to techniques for reducing interference between satellite networks, though they are not limited to that context. Techniques that are specially suitable for reducing interference between earth stations and terrestrial stations are discussed in [109]. We also note that there are other techniques, which have not yet found widespread use, that do not fall into the category of this chapter. For example, companding, DSI, error-correcting codes are all capacity-increasing measures for a given level of interference, and hence they are interference reducing in some sense. However, such techniques are better viewed as "signal processing" rather than as affecting the level of interference. A brief look at the benefits of coding was given in Chapter 4, but in general further study is needed to systematically quantify the possible advantages of signal processing techniques.

done at RF or at a suitable intermediate frequency stage. Some of the approaches to depolarization in dual-polarized networks can be considered a subset of this idea, but this specific situation will be treated separately in Section 8.5. We should note that one attraction of this method is that it is independent of the details of wanted and interfering signals, and in this sense is the most general of the interference reduction methods.

The replica of the interference is obtained from an "auxiliary source," which is commonly envisioned as a separate small antenna at the earth station site. However, in the orbit utilization context, there are generally several interference entries which may be of comparable magnitude. In order to obtain a clean sample of each of these it would likely be impractical to erect multiple antennas, but it should be quite feasible to install a feed cluster either on the main or auxiliary antenna, thus creating a series of subsidiary beams towards the interfering satellites. (A physically separate antenna would probably be required for terrestrial interference.) To minimize the complexity of the auxiliary source its beamwidth should be so chosen that it is relatively large with respect to satellite drift to avoid tracking, and its gain sufficiently higher than the sidelobe gain of the main antenna. These objectives are achievable with moderately small antennas. Of course, without tracking, the amplitude, at least, will fluctuate, but this can be taken care of with an adaptive system, as described later.

The principle of operation can be illustrated, assuming, initially, undistorted waveforms. Let the signal received at the main antenna be given by

$$r(t) = w(t) + i(t)$$

where the (single) interferer $i(t)$ may be taken as

$$i(t) = R(t) \cos [\omega_2 t + \psi(t)] \tag{8-1}$$

Now let

$$i_a(t) = KR (t\text{-}\tau) \cos [\omega_2 t + \psi(t\text{-}\tau) + \theta] \tag{8-2}$$

represent the auxiliary interference signal as it appears at a point in the system [see Figure 8-1] where it is available for processing.* The

*Note that the very form (8-2) implies that we are ignoring the corruption of the interference replica with the wanted signal itself, which is inevitable. However, this effect should be second-order with a properly designed auxiliary source.

relative amplitude K, time delay τ, and phase shift θ are induced by the differences that the interference encounters over the two paths to point ① i.e., differences in gain and electrical path lengths.

Thus, if we have a variable attenuator and variable delay element, a suitable interference cancelling system would appear as schematically shown in Figure 8-1. Note, in addition, that the figure shows an additional device, a variable phase shifter. Of course, this is also a delay element. We make the distinction here only to point out that in practice one may need a "coarse" adjustment (i.e., the variable τ) and a "vernier" adjustment (the phase shifter) because, since $\theta = \omega_2 \tau$, a very small error in τ can result in a comparatively large error in θ, depending on ω_2. In some cases it may only be necessary to compensate for the phase shift, rather than the absolute delay [note that θ depends on the latter only *modulo 2π*], depending on the signal's reciprocal bandwidth in relation to that delay. We note that there will generally have to be a delay element both in the $r(t)$ and $i_a(t)$ lines since the sign of τ may be positive or negative. For multiple interference sources the $i_a(t)$ circuitry will generally have to be replicated as many times as there are such sources to be cancelled. With only one interferer, manual adjustment of the variable parameters might be achieved without too much difficulty in actual operation, e.g., by observing the output of normally unoccupied slot of frequencies, such as between groups of telephone channels. However, the practicality of this evidently depends on the stability of the triplet (K, τ, θ).

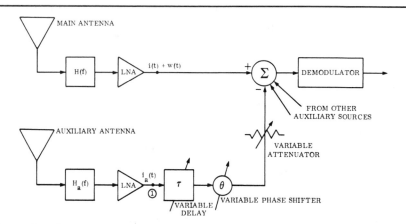

Figure 8-1 Block Diagram of Simplified Interference Canceller

For reasons to be discussed shortly, it may be more practical to consider adaptive adjustment of the variable parameters.

The preceding simplified discussion suffers two types of shortcomings. One relates to uplink interference and the other to distortion and filtering effects. On the first of these, it is obvious that one cannot obtain an auxiliary source in the sense we have been discussing. However, in the absence of distortion, the downlink auxiliary interference is in fact sufficient since the uplink interference as seen at the receiver input and the downlink interference would differ only by a scale factor, time delay, and phase shift. Hence, by taking $i_a(t)$ and passing it through two lines with separate adjustable elements we would be able to cancel both uplink and downlink interference.

As mentioned, the second simplifying assumption in our initial discussion was the absence of distortion, which arises from filtering or nonlinear elements. In actuality, the presence of these impairments means that the envelope and phase of $i(t)$ and $i_a(t)$ need not be identical, which means that operating only on the magnitude, delay, and phase of $i_a(t)$ would not be sufficient to effect cancellation of $i(t)$. On the downlink, for example, $i(t)$ and $i_a(t)$ could differ if the front-end filtering of their respective receivers were not identical. This might not be too significant if, say, $i_a(t)$ were not wider in bandwidth than $w(t)$ and wholly contained within the passband of the latter. However, if there were an appreciable carrier-frequency offset between $i(t)$ and $w(t)$ so that, the spectrum of $i(t)$ straddled the band-edge of the receiver transfer function then $i(t)$ would undergo considerable modification in passing from the input to the antenna to the output of the receiver. In particular, the receiver output interference, say $i'(t)$, would not have the same envelope and phase functions as the output of the receiver front-end of the auxiliary antenna if this receiver were tuned to the center frequency of $i(t)$. One way of correcting for this, of course, would be to have duplicate front-ends at the main and auxiliary antennas.

Another factor which should be considered is that in general the uplink and downlink interference will not have passed through the identical sets of elements. Therefore, modification of the amplitude, delay, and phase shift of the downlink interference may not be sufficient to obtain a true replica of the uplink interference.

With the foregoing considerations in mind, a more general interference cancelling structure may be advanced, as shown in Figure 8-2. The main additional feature is the variable or adaptive filter* in the path of the auxiliary interference signal, which can be used to compensate to some extent for differences in the waveforms of the actual interference and the auxiliary signal. A suitable structure for the filter might be a tapped delay line (TDL) with complex weights $W_k = \alpha_k \, exp \, (j\beta_k)$. Thus, for example, in the absence of distortion the TDL would have two appropriately separated taps, one for the uplink and one for the downlink.

An important consideration in this context is how to adjust the variable filter. Because of the possibly random variations in (K, τ, θ) for each of the interference entries, as well as the complex nature of the distortions, it is likely that a practical structure will need to be adaptive. There are potentially a number of suitable adaptation algorithms to control the variable filter. The control variable might be at baseband: for example, in an FDM/FM wanted signal it might be the noise in a test channel, or in a digital signal it might be the BER in a sync word. However, the technique most often considered [110, 115], which does not depend upon the particulars of the wanted signal, involves a comparison of the auxiliary intereference to the

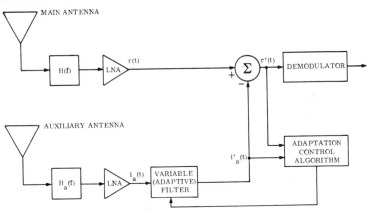

Figure 8-2 Simplified Block Diagram of Adaptive Interference Cancelling System with Specific Example of Feedback Loop for Adaptation Algorithm

*The adaptive filter block can be assumed to contain any fixed coarse corrections.

signal-plus-interference at the output of the main channel, i.e., after cancellation. The basic idea is that the wanted signal and the interference are uncorrelated (actually, independent in general), while the two samples of interference, $i(t)$ and $i_a(t)$ are completely correlated when properly "lined up" in phase and delay. When the latter condition holds, the correlation between $i'_a(t)$ and $r'(t)$ [see Fig. 8-2] can serve as the control variable, the desired end state being to drive this correlation to zero. This type of system has been implicitly assumed in the block diagram of Figure 8-2. Care must be taken, however, that the time delay be well calibrated for otherwise an erroneous null may be observed since a signal is uncorrelated with a sufficiently delayed replica of itself. This pitfall could be corrected by first trying to maximize the correlation between $i'_a(t)$ and $r(t)$, which would ensure that $i'_a(t)$ and $i(t)$ were as identical as possible. In specific implementations of the type of adaptive system being discussed, good results have been reported [110, 115]. The effect of certain implementation errors has also been discussed [112].

8.2 Baseband Interference Cancellation for Angle-Modulated Signals

If specific assumptions can be made as to the nature of the signals involved, then it may be possible to derive interference cancellers which exploit the properties assumed. Of course, the motivation for considering such ad hoc structures, rather than the more general one of the preceding section, would be the possibility of simpler implementation.

Two suggestions have been made [113, 114], for interference cancellers when the wanted signal can be assumed to be an analog angle-modulated carrier. In both schemes the subtraction is effected at baseband.

In the first proposal [113], it is assumed that an independent auxiliary interference source is available. The idea then is to mix (multiply) the auxiliary interference with the signal-plus-intereference. The resulting lowpass output approximately reproduces the interference at the output of the angle demodulator, which it can then cancel by appropriate scaling. We outline here a somewhat different approach than that originally described in [113]. Let the wanted signal be given by

$$w(t) = A \cos [\omega_1 t + \phi(t)] \qquad (8\text{-}3)$$

and let the interference accompanying $w(t)$ and the auxiliary interference be given by (8-1) and (8-2), respectively. As with the RF method, we first need to align $i(t)$ and $i_a(t)$ in delay and phase, and in this respect the current method does not seem to offer any particular advantage. Assuming this alignment has been done the auxiliary source can be expressed as

$$i'_a(t) = KR(t) \cos [\omega_2 t + \psi(t)] \tag{8-4}$$

For reasons which will soon become apparent, it is necessary to induce a 90° phase shift in $i'_a(t)$ without otherwise altering the synchronism with $i(t)$. This can be done, as shown in Figure 8-3, by downconverting both $r(t)$ and $i_a(t)$ coherently to some convenient IF frequency, but using quadrature local oscillators. Alternatively this might be (more simply) accomplished by adjusting the phase shifter an additional 90° rotation. However, this will cause some delay in the modulation which may or may not be significant. (For convenience we now assume that ω_1 and ω_2 represent the intermediate frequencies.) Hence, at this point the auxiliary interference is given by

$$i''_a(t) = KR(t) \sin [\omega_2 t + \psi(t)] \tag{8-5}$$

The mixer then produces at its output

$$e_m(t) = [w(t) + i(t)]i''_a(t) \tag{8-6}$$

the lowpass portion of which is

$$e_{m,l}(t) = \tfrac{1}{2} AKR(t) \sin [(\omega_2 - \omega_1)t + \psi(t) - \phi(t)] \tag{8-7}$$

It should be noted that the $i(t) \cdot i''_a(t)$ product has no baseband component because the terms are in quadrature.

Now, the output of the angle demodulator, the input to which is $r(t) = w(t) + i(t)$, has an interference component given by

$$\lambda(t) = \frac{R(t)}{A} \sin [(\omega_2 - \omega_1)t + \psi(t) - \phi(t)] \tag{8-8}$$

which is the first term of the series (6-14), appropriate as an approximation for "small" interference, assumed to be the case here. [The angle μ in (6-14) can be considered absorbed here in $\psi(t)$]. Obviously, (8-7) and (8-8) are identical in form. Hence, a scale factor applied to the former is all that is necessary to effect a subtraction of the latter, as shown in Figure 8-3. If the demodulator is an FM demodulator, as is more likely to be the case, the actual interference output is $\lambda'(t)$; hence to cancel this signal we need first to differentiate $e_{m,i}(t)$ as shown. If there are multiple interferers, the equipment shown in the figure must be replicated for each entry. Laboratory trial of this technique (with a different implementation than that in Fig. 8-3) has shown about 15 dB of improvement [113.]

The second method does not depend upon an auxiliary source, and for this reason is very attractive though its application is limited. The method makes use of the fact that for a wanted (constant-envelope) angle-modulated carrier, the actual received signal $r(t)$ has an envelope variation induced by the interference. Hence an AM detector can be used to extract information, as will be shown, which can be processed to cancel the demodulator output interference.

Consider the interference to have the form (8-1), which can be a single interferer or represent the aggregate interference as seen at the demodulator input. The wanted signal is given by (8-3). The envelope-squared of $r(t) = w(t) + i(t)$ is easily shown to be

$$E^2(t) = A^2 + R^2(t) + 2AR(t) \cos\left[(\omega_2 - \omega_1)t + \psi(t) - \phi(t)\right]$$

whence, for $R(t)/A \ll 1$, the envelope is approximately given by

$$E(t) \simeq A + \frac{R(t)}{A} \cos\left[(\omega_1 - \omega_2)t + \psi(t) - \phi(t)\right]$$

Comparing this to Eq. (8-5) we see that, after filtering out the dc term, it is in phase quadrature to $\lambda(t)$. In order to effect cancellation, then,

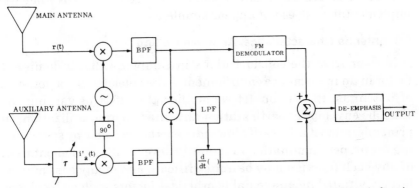

Figure 8-3 A Baseband Interference Canceller for Wanted Angle-Modulated Signals

the $cos(\cdot)$ term must be transformed to $sin(\cdot)$. Let the bandwidth of $\psi(t) - \phi(t)$ be denoted by B. If $|f_2 - f_1| >> B$, then a conventional $90°$ phase shifter (hybrid) will suffice. If only $|f_2 - f_1| \geq B$, then a wideband $90°$ phase shifter (a "Hilbert transformer"), is required, though its feasibility will depend on the magnitude of B.

The approach actually considered by Pontano[114] and tested in the laboratory utilized constant envelope carriers and an FM detector. Thus, the (filtered) envelope could be written as

$$\widetilde{E}(t) = r \cos\left[(\omega_2 - \omega_1) + \psi(t) - \phi(t)\right]$$

while the FM demodulator output, $\lambda'(t)$, is

$$\lambda'(t) = r\left[(\omega_2 - \omega_1)t + \psi'(t) - \phi'(t)\right] \cos\left[(\omega_2 - \omega_1)t + \psi(t) - \phi(t)\right]$$

If $(\omega_2 - \omega_1)$ is "large" compared to the instantaneous frequency of the modulation term, then

$$\lambda'(t) \approx r (\omega_2 - \omega_1) \cos\left[(\omega_2 - \omega_1)t + \psi(t) - \phi(t)\right]$$

and amplifying $\widetilde{E}(t)$ by the factor $(\omega_2 - \omega_1)$ will enable approximate cancellation of $\lambda'(t)$. As expected, and as measured by Pontano[114], the degree of cancellation depends on the carrier separation and becomes ineffective for small separations.

Thus, despite the attractiveness of this approach, its applicability appears to be limited to adjacent-channel interference which meets the conditions on carrier separation laid out above. It may be possible to remove this limitation if practical means can be found to effect the quadrature transformation for an analog wave $cos[x(t)]$; mathematically, of course, it can be done, e.g., from $\sqrt{1 - cos^2}$, but the implementation does not appear simple.

8.3 Antenna Characteristics: Adaptive Nulling

It is clear from the considerations in Chapter 4 that a dominant factor in an interference environment is the sidelobe characteristics of antennas. In Section 4.4 we studied the effect of varying the sidelobe envelope of earth station antennas on orbit utilization. In principle, almost arbitrarily low sidelobes are possible by synthesizing the proper illumination function.* However, the implementation of any such function may be quite difficult, and in any case may be largely vitiated by unavoidable practical factors such as blockage,

---

*But the main beam characteristics will also be affected in the process.

surface irregularities, and edge diffraction. These factors (which apply to conventional reflectors) limit the improvement that can be expected with respect to the standardly used CCIR reference pattern *32-25 log θ*. Harris et al [116] have examined the individual contribution of these factors and have concluded that a sidelobe envelope of the form *29-25 log θ* is a reasonable short-term objective for axisymmetric Cassegrain antennas. For offset-fed antennas these same authors suggest that the pattern *26-33 logθ* could be achieved. Further improvements in specific planes, e.g., that joining an earth station and the geostationary orbit, may be possible by judicious placement of microwave absorbers [117]. While the improvements are considerable they are not dramatic at small angles. We note that in one sense the effort involved in obtaining uniformly low sidelobes achieves more than is required. For the existence of low sidelobes in all directions but the ones whence the interference comes is over-achievement not put to good use. Conversely, what is of interest is to obtain very low response ("nulls" if possible), in just a few discrete directions corresponding to the most significant interfering satellites.

In order to realize this aim, more "active" measures (in more than one sense) must be entertained than has been usual in conventional antenna designs. There is an extensive literature on this subject, generally under the heading of adaptive antenna systems [see, e.g., 118-121]. However, the theory and the applications have almost exclusively been limited to phased array antennas, which to this stage have not been practical for commercial applications. As is well known, the basic approach to synthesizing a pattern (possibly with nulls in prescribed directions) with phased arrays consists of a suitably weighted combination of the responses of spatially separated elementary antennas. Several criteria for setting the weights, and procedures (algorithms) for adaptively determining them, have been examined in the literature in a formal mathematical context. Assuming just two antennas, it can then be shown [120] that one of these criteria (minimum mean-square error estimate of the desired signal) implemented with the least mean-square (Widrow-Hoff) algorithm leads to the "adaptive noise canceller" illustrated in Figure 8-4 [120]. The narrowband condition, namely that the ratio of center frequency-to-bandwidth be much greater than unity, is implicit in the derivation. It can be shown [120] that the very same system of Figure 8-4 can be considered as performing adaptive null steering, i.e., one can replace (mathematically) this system by one having a single antenna with an equivalent gain for the desired signal and another equivalent gain, which tends ideally to zero (a null) in the direction of the interferer.

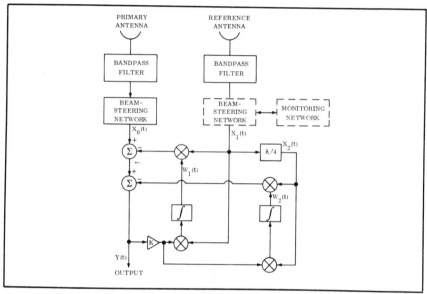

Figure 8-4   Interference Canceller Derived from Adaptive Antenna Theory

The key point of the preceding discussion is that the block diagram of Figure 8-4 is structurally identical to that of Figure 8-2, though the exact workings of the boxes need not be the same. In a sense the adaptive antenna approach validates the latter figure, which was arrived at in a more intuitive fashion, one which begins with the assumption of two antennas and proceeds to develop reasonable circuitry on a heuristic basis. Thus, whether we choose to label an installation as an adaptive interference cancellation system or an adaptive antenna system is largely a matter of where we choose to draw the interface between the "antenna" and the "receiver." Starting from either point of view we can arrive at more or less the same configuration depending on the mathematical conditions imposed. So, in effect, the more "active" measures we sought have already been considered in Section 8.1.

We have been considering up to this point measures that can be applied to a receiving station. Similar techniques can be applied to transmitting stations, though the operational situation is not quite the same and the need for adaptive equipment is less strong. Such techniques are probably least practical for transmitting earth stations since placing a null in the transmit pattern toward a multiplicity of other independently drifting satellites will not be a simple matter. However, it should be quite feasible to apply the same methods to area coverage satellite antennas, as discussed in the next subsection.

8.3.1 Satellite Antenna Area Coverage Nulling (Beam Shaping)

Here, the idea is to maximize the radiated power towards the service area and minimize it elsewhere. In particular it is desirable to place a "null" only in given directions, namely those where significant density of receiving stations of other satellites are to be found. In effect, we are speaking here of "beam shaping" in the same sense that it has been used elsewhere in this book. However, placing it in this context will clarify its relationship to nulling or interference cancellation. That relationship derives from reciprocity, the notion of which tells us that if we can generate a null on reception using an auxilliary antenna, then it should be possible to synthesize a "shaped" transmit beam using the same approach. A technique to achieve this aim makes use of contiguous beam coverage in order to minimize radiated fields in a multiplicity of directions around the antenna main beam coverage [122, 123]. Contiguous multiple beams, in addition, have the flexibility to permit reconfigurable regional coverage, with the ability to achieve large bandwidths, high efficiency for the desired coverage, and low depolarization and interference levels. Auxiliary beams with low level excitation can be used between the adjacent coverage areas or adjacent coverage directions in order to cancel undesired high sidelobe levels to the unintended coverage directions. This technique can create wideband sizeable reductions in the pattern in prescribed directions. The generic reflector configuration in this approach is shown in Figure 8-5 [123]. Also shown in this figure, as an example, is a beam topology plan for coverage of North American and Caribbean "regions." The auxiliary beams, which are used to control sidelobe levels from each

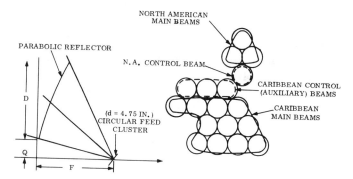

Figure 8-5 Reflector Configuration and Beam Topology (126 in. reflector)
Source: Ref. [123], IEEE © 1982

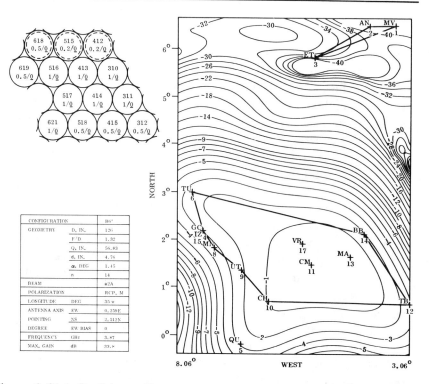

Figure 8-6(a) Caribbean/South American Coverage, 126 in. reflector, 3.87 GHz, 35° Sync. Orbit.
Source: Ref. [123], IEEE © 1982

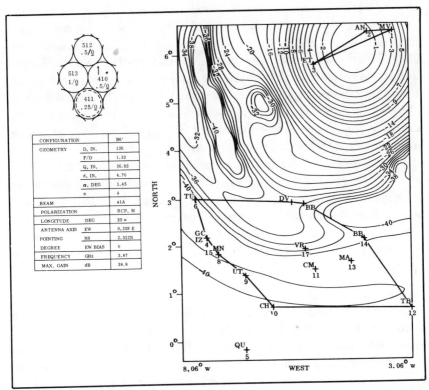

Figure 8-6(b) North American Coverage, 126 in. reflector, 3.87 GHz, 35° West Sync. Orbit.

Source: Ref. [123], IEEE © 1982

regional cluster in the direction of the other region, are indicated by dotted lines. With the specific reflector size, frequency, and satellite position shown in the caption, Figure 8-6, parts (a) and (b), show the detailed radiation patterns emanating from each beam cluster [123]. Examination of these patterns shows better than 30 dB isolation between the two coverage areas.

8.4 Receiver Structure

The demodulators and detectors normally implemented are typically long-standing designs derived initially to combat thermal noise. In earlier chapters we examined the performance of these traditional receivers when the signal was corrupted with interference as well as thermal noise. The question naturally arises, then, as to what the receiver structure would be and how well it would perform if it were specifically designed to combat noise-plus-interference. Clearly the answer depends on the nature of the wanted signal and the manner in which we choose to characterize the interference. But in any case very little work seems to have been done in this area. One might consider the interference cancelling circuitry in the several previous figures to constitute this type of receiver structure augmentation. However, these structures are based on the availability of an auxiliary signal. The traditional derivations of receiver structure are based on a single input signal (and that is the sense we give it here), so that any useful result obtained from this approach would have the further advantage of not requiring an auxiliary antenna. In effect, then, the second of the interference cancelling methods in Section 8.2 does belong to this present category. We now describe briefly some other work which has been done in this area.

Wilmut and Campbell [124] have obtained the form of the optimum receiver for a binary PSK signal in the presence of an interfering sinusoid and of a suboptimum receiver (called a "correlation" receiver) which is a correlation receiver whose reference reflects knowledge of the interference. They show that this receiver is superior to the standard correlation receiver (one which is matched to the desired signal only), but not greatly so when the intereference is at the relatively low level expected in most cases. In fact, Goldman [125] has shown that the standard correlation receiver is indeed optimum for a binary PSK signal when the interference consists of any number of independent co-channel signals whose collective peak value is less than the desired signal envelope. It is further stated in [125] that, when the interference frequencies are unknown and the peak condition is satisfied, the standard correlation receiver is minimax in that it minimizes the worst performance over all receivers designed on the assumption of a specific set of interference frequencies. The results discussed so far were obtained for an idealized channel, and we are not aware of comparable derivations for a realistic channel. However, we have seen in the discussion of Chapter 7 that the characteristics of the detection ("matched") filter can have a significant effect on the BER of a filtered-channel MSK system subjected to interference. Therefore, the implication is that digital system performance can be improved using an appropriate receiver structure, but a systematic investigation has yet to be performed.

Another approach to the receiver structure has been considered by Kowalski [111]. He proposes a dual tapped-delay-line configuration which is intended specifically to reject multiple narrowband interferences into a wideband signal. Conceivably this scheme might be used where a wideband telephony signal, for example, is lined up in frequency with an SCPC transponder.

As a final example of receiver structure designed to combat interference, Stojanović et al. [42] have proposed and analyzed a new FM demodulator (FMD). The proposed structure is ad hoc but is motivated by the very same observation that led to the second baseband cancelling scheme of Section 8.2, namely that the envelope of signal plus interference, when the former is constant envelope, carries information about the latter which presumably can be put to good use. Actually, the structure in [42] can be shown to be a more general version of that in [114]. Rather good NPR improvement has been computed for the proposed FMD although all the computations shown have assumed adjacent-channel interference; its usefulness for co-channel interference remains to be determined.

8.5 Depolarization Compensation

As discussed in Chapter 4, a common approach for increasing the traffic density of a satellite network is to transmit information on carriers with orthogonal polarizations. As is well-known the major obstacle to the successful implementation of a dual-polarized system is the existence of "depolarization," i.e., the reception on one polarization of a fraction of the signal intended to be purely of the orthogonal polarization. Thus, a dual-polarized network exhibits self-interference, which must be kept to a tolerable level if it is to operate satisfactorily. Depolarization stems from two sources; the antenna system (including feeds and other components), and the propagation medium. The first of these can presumably be designed so that the cross-polarization discrimination (or XPD, to be defined shortly) is adequately low — otherwise dual-polarized transmission would not be possible even in an ideal medium. The actual medium, in fact, induces depolarization to a degree that depends on propagation conditons, i.e., in a randomly time-varying manner.* Therefore, adaptive means are called for to compensate for the depolarization.

One possibility for compensation stems from the fact that at a given dual-polarized receiver both of the signals sent on the orthogonal polarizations are available and hence can be used as one another's auxiliary source to effect cancellation. Thus, depolarization compensation schemes can be thought of as a subset of the class considered in Section 8.1, but are more conveniently addressed separately because of the specialized nature of the phenomenon. There are other physical means to correct for depolarization, but these are not naturally described in terms of interference cancelling. We adopt here the latter viewpoint because it is conceptually straightforward, it enables us to view the process within a larger context, and perhaps more importantly, it leads to a natural generalization not possible with other implementations. We shall here only map out the problem in broad terms since the main idea has already been discussed in Section 8.1, and numerous specific approaches have been discussed in detail in the literature (see, e.g., [127-135]). Of course, the effect of residual interference after cancellation should be treated by the methods discussed earlier in Chapters 6 and 7, according to the nature of the signals involved.

Let the transmitted signals on polarization 1 and 2, respectively, be given by

$$s_1(t) = \rho_1(t) \cos [\omega_1 t + \phi_1(t)]$$
$$s_2(t) = \rho_2(t) \cos [\omega_2 t + \phi_2(t)],$$

where ω_1 and ω_2 may be the same. The received signals on polarizations 1 and 2, respectively, can then be written as

$$r_1(t) = a_{11}\rho_1(t\text{-}\tau_{11}) \cos [\omega_1(t\text{-}\tau_{11}) + \phi_1(t\text{-}\tau_{11})]$$
$$+ a_{21}\rho_2(t\text{-}\tau_{21}) \cos [\omega_2(t\text{-}\tau_{21}) + \phi_2(t\text{-}\tau_{21})] \qquad (8\text{-}9)$$

$$r_2(t) = a_{22}\rho_2(t\text{-}\tau_{22}) \cos [\omega_2(t\text{-}\tau_{22}) + \phi_2(t\text{-}\tau_{22})]$$
$$+ a_{12}\rho_1(t\text{-}\tau_{12}) \cos [\omega_1(t\text{-}\tau_{12}) + \phi_1(t\text{-}\tau_{12})] \qquad (8\text{-}10)$$

The *XPD* is defined as

$$XPD = 20 \log (a_{11}/a_{12})$$

---

*The mechanisms causing depolarization, as well as its magnitude as a function of the relevant parameters, have been extensively discussed in the literature. A good summary can be found in [126].

for polarization 1, and as

$$XPD = 20 \log (a_{22}/a_{21})$$

for polarization 2. Note that what is actually of interest is the cross-polarization isolation, *XPI*, defined as

$$XPI = 20 \log (a_{11}/a_{21})$$

for polarization 1, and

$$XPI = 20 \log (a_{22}/a_{12})$$

for polarization 2. However, only the *XPD* is amenable to measurement, which can be done by alternately transmitting the two polarizations and observing the receiver outputs. Furthermore, it has been pointed out [126] that if the depolarization effects stem primarily from the medium the *XPD* and *XPI* are essentially identical. The pairs (a_{ij}, τ_{ij}) are functions of frequency, elevation angle, and other details characterizing the medium. For rain-induced depolarization it has been observed [126] that the *XPD* generally decreases (worsens) as the attenuation increases (a_{11} and a_{22} decrease).

If the delays τ_{ij} are small compared to the signal decorrelation time, then (8-9) and (8-10) reduce approximately to*

$$r_1(t) = a_{11}\rho_1(t) \cos [\omega_1 t + \phi_1(t) + \theta_{11}] + a_{21}\rho_2(t) \cos [\omega_2 t + \phi_2(t) + \theta_{21}]$$
$$r_2(t) = a_{22}\rho_2(t) \cos [\omega_2 t + \phi_2(t) + \theta_{22}] + a_{12}\rho_1(t) \cos [\omega_1 t + \phi_1(t) + \theta_{12}]$$

with an obvious identification for the θ_{ij}. These equations reflect the more standard description of the depolarization effect, namely in terms of attenuation and phase shift. This permits us a more compact representation in terms of complex coefficients $A_{ij} = a_{ij} \exp (j\theta_{ij})$. If we let $S_i(t)$, $R_i(t)$ stand for the complex envelopes, then the matrix notation

$$\begin{bmatrix} R_1(t) \\ R_2(t) \end{bmatrix} = \begin{bmatrix} A_{11} & A_{21} \\ A_{12} & A_{22} \end{bmatrix} \begin{bmatrix} S_1(t) \\ S_2(t) \end{bmatrix}$$

results, which has been commonly used in this context. Note, however, that it is an approximation in the sense earlier shown. Many polarization compensation schemes have been proposed and all of them, except [131], rely on this assumption.

---

*In fact, the data shown in [126] indicate that this should be almost always true even for very high rain rates and bandwidths.

As can be seen, then, compensation can take place by appropriately phase-shifting and attenuating each channel and feeding it back to the other to produce approximate cancellation. Note that the *XPD* cannot be too low otherwise the requirement for an uncorrupted auxiliary signal will not be met.

For certain sets of conditions the phase shifts (or their differences) are approximately predictable. In such cases, simplified compensation methods may be feasible [129]. However, it is difficult to say whether the loss of flexibility thereby incurred is justified by the cost savings presumably resulting from the simplified design.

Most of the depolarization compensation schemes that have been discussed or implemented depend on the transmission of "pilot" signals from the satellite to determine the values of the depolarization parameters. The possible frequency dependence of these parameters across a large bandwidth is a limitation of this approach (although results reported in [126] indicate that this should not be a major effect). Makino et al. [130] have proposed a method for TDMA which does not depend on pilots.

One problem not directly addressed in the preceding discussion is the fact that depolarization will also occur on the uplink, while the formulation presented so far implicitly assumes that the transmitted satellite signals are uncorrupted. One possible solution that has been mentioned [129, 132, 133] is to pre-compensate the uplink signal at the transmitting station. Another possibility, which appears to be the most general approach offerred to date, and is particularly suitable for digital signals, is to pass the received signals through a 2-input, 2-output tapped-delay-line structure with cross-coupling [131]. This structure is identical in form to a baseband equalizer for quadrature signals, and indeed is suggested by the fact that the crosstalk in the latter case leads to received signals which can be formulated in the manner of Eq. (8-9) and (8-10), except that the main signals in that instance are in phase quadrature as well as co-frequency. The equalizer structure, with sufficient taps, can also compensate for the fact that the uplink depolarized signals are distorted differently from the main signals from which they arose. In effect, the standard compensation circuits can be considered degenerate cases of this structure.

REFERENCES / PART THREE

1. Takahashi, K., "Collision Between Satellites in Stationary Orbits," *IEEE Trans. Aerosp. & Electronic Syst.,* Vol. AES-17, No. 4, pp. 591-596, July 1981 (Additions and Corrections, Sept. 1981).

2. Middleton, D., and A. D. Spaulding, "Deterministic Signals and Interference in FM Reception: The 'Instantaneous' Approach, with Undistorted Inputs," *Proc. IEEE,* Vol. 68, No. 2, Dec. 1980, pp. 1522-1536.

3. Furuya, T., A. Fujii, and K. Tezuka, "Impulse noise in FM receivers in the presence of adjacent channel interference and thermal noise," presented at Int. Conf. Communications, Philadelphia, PA, June 19-21, 1972.

4. Wachs, M., "Analysis of adjacent channel interference in a multicarrier FM communication system" *COMSAT Tech, Rev.,* Vol. 1, No. 1, pp. 139-170, Fall 1971.

5. Bennett, W. R., H. E. Curtis, and S. O. Rice, "Interchannel interference in FM and PM systems under noise loading conditions," *Bell Syst. Tech. J.,* Vol. 34, pp. 601-636, May 1955.

6. Abramson, N., "Bandwidth and spectra of phase and frequency-modulated waves," *IEEE Trans. Commun. Syst.,* Vol. CS-11, pp. 407-414, Dec. 1963.

7. Ferris, C. C., "Spectral characteristics of FDM-FM signals," *IEEE Trans. Commun. Technol.,* Vol. COM-16, pp. 233-238, Apr. 1968.

8. Lundquist, L., "Digital PM spectra by transform techniques," *Bell Syst. Tech. J.,* Vol. 48, No. 2, pp. 397-411, Feb. 1969.

9. Middleton, D., *An Introduction to Statistical Communication Theory,* New York: McGraw-Hill, 1960.

10. Rowe, H. E., *Signals and Noise In Communications Systems,* New York: Van Nostrand, 1965.

11. Rowe, H. E., and V. K. Prabhu, "Power spectrum of a digital, frequency-modulation signal," *Bell Syst. Tech. J.,* Vol. 54, No. 6, pp. 1095-1125, July-Aug. 1975.

12. Koh, E. K., and O. Shimbo, "Computation of interference into anle-modulated systems carrying multichannel telephone signals," *IEEE Trans. Commun.,* Vol. COM-24, pp. 259-263, Feb. 1976.

13. Boroditch, S. Y., "Calculating the permissible magnitude of radio interference in multi-channel radio relay systems," *Electrosviaz,* Vol. 1, No. 1, pp. 13-24, Jan. 1962.

14. Hamer, R., "Radio-frequency interference in multi-channel telephony FM radio systems," *Proc. Inst. Elect. Eng.,* Vol. 108B, pp. 75-89, January 1961.

15. Hayashi, S., "On the interference characteristics of the phase modulation receiver for multiplex transmission," Proc. IECE (Jap), Vol. 35, pp. 522-528, Nov. 1952.

16. Johns, P. B., "Graphical method for the determination of interference transfer factors between interfering frequency-modulated multichannel telephony systems," *Electron. Lett.,* Vol. 2, No. 3, pp. 84-86, Mar. 1966.

17. Johns, P. B., "Interference between terrestrial line-of-sight radio-relay systems and communication-satellite systems," *Electron. Lett.,* Vol. 2, No. 5, pp. 177-178, May 1966.

18. Medhurst, R. G., E. M. Hicks, and W. Grossett, "Distortion in frequency-division-multiplex FM systems due to an interfering carrier," *Proc. Inst. Elec. Eng.,* Vol. 105B, pp. 282-292, May 1958.

19. Medhurst, R. G., "RF Spectra and interfering carrier distortion in FM trunk radio systems with low modulation ratios," *IRE Trans. Commun. Syst.,* Vol. CS-9, pp. 107-115, June 1961.

20. Medhurst, R. G., "FM interfering carrier distortion: general formula," *Proc. Inst. Elec. Eng.,* Vol. 109B, pp. 149-150, Mar. 1962.

21. Medhurst, R. G., and J. H. Roberts, "Expected interference levels due to interactions between line-of-sight radio relay systems and broadband satellite systems," *Proc. Inst. Elec. Eng.*, Vol. 111, No. 3, pp. 519-523, Mar. 1964.

22. Pontano, B. A., J. C. Fuenzalida, and N. K. M. Chitre, "Interference into angle-modulated systems carrying multichannel telephony signals," *IEEE Trans. Commun.*, Vol. COM-21, pp. 714-727, June 1973.

23. Prabhu V. K., and L. H. Enloe, "Interchannel interference consideration in angle-modulated systems," *Bell Syst. Tech. J.*, Vol. 48, pp. 2333-2358, Sept. 1969.

24. Beyer, J. P., R. D. Briskman, D. W. Lipke, and K. Manning, "Some Orbital Spacing Considerations for Geostationary Communications Satellites, (Using the 4 and 6 GHz Frequency Bands)," *COMSAT Corp. Tech. Memo DS-1-68*, Oct. 25, 1968.

25. Jeruchim, M. C. and D. A. Kane, "Orbit/Spectrum Utilization Study" Vol. IV, General Electric Co., Phila., PA, Doc. No. 705D4293, 31 Dec. 1970. (Available through NTIS: PB-203-389.)

26. Prasanna, S., G. Sharp, and S. Das, "Interference Between FM Carriers and a Digitally Modulated SCPC Circuit," *IEEE Trans. Areosp. & Electronic Syst.*, Vol. AES-13, No. 4, pp. 427-430, July 1977.

27. Biernson, G., "A feedback-control model of human vision," *Proc. IEEE*, Vol. 54, pp. 858-872, June 1968.

28. Budrikis, Z. L. "Visual Fidelity criterion and modeling," *Proc. IEEE*, Vol. 60, pp. 771-779, July 1972.

29. Goodman, J. S., and D. E. Pearson, "Multidimensional Scaling of Multiply-Impaired Television Pictures," *IEEE Trans. Syst., Man, Cybern.*, Vol. SMC-9, June 1979, pp. 353-356.

30. Granrath, D. J., "The role of human visual models in image processing," *Proc. IEEE, Vol. 69, pp. 552-561, May 1981.*

31. Pearson, D. E., "Methods for scaling television picture quality: A survey," presented at Symp. Picture Bandwidth Compression, M.I.T., Cambridge, MA, Apr. 2-4, 1969.

32. Franks, L. E., "A model for the random video process," *Bell Syst. Tech. J.,* Vol. 45, pp. 609-630, Apr. 1966.

33. CCIR: XIVth Plenary Assembly, Kyoto, Japan, Vol. XI, *Report 405* (Subjective Assessment of the Quality of Television Pictures), Geneva, Switzerland, 1978.

34. CCIR: XIVth Plenary Assembly, Kyoto, Japan, Vol. XI, *Report 634* (Broadcasting-satellite service (television): subjectively measured interference protection ratios for planning television broadcasting systems), Geneva, 1978.

35. CCIR: XIVth Plenary Assembly, Kyoto, Japan, Vol. IX, *Report 449-1* (Measured interference into frequency-modulation television systems using frequencies shared within systems in the fixed satellite service between these systems and terrestrial systems), Geneva, 1978.

36. Sampaio-Nato, R., J. P. A. Albuquerque, "Intermodulation Effects in the Transmission of Voice-Activated SCPC/FM Carriers through a nonlinear repeater," *IEEE Trans. Commun.,* Vol. COM-29, pp. 1537-1547, October 1981.

37. Shimbo, O., "Effects of Intermodulation, AM-PM Conversion, and Additive Noise in Multicarrier TWT Systems," *Proc. IEEE,* Vol. 59, No. 2, Feb. 1971.

38. Berman, A. L., and C. E. Mahle, "Nonlinear Phase Shift in Traveling-Wave Tubes as Applied to Multiple Access Communications Satellites," *IEEE Trans. Com. Tech.,* Vol. COM-18, No. 1, Feb. 1970.

39. Blachman, N. M., "Detectors Bandpass Nonlinearities, and their Optimization: Inversion of the Chebyshev Transform," *IEEE Trans. Inform Theory,* Vol. IT-17, No. 4, July 1971, pp. 398-404.

40. Jeruchim, M. C., "Interference in angle-modulated systems with predetection filtering," *IEEE Trans. Commun. Technol.,* Vol. COM-19, pp. 723-726, Oct. 1971.

41. Khatri, R. K., and J. E. Wilkees, "Convolution noise and distortion in FDM/FM systems," presented at Int. Conf. Communications, Philadelphia, PA, June 19-21, 1972.

42. Stojanović, Z. D., M. L. Dukić, and I. S. Stojanović, "A New Demodulation Method Improving FM System Interference Immunity," *IEEE Trans. Commun.*, Vol. COM-29, No. 7, pp. 1001-1011, July 1981.

43. Eng, K. Y., and T. E. Stern, "The Order-and-Type Prediction Problem Arising from Passive Intermodulation Interference in Communications Satellites," *IEEE Trans. Commun.*, Vol. COM-29, No. 5, pp. 549-555, May 1981.

44. Eng. K. Y. and O.-C. Yue, "High-Order Intermodulation Effects in Digital Satellite Channels," *IEEE Trans. Aerospace & Elec. Syst.*, Vol. AES-17, No. 3, pp. 438-445, May 1981.

45. Chang, P. Y., and R. J. Fang, "Intermodulation in Memoryless Nonlinear Amplifiers Accessed by FM and/or PSK Signals," *COMSAT Tech. Rev.*, Vol. 1, No. 11, Spring 1978, pp. 89-139.

46. Chitre, N. K. M., and J. C. Fuenzalida, "Baseband distortion caused by intermodulation in multicarrier FM systems," *COMSAT Tech. Rev.*, Vol. 2, No. 1, pp. 147-172, Spring 1972.

47. Fuenzalida, J. C., O. Shimbo, and W. L. Cook, "Time-Domain Analysis of Intermodulation Effects Caused by Nonlinear Amplifiers," COMSAT Tech. Rev., Vol. 3, No. 1, Spring, 1973, pp. 89-143.

48. Kahverad, M., "Multiple FM/FDM Carriers Through Nonlinear Amplifiers," *IEEE Trans. Commun.*, Vol. COM-29, No. 5, pp. 751-756, May 1981.

49. Westcott, R. J., "Investigation of multiple FM/FDM carriers through a satellite TWT operating near to saturation," *Proc. Inst. Elec. Eng.*, Vol. 114, No. 6, pp. 726-740, June 1967.

50. Jeruchim, M. C., "Digital computer simulation of satellite quadrature communication systems," presented at Nat. Telecommunications Conf., New Orleans, LA, December 1975.

51.   Devieux, C. L., and M. E. Jones, "A Practical Optimization Approach for QPSK/TDMA Satellite Channel Filtering," *IEEE Trans. Commun.,* Vol. COM-29, No. 5, pp. 556-566, May 1981.

52.   Fang, R. J. F., "Quaternary Transmission over Satellite Channels with Cascaded Nonlinear Elements and Adjacent Channel Interference," *IEEE Trans. Commun.,* Vol. COM-29, No. 5, pp. 567-581, May 1981.

53.   Goldman, J., "Multiple error-performance of PSK systems with cochannel interference and noise," *IEEE Trans. Commun., Technol.,* Vol. COM-19, pp. 420-430, Aug. 1971.

54.   Koerner, M. A., "Effect of interference on a binary communication channel using known signals," Jet Propulsion Laboratory, Pasadena, CA, Tech. Rep. 32-1281, Dec. 1, 1968.

55.   Prabhu, V. K., "Error rate consideration for coherent phase-shift keyed systems with co-channel interference," *Bell Syst., Tech. J.,* Vol. 48, No. 3, pp. 743-767, Mar. 1969.

56.   Rosenbaum, A. S., "PSK error performance in Gaussian noise and interference," *Bell Syst. Tech. J.,* Vol. 48, No. 2, pp. 413-442, Feb. 1969.

57.   Rosenbaum, A. S., "Binary PSK error probabilities with multiple cochannel interferences," *IEEE Trans. Commun. Technol.,* Vol. COM-18, pp. 241-253, June 1970.

58.   Shimbo, O., and R. Fang, "Effects of cochannel interference and Gaussian noise in M-ary PSK systems," COMSAT Tech. Rev., Vol. 3, No. 1, pp. 183-207, Spring 1973.

59.   Benedetto, S., E. Biglieri, and V. Castellani, "Combined effects of intersymbol, interchannel, and co-channel interference in M-ary CPSK systems," *IEEE Trans. Commun.,* Vol. COM-21, pp. 997-1008, Sept. 1973.

60.   Celebiler, M. I., and G. M. Coupé, "Effects of Termal Noise, Filtering and Co-channel Interference of the Probability of Error in Binary Coherent PSK Systems," *IEEE Trans. Commun.,* Vol. COM-26, No. 2, pp. 257-267, Feb. 1978.

61. Cruz, J. R., and R. S. Simpson, "Cochannel and Intersymbol Interference in Quadrature-Carrier Modulation Systems," *IEEE Trans. Commun.*, Vol. COM-29, No. 3, pp. 285-297, March 1981.

62. Fang, R., and O. Shimbo, "Unified analysis of a class of digital systems in additive noise and interference," *IEEE Trans. Commun.*, Vol. COM-21, pp. 1075-1091, Oct. 1973.

63. Kennedy, D. J., and O. Shimbo, "Cochannel Interference in Nonlinear QPSK Satellite Systems," *IEEE Trans. Commun.*, Vol. COM-29, No. 5, pp. 582-592, May 1981.

64. Matthews, N. A., and A. H. Aghvami, "Binary and Quaternary CPSK Transmission Through Nonlinear Channels in Additive Guassian Noise and Cochannel Interference," *Proc. IEE* (Part F), Vol. 128, pp. 96-103, April 1981.

65. Oka, I., S. Kabasawa, N. Morinaga, and T. Namekawa, "Interference Immunity Effect in CPSK Systems with Hard-Limiting Transponders," *IEEE Trans. Aerosp. & Electronic Syst.*, Vol. 17, No. 1, pp. 93-99, Jan. 1981 (Corrections and Addendum, Vol. 17, March 1981).

66. Huang, T.-C., J. K. Omura, and W. C. Lindsey, "Analysis of Coherent Satellite Communication Systems in the Presence of Interference and Noise," IEEE Trans. Commun., Vol. COM-29, No. 5, pp. 593-604, May 1981.

67. Chang, P., R. J. Fang, and M. E. Jones, "Performance over Cascaded and Band-Limited Nonlinear Satellite Channels in the Presence of Interference," Vol. 3, pp. 47.3.1-37.3.10, *Proc. Internat. Conf. Commun.*, (ICC '81), Denver, CO, June 14-18, 1981.

68. Kabasawa, S., N. Morinaga, and T. Namekawa, "M-ary CPSK Detection with Noisy Reference and Interferences," *IEEE Trans. Aerosp. & Electronic Syst.*, Vol. 16, No. 5, pp. 712-719, Sept. 1980.

69. Wachs, M. R., and D. E. Weinreich, "A laboratory study of the effects of CW interference on digital transmission over non-

linear satellite channels," in *Proc. INTELSAT 3rd Int. Conf. on Digital Communications* (Kyoto, Japan), pp. 65-72, Nov. 11-13, 1975.

70. Weinreich, D. E., and M. R. Wachs, "A laboratory simulation of multiple unmodulated interference sources on digital satellite channels," presented at Int. Conf. Communications, Philadephia, PA, June 1976.

71. Kliger, I. E., and C. F. Olenberger, "Phase-lock loop jump phenomenon in the presence of two signals," *IEEE Trans. Aerosp. Electron. Syst.*, Vol. AES-12, pp. 55-64, Jan. 1976.

72. Levitt, B. K. "Carrier Tracking-Loop Performance in the Presence of Strong CW Interference," *IEEE Trans. Commun.*, Vol. COM-29, No. 6, pp. 911-916, June 1981.

73. Ziemer, R. E., "Perturbation analysis of the effect of CW interference in Costas loops," in *Proc. Nat. Telecommunications Conf.* (Houston, TX), pp. 20G-1-20G-6, Dec. 1972.

74. Rosenbaum, A. S., "Error performance of multiphase DPSK with noise and interference," *IEEE Trans. Commun. Technol.*, Vol. COM-18, pp. 821-824, Dec. 1970.

75. Kostić, I. M., "Error rates of DCPSK Signals in hard-limited multilink system with cochannel interference and noise," *IEEE Trans. Commun.*, Vol. COM-30, pp. 222-230, Jan. 1982.

76. Bodner, H. A., "Error rate considerations of single sideband amplitude modulation systems with cochannel interference," presented at Int. Conf. Communications, June 14-16, 1971, Montreal, Canada.

77. Cohen, S. A., "Interference effects of pseudo-random frequency hopping signals," *IEEE Trans. Aero sp. Electron. Syst.*, Vol. AES-7, pp. 279-287, Mar. 1971.

78. Kullstam, P. A., "Spread Spectrum Performance Analysis in Arbitrary Interference," *IEEE Trans. Commun.*, Vol. COM-25, No. 8, pp. 848-853, Aug. 1977.

79. Massaro, M. J., "Error performance of M-ray nonocoherent FSK in the presence of CW tone interference," *IEEE Trans. Commun.*, Vol. COM-23, pp. 1367-1369, Nov. 1975.

80. Prabhu, V. K., "Cochannel Interference Immunity of High Capacity QAM," *Conf. Red.*, Vol. 4, pp. 68.1.1-68.8.4, Intern. Conf. Commun., (ICC 81) Denver, CO, June 14-18, 1981.

81. Wang L., "Error probability of a binary noncoherent FSK system in the presence of two CW tone interferers," *IEEE Trans. Commun.*, Vol. COM-22, pp. 1948-1949, Dec. 1974.

82. Spilker, J. J. "Digital Communications by Satellite," Englewood Cliff, N. J., Prentice-Hall, 1977.

83. Benedetto, S., G. Devincentiis, and A. Luvison, "Error probability in the presence of intersymbol interference and additive noise for multilevel digital signals," *IEEE Trans. Commun.*, Vol. COM-21, pp. 181-190, Mar. 1973.

84. Shimbo, O., R. J. Fang, and M. Celebiler, "Performance of M-ary PSK systems in Gaussian noise and intersymbol interference," *IEEE Trans. Information Theory*, Vol. IT-19, pp. 44-58, Jan. 1973.

85. Bennett, W. R. "Distribution of the Sum of Randomly Phased Components," *Quart. App. Math.*, April 1948.

86. Jeruchim, M. C., and F. E. Lilley, "Spacing limitations of geostationary satellites using multilevel PSK signals," *IEEE Trans. Commun.*, Vol. COM-20, Oct. 1972.

87. Kostić, I. M., "Fourier-Series Representation for Phase Density of Sum of Signal, Noise, and Interference," *Electronic Letters,* Vol. 15, pp. 541-542 (errata, p. 701), Aug. 1979.

87a. J. Goldman, "Statistical properties of a sum of sinusoids and Gaussian noise and its generalization to higher dimensions," *Bell Syst. Tech. J.,* Vol. 53, No. 4, pp. 557-580, Apr. 1974.

87b. Bendetto, S., G. De Vincentiis, and A. Luvison, "Application of Gauss Quadrature Rules to Digital Communication Problems," *IEEE Trans. Commun.,* Vol. COM-21, No. 10, pp. 1159-1165, Oct. 1973.

88. Prabhu, V. K., "Bandwidth occurancy in PSK systems," *IEEE Trans. Commun.,* Vol. COM-24, pp. 456-462, Apr. 1976.

89.   Aein, J. M., "On the effects of undesired signal interference to a coherent digital carrier," Inst. for Defense Analyses, Paper P-812, Feb. 1972.

90.   Aein, J. M., and R. D. Turner, "Effect of cochannel interference on CPSK carriers," *IEEE Trans. Commun.,* Vol. COM-21, pp. 783-790, July 1973.

91.   Frenkel, G., "The Error Rate Caused by Mutual Interference Between Digital Transmission," unpublished notes.

92.   Glave, F. E., and A. S. Rosenbaum, "An upper bound analysis, for coherent phase-shift keying with cochannel, adjacent-channel, and intersymbol interference," *IEEE Trans. Commun.,* Vol. COM-23, pp. 586-597, June 1975.

93.   McLane, P. J., "Error rate lower bounds for digital communication with multiple interference," *IEEE Trans. Commun.,* Vol. COM-23, pp. 539-543, May 1975.

94.   McLane, P. J., "On Multiple Interference Lower Bounds," *Proc. IEEE,* Vol. 65, No. 2, pp. 273-275, Feb. 1977.

95.   McLane, P. J., and D. F. Boyko, "Error Bounds for Digital PAM Transmission with Phase Jitter and Intersymbol or Cochannel Interfernce," IEEE Trans. Commun., Vol. COM-25, No. 5, pp. 536-541, May 1977.

96.   McLane, P. J., "Additions and Corrections to 'Error Rate Lower Bounds for Digital Communication with Multiple Interferences'," *IEEE Trans. Commun.,* Vol. COM-26, no. 1, pp. 178-179, Jan. 1978.

97.   Prabhu, V. K., "Error-probability upper bound for coherently detected PSK signal with cochannel interference," *Electron. Lett.,* Vol. 5, No. 16, pp. 383-386, Aug. 1969.

98.   Rosenbaum, A. S., and F. E. Glave, "An error-probability upper bound for coherent phase-shift keying with peak limited interference," *IEEE Trans. Commun.,* Vol. COM-22, pp. 6-16, Jan. 1974.

99.   Spowage, K. E. A., "An Upper Bound on the Error Probability for Differential Phase-Shift Keying in the Presence of

Additive Gaussian White Noise and Peak-Limited Interference," *IEEE Trans. Commun.*, Vol. COM-24, No. 11, pp. 1276-1279, Nov. 1976.

100. Tobin, R. M., and K. Yao, "Upper and Lower Bounds for Coherent Phase-Shift-Keyed (CPSK) Systems with Cochannel Interference," *IEEE Trans. Commun.*, Vol. COM-25, No. 2, pp. 281-287, Feb. 1977.

100a. Krishnamurthy, J., "Bound on probability of error due to cochannel interference," *IEEE Trans. Aerosp. Electron. Syst.*, Vol. AES-11, pp.1373-1377, Nov. 1975.

101. CCIR: XIVth Plenary Assembly, Kyoto, Japan, Vol. IX, *Report 388-3* (Methods for determining interference in terrestrial radio-relay systems and systems in the fixed satellite service), Geneva, 1978.

102. Yao, K., and E. Biglieri, "Multidimensional Moment Error Bounds for Digital Communication Systems," *IEEE Trans. Inform. Theory,* Vol. IT-26, No. 4, pp. 454-464, July 1980.

103. M. R. Wachs and D. Kurjan, "Interference into the SPADE system by a cochannel TV signal," *COMSAT Tech. Rev.,* vol. 6, No. 2, Fall 1976.

104. Wachs, M. "Energy Dispersal as Applied to Interference Reduction," *Conf. Rcd.,* Intern. Conf. Commun. (ICC '77), Chicago, IL, June 12-15, 1977.

105. Yam, S. E. "New TV Energy Dispersal Techniques for Interference Reduction," *COMSAT Tech. Rev.,* Vol. 10, No. 1, Spring 1980, pp. 103-150.

106. Steinberger, M. L., P. Balaban, and K. S. Shanmugam, "On the Effect of Uplink Noise on a Nonlinear Digital Satellite Channel," *Conf. Rcd.,* Vol. 1, pp. 20.2.1-20.2.5, Interm. Conf. Commun. (ICC 81), Denver, CO, June 14-18, 1981.

107. CCIR: XIVth Plenary Assembly, Kyoto, Japan, Vol. IV, *Rec. 522* (Allowable bit error rates at the output of the hypothetical reference circuit for systems in the fixed satellite service using pulse-code modulation for telephony), Geneva, 1978.

108.   CCIR: XIVth Plenary Assembly, Kyoto, Japan, Vol. IV, *Rec. 523* (Maximum permissible levels of interference in a geostationary satellite network in the fixed satellite service using 8-bit PCM encoded telephony, caused by other networks of this service), Geneva, 1978.

109.   Gould, R. G., and C. Schmitt, "Interference Reduction Techniques for Satellite Earth Stations," 2nd Symposium and Technical Exhibition on Electromagnetic Compatibility, Montreux, Switzerland, June 28-30, 1977.

110.   Horton, E. D., "An Adaptive Co-Channel Interference Suppression System to Suppress High Level Interference in Satellite Communication Earth Terminals," *Conf. Rcd.*, Vol. I, pp. 13.4.1-13.4.5, Nat'l Telecommun. Conf., Dallas, TX, Nov.-Dec., 1976.

111.   Kowalski, A. M., "Adaptive Filter for Interference Suppression," Conf. Rcd., Vol. I, pp. 04.5.1-04.5.6, Nat'l Telecommun. Conf., Los Angeles, CA, Dec. 1977.

112.   Lubell, P. D., and F. D., Rebhun, "Suppression of Co-Channel Interference with Adaptive Cancellation Devices at Communications Satellite Earth Stations," *Proc. 1977 Intern. Conf. Commun.* (ICC), Vol. 3, Chicago, June 12-15, 1977.

113.   Pontano, B. A., "A Method of Baseband Interference Cancellation for Angle-Modulated Carriers," 1980 IEEE Electromagnetic Compatiblity Conf. Baltimore, MD, Oct. 1980.

114.   Pontano, B. A., "Methods of Interference Cancellation for Improved Orbit and Spectrum Utilization," *Conf. Rcd.,* Nat'l Telecommun. Conf., Houston, TX, Dec. 1980.

115.   Sauter, W. A., "Interference Cancellation Systems for Colocated Receivers and Sources," *Conf. Rcd.,* Vol. I, pp. 04.1.1-0.4.1.5, Nat'l Telecommun. Conf., Los Angeles, CA, Dec. 1977.

116.   Harris, A. B., B. Claydon, and K. M. Keen, "Reducing the Sidelobes of Earth-Station Antennas," *Conf. Rcd.,* Vol. I, pp. 6.6.1-6.6.5, Intern. Conf. Commun., (ICC), Boston, MA, June 10-14, 1979.

117. Janky, J. M., B. B. Lusignan, L.-S. Lee, E. C. Ha, and E. E. Reinhart, "New Sidelobe Envelopes for Small Aperture Earth Stations," *IEEE Trans. Broadcast.*, Vol. BC-22, June 1976, pp. 39-44.

118. Mailloux, R. J., "Phased Array Theory and Technology," *Proc. IEEE,* Vol. 70, No. 3, March 1982, pp. 246-291.

119. Takao, K., and K. Korniyama, "An Adaptive Antenna for Rejection of Wideband Interference," *IEEE Trans. Aerosp. and Electronic Syst.,* Vol. AES-16, July 1980, pp. 452-459.

120. Torrieri, D. J. *Principles of Military Communication Systems,* Dedham, MA, Artech House, 1981, (Chapter 5, Adaptive Antenna Systems).

121. Widrow, B., P. E. Mantey, L. J. Griffiths, and B. B. Goode, "Adaptive Antenna Systems," *Proc. IEEE,* Vol. 55, Dec. 1967, pp. 2143-2159.

122. Afifi, M., and P. Foldes, "Optimum Contiguous Multibeam Antenna Coverage," *IEEE AP-S Intern Symp. Digest, Quebec, Canada, June 1980.*

123. Afifi, M., and P. Foldes, "Continguous Multibeam Regional Coveragfe with Low Depolarization and Interference Levels," *IEEE AP-S Symposium,* Albuquerque, NM, May 1982.

124. Wilmut, M. J., and L. L. Campbell, "Signal detection in the presence of cochannel interference and noise," *IEEE Trans. Commun.,* Vol. COM-20, pp. 1153-1158, Dec. 1972.

125. Goldman, J., "Detection in the presence of spherically symmetric random vectors," *IEEE Trans. Inform. Theory,* Vol. IT-22, pp. 52-59, Jan. 1976.

126. Ippolito, L. J., "Radio Propagation for Space Communications Systems," *Proc. IEEE,* Vol. 69, No. 6, June 1981, pp. 697-727.

127. Baird, C. A., and G. G. Rassweiler, "Measured Results of Polarization Crosstalk cancellation Using LMS Control," *Conf. Rcd.,* Nat'l Telecommun. Conf., Vol. I, pp. 04.3.1-04.3.6, Los Angeles, CA, Dec. 1977.

128. DiFonzo, D. F., W. S. Trachtman, and A. E. Williams, "Adaptive Polarization Control for Satellite Frequency Reuse Systems," *COMSAT Tech. Rev.,* Vol. 6, No. 2, Fall 1976, pp. 253-283.

129. Lee, L.-S., "The Feasibility of Two One-Parameter Polarization Control Methods in Satellite Communications," *IEEE Trans. Commun.,* Vol. COM-29, May 1981, pp. 735-743.

130. Makimo, H., Y. Orino, S. Orui, H. Fuketa, and T. Inoue, "A New Adaptive Control System for Compensating Cross-Polarization-Couplings on the Up and/or Down Path in Frequency Reuse Satellite Communication Systems," *Proc. 8th AIAA Commun. Satellite Systems Conf.,* Orlando, FL, April, 20-24, 1980.

131. Namiki, J., and S. Takahara, "Adaptive Receiver for Cross-Polarized Digital Transmission," Vol. 3, pp. 46.3.1-46.3.5, *Proc. Internat. Conf. Commun.,* (ICC '81), Denver, CO, June 14-18, 1981.

132. Nouri, M., and M. R. Braine, "An Adaptive Interference Control System for Earth-Satellite Links Above 10 GHz," *Marconi Review,* Vol. XLII, No. 216, First Quarter, 1980.

133. Persinger, R. R., R. W. Gruner, J. E. Effland, and D. F. DiFonzo, "Operational Measurements of a 4/6 GHz Adaptive Polarization Compensation Network Employing Up/Down Link Correlation Algorithms," IEE Conference, London, April 1981.

134. Tseing, F. T., and L. S. Lee, "Accurate Detection of Satellite Beacon Polarization States Using Cascaded Heterodined Phase-Locked Loops," *IEEE Trans. Commun.,* Vol. COM-30, No. 1, Jan. 1982, pp. 283-293.

135. Weber, III, W. J., "A Decision-Directed Network for Dual-Polarization Crosstalk," Vol. 3, pp., 40.4.1-40.4.7, *Proc. Internat. Conf. Commun.,* (ICC '79), Boston, MA, June 10-14, 1979.

APPENDICES

---------------------------**A**---------------------------

Some Signal-to-Noise Ratio Relationships for Analog Signals

In this appendix we set down for convenience some standard formulas relating the signal-to-noise and carrier-to-noise ratios to basic design parameters. In particular, we shall obtain expressions for the receiver transfer characteristic (RTC) in a thermal noise environment for FDM/FM multichannel telephony, FM television, and FM single channel per carrier (SCPC) voice transmission. We shall introduce the concepts of demodulator transfer characteristics (DTC) and channelizing transfer characteristic (CTC) into which the RTC can be decomposed, and which are necessary to obtain the end-to-end performance of a satellite link with a processing satellite. Finally, for ready reference we shall express the carrier-to-noise ratio in terms of the standard design parameters.

A.1 Receiver Transfer Characteristic for FDM/FM Multichannel Telephony

The process of recovering a particular telephone channel from a modulated carrier can be thought of as occurring in two distinct steps: (1) demodulation of the carrier, and (2) processing of the demodulated multichannel baseband and extraction therefrom of the wanted channels [1]. These steps yield respectively, the DTC and the CTC.

The well-known equation for output SNR in an FM system is

$$\frac{S}{N} = 3\beta^2 \frac{S}{2N_o f_m} \tag{A-1}$$

where

β = peak modulation index
S = signal (carrier) power; (the symbol C will be used)
N_o = receiver noise spectral density
f_m = top baseband frequency

An equivalent from better suited to our purpose is

$$\frac{S}{N} = 6M_o^2 \ \frac{S}{N_oB} \ \frac{B}{2f_m} \qquad \text{(A-2)}$$

where

M_o = rms (multichannel) modulation index
B = occupied radio-frequency bandwith

By definition the CNR is

$$(C/N) \triangleq (S/N_oB) \qquad \text{(A-3)}$$

and using Carson's rule, i.e.,

$$B = 2f_m \ (\alpha M_o + 1) \qquad \text{(A-4)}$$

where α = peak-to-rms ratio of the multichannel baseband, we find upon substitution of (A-3) and (A-4) into (A-2), that

$$\frac{S}{N} = 6M_o^2 \ (\alpha M_o + 1) \ (C/N) \qquad \text{(A-5)}$$

The *demodulator transfer characteristic* (DTC) is defined as

$$DTC \triangleq 6M_o^2 \ (\alpha M_o + 1) \qquad \text{(A-6)}$$

Equation (A-5) represents the SNR for the entire multi-channel baseband. In order to find the SNR for a particular channel, we need to account for

(a) the ratio of the noise in the entire baseband to that in the channel of interest;

(b) the ratio of signal power in the channel of interest to that in the entire baseband;

(c) the effect of any filtering and noise weighting that may take place.

It is well-known that the demodulator output noise spectral density is of the form

$$G_o(f) = kf^2$$

where $k = k_r N_o/C$ and k_r is a receiver constant. Hence, the total output baseband noise is

$$\int_0^{fm} kf^2 df = \frac{k}{3} f_m^3$$

while that in a particular telephone channel of width b Hz and centered at f_{ch} is given by

$$\int_{f_{ch}-b/2}^{f_{ch}+b/2} kf^2 df = \frac{k}{3} \left[(f_{ch} + b/2)^3 - (f_{ch} - b/2)^3 \right] = kbf_{ch}^2 \left[1 + \frac{b^2}{12 f_{ch}^2} \right]$$

which, for $f_{ch} \gg b$, the case of usual interest, reduces to bf_{ch}^2. Hence, factor (a) above is

$$\frac{baseband \ noise}{channel \ noise} = \frac{f_m^3}{3bf_{ch}^2} \tag{A-7}$$

To obtain the second factor, it is necessary to recall [2] that the multichannel baseband power at a point of zero relative level is given by

$$10 \log S_n = \begin{cases} -15 + 10 \log n, \ dBm0, \ n \geq 240 \ channels \\ \\ -1 + \ 4 \log n, \ dBm0, \ 12 \leq n < 240 \ channels \end{cases} \tag{A-8}$$

Now, the "signal" power in a channel is often defined in two different (but relatable) ways. One common practice is to define a "test-tone," whose power is conventionally defined as

$$10 \log S_{tt} = 0 \ dBm0 \tag{A-9}$$

The second method is to define the power as that existing in the channel of interest when P_n is uniformly distributed over the baseband. Thus, this power is

$$S_{ch} = S_n \ (b/F), \tag{A-10}$$

where F is the baseband bandwith, and this definition lead to the SNR measure called the noise-power ratio (NPRF).*

---

*Note that F is the difference between the highest and lowest baseband frequency and is not necessarily equal to f_m.

For the moment we pursue only the test-tone SNR derivation. Thus, factor (b) above is given by

$$\frac{test\text{-}tone\ power}{baseband\ power} = (5_n)^{-1}$$

$$= 31.6/n \ , \ n \geq 240 \ channels \qquad (A\text{-}11)$$

$$= 1.26/n^{0.4}, \ 12 \leq n < 240 \ channels$$

The third factor, (c), listed above we account for symbolically be defining

$p(f_{ch})$ = the pre-emphasis/de-emphasis improvement at f_{ch}

w = psophometric weighting factor

Combining the previous factors we obtain the *channelizing transfer characteristic* [1], or the CTC, as

$$CTC = \left(\frac{f_m^3}{3\ b f_{ch}^2}\right)\ S_n^{-1} p\ (f_{ch}) \cdot w \qquad (A\text{-}12)$$

Typically, we can write $f_m = \beta_n$, where β depends on the channel arrangements. Thus,

$$CTC = \begin{cases} \dfrac{1}{3} \left(\dfrac{f_m}{f_{ch}}\right)^2 31.6\ \dfrac{Y}{b} \cdot p\ (f_{ch}) \cdot w, \ n \geq 240 \\[3mm] \dfrac{1}{3} \left(\dfrac{f_m}{f_{ch}}\right)^2 1.26\ \dfrac{Y}{b} n^{0.6} \cdot p\ (f_{ch}) \cdot w, \ 12 \leq n < 240 \end{cases} \qquad (A\text{-}13)$$

Hence, the receiver transfer characteristic (relative to a standard test-tone) is

$$RTC = DTC\ x\ CTC$$

$$= 63.2\ M_0^2 (\alpha\ M_0 + 1) \left(\frac{f_m}{f_{ch}}\right)^2 \cdot w, \ 12 \leq n < 240$$

$$\qquad\qquad\qquad\qquad\qquad\qquad\qquad\qquad\qquad (A\text{-}14)$$

$$= 2.52\ n^{0.6}\ M_0^2\ (\alpha\ M_0^2 + 1) \left(\frac{f_m}{f_{ch}}\right)^2 \frac{Y}{b} p\ (f_{ch}) \cdot w, \ 12 \leq n < 240$$

and the test-tone SNR is

$$\left(\frac{S}{N}\right)_{tt} = R_t \left(\frac{C}{N}\right)$$

where $R_t \triangleq RTC$.

We are normally interested in the worst channel, which is usually the top channel, so that $f_{ch} \cong f_m$. Further, one standardly takes $\beta = 4200$, $b = 3100$, $w = 1.78$ (2.5 db) and $p(f_m) = 2.5$ (4 db). We then get the simplified form

$$R_t = \begin{cases} 382.75 \, M_o^2 \, (\alpha \, M_o + 1) \quad , & n \geq 240 \\ 15.24 \, n^{0.6} \, M_o^2 \, (\alpha \, M_o + 1), & 12 \leq n < 240 \end{cases} \tag{A-15}$$

which is used elsewhere in the text with the coefficients rounded off for convenience.

Since the test-tone power is by definition $1\,mW$, it is evident that the noise in the channel, measured in picowatts, is simply

$$N = 10^9 \, / \, (S/N)_{tt} \tag{A-16}$$

which is the form we have most frequently used.

The NPR is the channel SNR using (A-10) as the definition of signal power, rather than (A-9). Since $S_{tt} = 1\ mw$, the ratio

$$S_{ch}/S_{tt} = S_n(b/F)$$

is the factor necessary to convert from SNR to NPR. Denoting with a prime the CTC appropriate to the latter, we get from (A-12)

$$CTC' = \left(\frac{f_m^3}{3 \, bf_{ch}^2} \right) \frac{b}{F} \, p(f_{ch}) \, w \tag{A-17}$$

We can write

$$F = f_m - f_1 = f_m \, (1 - \varepsilon)$$

where f_1 is the lowest baseband frequency and $\varepsilon = f_1/f_m$. Thus, (A-17) becomes

$$CTC' = \frac{1}{3} \left(\frac{f_m}{f_{ch}} \right)^2 \frac{1}{(1 - \varepsilon)} p(f_{ch}) \, w$$

Assuming, as before that the worst channel is at $f_{ch} = f_m$ we find

$$\frac{NPR}{(C/N)} = \frac{8.93}{(1 - \varepsilon)} M_0^2 (\alpha \, M_0 + 1) \tag{A-18}$$

which, it may be noted, is independent of the number of channels.

Still another form for the SNR is sometimes used. This calls upon the idea of the test-tone deviation (or modulation index). Specifically, combining (A-2) and (A-12) we have

$$\left(\frac{S}{N}\right)_{tt} = b\, M_0^2\, (C/N)\, \frac{B}{2\, f_m}\, \frac{f_m^3}{3\, b\, f_{ch}^2} S_n^{-1}\, p\, (f_{ch}) \cdot w$$

Making the usual assumption $f_{ch} = f_m$, this reduces to

$$\left(\frac{S}{N}\right)_{tt} = (C/N)\, (B/b)\, \frac{\Delta f_{rms}^2\, S_n^{-1}}{f_m^2} p \cdot w$$

since $M_0 = \Delta f_{rms}/f_m$, and setting $p\, (f_m) \triangleq p$. Since S_n^{-1} is the ratio of test-tone power to baseband power, we can define

$$\Delta f_{tt}^2 = \Delta f_{rms}^2\, S_n^{-1}$$

as the *rms* test-tone deviation, and

$$M_{tt} = \Delta f_{tt}/f_m$$

as the rms test-tone modulation index. Hence,

$$\left(\frac{S}{N}\right)_{tt} = (C/N)\, M_{tt\,(B/b)\,p}^2 \cdot w \tag{A-19}$$

which is the alternative form we are seeking.

A.2 Receiver Transfer Characteristic for FM Television

In a single channel system, as with television, there is no channelizing transfer characteristic, except for pre-emphasis and weighting, since the channel itself forms the entire baseband. Thus,

$$CTC = I\, (p,w)$$

where p and w represent, respectively, the application of pre-emphasis and noise weighting. The pre-emphasis/weighting improvement, I, is not the product of the individual factors acting alone. Table A-1 shows I for noise weighting and noise weighting + pre-emphasis for several video standards, where the pre-emphasis/de-emphasis characteristics recommended by the CCIR [3] for radio-relay systems has been used.

TABLE A-1

System	Nominal Upper Frequency (MHz)	Number of Lines	I(dB) Noise-Weighting Only	I(dB) Weighting Plus Pre-emphasis
M (U.S. & Canada)	4	525	10.2	12.5
B, C, G, H	5	625	16.3	16.4
D, K, L	6	625	17.8	18.25
E	10	819	16.3	20.65

The SNR for television is defined as the ratio of the peak-to-peak picture signal to rms weighted noise. From this definition it follows readily that the RTC for an FM television signal is given by

$$R_t = 12M^2 (M + 1) I (p,w) \tag{A-20}$$

where

$M = \Delta f/f_v$
$2\Delta f$ = peak-to-peak frequency deviation
f_v = nominal upper video bandwith,

and in obtaining Carson's rule bandwith

$$B = 2f_v (M + 1)$$

has been used.

A.3 Receiver Transfer Characteristic for SCPC-FM Signals

For SCPC-FM transmission, the standard reference waveform is taken to be a test-tone of frequency f_m (either 1000 Hz or 800 Hz), with peak voltage T_{max} corresponding to a peak deviation Δf. The input to the modulator is peak-limited so that input voltages larger than T_{max} are clipped and the calibration is such that under actual (speech) loading, only a specific (small fraction) of talkers suffer clipping of peaks higher than a certain level above their rms value. With $\beta = \Delta f/f_m$, the test-tone-to-noise ratio well above threshold is given by (A-1), which with a slight manipulation can be written as

$$\left(\frac{S}{N}\right)_{tt} = 3\beta^2 \left(\frac{B}{2 f_m}\right) (C/N)$$

Denoting by I_w, I_p, and I_c the noise-weighting, pre-emphasis, and compander improvements, respectively, we can express the RTC as

$$R_t = 3\beta^2 (B/2 f_m) I_w I_p I_c \tag{A-21}$$

The value and nature of these improvements has been extensively discussed in the literature [4,5].

A.4 Signal-to-Noise Ratio for a Processing Satellite

In this book we deal exclusively with satellites having translating repeaters. It is of interest, however, to see in what way our considerations would need to be altered when a "processing" satellite is used. By "processing" we mean here demodulation to baseband and remodulation onto an RF carrier, not necessarily employing the same modulation characteristics as on the uplink. The concepts of DTC and CTC introduced earlier are necessary to arrive at the desired conclusion. This is because the CTC is a factor which is applied only once in an n-hop regenerative system, namely on the last hop when the desired "channel" is extracted.

Let X represent an arbitrary unwanted signal, which may be noise or interference. Let $(C/X)_u$ and $(C/X)_d$ represent the uplink and downlink carrer-to-unwanted signal ratios, and DTC_u and DTC_d the associated demodulator transfer characteristics. If the CTC were unity, the SNR at the output of the second (ground) demodulator would be simply,

$$(S/N)^{-1} = [(C/X)_u \, RTC_u]^{-1} + [C/X)_d \, RTC_d]^{-1}$$

Now, applying the CTC to the second demodulator output, yields

$$(S/N)^{-1} = R_u^{-1} \, (C/X)_u^{-1} + R_d^{-1} \, (C/X)_d^{-1} \tag{A-22}$$

where
$$R_u = RTC_u \cdot CTC$$
$$R_d = RTC_d \cdot CTC$$

are the effective receiver transfer characteristics associated with each link. Thus, the equations of Section 4.1 of the text are straightforwardly generalized to the case where the network under scrutiny has a processing satellite. Note that equation (A-22) applies whether the unwanted uplink and downlink signals also originate from a network with a processing satellite or from one with a translating repeater.

In the particular instance that the modulation characteristics are identical on both uplink and downlink, and that the uplink (e.g., the interfering network has a repeater), equation (A-22) reduces to

$$(S/N)^{-1} = R^{-1} \, [(C/X)_u^{-1} + (C/X)_d^{-1}] \tag{A-23}$$

where R is the factor $DTC \cdot CTC$ which is common to both directions. This equation is identical in appearance to the one which applies for non-processing satellites. Under the conditions leading to (A-23) there is clearly no advantage, in terms of SNR, to complicating the satellite. Of course, processing could be required for signal routing purposes, irrespective of SNR considerations.

The preceding discussion has applied, implicitly, to analog signals. For digital signals there will usually be a clear advantage, in terms of BER, to the use of regenerative repeating.

A.5 Carrier-to-Noise Ratio "Nomogram"

The carrier-to-(thermal) noise ratio (CNR) enters in many of the orbit utilization relationships. The relationships between the CNR and the terminal parameters is so well-known as to hardly need recounting. However, for the reader's convenience in facilitating the use of this book we give here a nonogram-like figure, which may prove helpful.

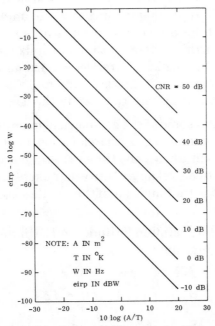

Figure A-1 Relationship Between Carrier-to-Noise Ratio and Link Parameters

The CNR may be written

$$CNR = eirp - 10 \log W + 10 \log (A/T) + 66, (dB) \qquad (A-24)$$

which is a slightly different form from the usual exposition, but which has the advantage of being frequency independent. In (A-24), W is the receiver bandwidth, T is the receiver noise temperature, and $A = \eta \pi (D/2)^2$ is the effective antenna receiving area, with η its efficiency and D the diameter. The result is plotted in Figure A-1, which should be applied separately for uplink and downlink to find the composite CNR.

B

Some Useful Geometrical Relationships

The homogeneous model is based upon certain simplifying assumptions, notably that the slant range from a given earth station to any satellite is essentially the same, and that the angle between two satellites as seen from an earth station is approximately that as viewed from the earth's center. In this appendix we present some exact geometrical data which the reader can examine to assess the magnitude of the approximations uses, and make adjustments if he wishes to solve some of the curves given in the text.

Figure B-1 illustrates some of the geometrical quantities of interest. They are S, the slant range; δ, the elevation angle; β, the angle between the subsatellite point and the earth station, as seen from the satellite; θ, the earth central angle between the subsatellite point and the earth station; ϕ_{sat}, the longitude of the satellite; ϕ_{sta}, the longitude of the earth station.

Denoting by λ the earth station latitude, and letting $\phi = \phi_{sta} - \phi_{sat}$, the following relationships apply:

$$\cos \theta = \cos \lambda \cdot \cos \phi \qquad (B-1)$$

$$\cos (\theta + \delta) = 0.15112 \cos \delta \qquad (B-2)$$

$$S = 22,763 \, \frac{\sin \theta}{\cos \delta} \, (n.\,m.) \qquad (B-3)$$

Eliminating θ allows us to plot the parametric relationship shown in

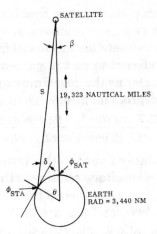

Figure B-1 Illustrating the Geostationary Satellite Geometry

Figure B-2. Among other things we see from this figure that the maximum ratio of slant ranges for any geometry is *22500/19323* ≈ *1.164*, or about 1.3 dB difference in received power level.

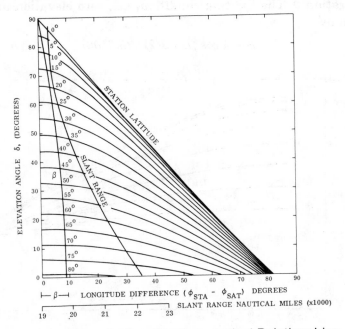

Figure B-2 Geostationary Satellite Geometrical Relationships

Another relationship of interest is that between θ, the earth central angle subtending two geostationary satellites, and θ_t, the angle between the same two satellite as viewed from an earth station. The angle θ is often referred to as the geocentric angle, and θ_t as the topocentric angle. Define the coordinate axes so that the two satellites are at $\pm\ \theta/2$ degrees longitude. Then, it can be shown that

$$\cos\ \theta_t = \frac{1 + 43.7\ \cos\ (l_2 - l_1) - 6.6106\ \cos\ \lambda\ (\ \cos\ l_1 + \cos\ l_2)}{\sqrt{(44.7 - 13.22\ \cos\ \lambda\ \cos\ l_1)\ (44.7 - 13.22\ \cos\ \lambda\ \cos\ l_2)}} \quad \text{(B-4)}$$

where λ is the absolute value of station latitude, $l_1 = L - \theta/2$, $l_2 = L + \theta/2$, and L is the station longitude relative to the mid-longitude between the two satellites; e.g., $L = 0$ means the station longitude is at zero degrees when the two satellites are at $\pm\ \theta/2$. It develops that for given λ and L, the ratio θ_t/θ is virtually independent of the actual value of θ, for θ at least up to 45°. (The difference as a function of θ is seen only in the third or fourth decimal place.) Hence we can present the results strictly as the ratio θ_t/θ which we do in Figure B-3, where the parameter on the curves is L. For a given station longitude and latitude, both satellites may not be visible depending on the satellite separation θ. The limiting condition, i.e., zero elevation angle, is given by

$$\cos\ \lambda\ \cos\ (L + \theta/2) = (6.6106)^{-1} \quad \text{(B-5)}$$

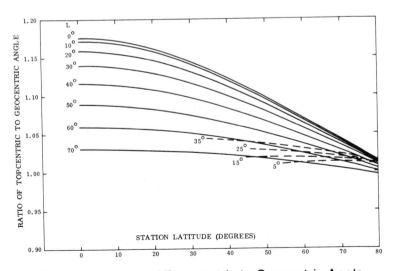

Figure B-3 Relationship of Topocentric to Geocentric Angle

The locus of *(λ, L)* values satisfying (B-5) for different *θ* has been superimposed on Figure B-3 as dashed lines, next to which are noted the values of *θ*. The region of validity of the solid lines thus lies above the dashed lines.

In summing the interference, we find the number N of interfering satellites occurring in most of the relationships derived earlier. For given intersatellite spacing $\Delta\theta$ and visible arc θ_v we have the limit $N \leq \theta_v / \Delta\theta$. Although the interference level is not very sensitive to N, one may wish to know exactly what it is. Thus, we present in Figure B-4 the length of the visible arc as a function of station latitude, for several elevation angles.

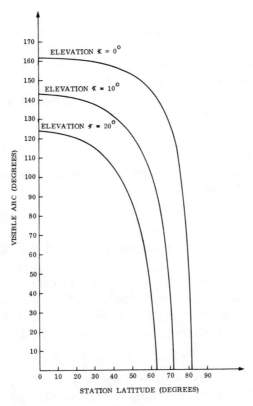

Figure B-4 Visible Geostationary Arc as Function of Station Latitude

C
Statistical Considerations for Antenna Sidelobes
and Implications on Orbit Utilization

The reference sidelobe antenna pattern $G(\theta) = A\theta^{-\beta}$ discussed in Section 4.4 is a convenient idealization for some purposes. But in actuality the sidelobe gain is not a smooth or simply characterizable function. In fact, it is not even a deterministic function. Because of random errors in the surface, alignment, fabrication, *et cetera,* the sidelobe gain is actually a random variable. In a probabilistic sense this means that at any given off-axis angle the measured gain from one antenna to another nominally identical antenna will vary, this variation being in principle describable by some distribution. The CCIR reference antenna pattern [6, 7] recognizes this statistical behavior and was empirically derived with the aim that only 10% of the sidelobe peaks of a number of measurements should exceed it. The resulting "envelope" *32 - 25 log θ* is conservative in that an interference entry at an arbitrary off-axis angle is likely to see a gain which is lower than the envelope.

When there are multiple interferers it is customary to use a reference antenna pattern, although it has been pointed out [8] that in this instance it is difficult to assign a probability level to the composite interference. In order to obtain a proper description of the (power) sum of multiple entries it is necessary to combine them statistically, as shown in the following.

In accordance with the homogeneous model the satellite antenna covers the earth stations of interest within its main beam so that statistical considerations do not enter. For narrow-beam satellite antennas it would in principle also be possible to account for sidelobe statistics but we shall not do so. Since the sidelobe gain is now assumed to be a random quantity we can write

$$(C/I)^{-1} = \Sigma\, G\,(i\Delta\theta)/G_r + \Sigma\, G_i\,(i\Delta\theta)/G_t$$

where, for simplicity, we have assumed co-channel, co-polarized operation. The first summation represents the downlink contribution and $G(\cdot)$ is the sidelobe gain of the wanted receiving antenna;

the second summation is the uplink interference and $G_i(\cdot)$ is the sidelobe gain of the uplink antenna accessing the ith satellite.

Generally we can expect that the behavior of the sidelobe gain will depend on the off-axis angle. Hence denote by $f_G(g, \theta)$ the probability density of the gain at angle θ, and let $X(\theta) = G(\theta)/G_r$, $Y(\theta) = G(\theta)/G_t$ with corresponding densities $f_X(x, \theta)$, $f_Y(y, \theta)$. Then the density of $(C/I)^{-1} \triangleq Z$, is

$$f_z(z) = f_X(x, \Delta\theta) * f_X(x, -\Delta\theta) * \cdots * f_X(x, N\Delta\theta) * f_X(x, -N\Delta\theta)$$
$$* f_Y(y, \Delta\theta) * f_Y(y, -\Delta\theta) * \cdots * f_Y(y, N\Delta\theta) * f_Y(y, -N\Delta\theta) \tag{C-1}$$

where $*$ stands for convolution, and assuming all the terms in the summations are statistically independent.

In order to solve Eq. (C-1) we need an explicit form for $f_G(g, \theta)$. The statistics of antenna sidelobes have been studied theoretically and experimentally [7, 9, 10]. It has been demonstrated [10] under reasonable assumptions that for a circular aperture the amplitude gain has a Rician density and, further, that under certain conditions this density reduces to the Rayleigh distribution. Measurements [9] appear to confirm, at least as an approximation, that the sidelobe gain is Rayleigh distributed. In any case, we shall suppose that this is true to illustrate the computational procedure and to see what consequences follow from the statistical approach.

If the amplitude gain is Rayleigh distributed, it follows straightforwardly that the power gain is exponentially distributed. Thus,

$$f_G(g, \theta) = \frac{1}{\gamma(\theta)} exp[-g/\gamma(\theta)], g \geq 0. \tag{C-2}$$

In order to establish $\gamma(\theta)$ explicitly, we assume, following [10], that the distribution (C-2) is such that the probability of exceeding the CCIR reference antenna gain is 10%, in approximate conformance with the derivation of that pattern. The cumulative distribution function (CDF) corresponding to (C-2) is

$$F_G(u, \theta) = \int_0^u f_G(g, \theta) \, dg = e^{-u/\gamma(\theta)} \tag{C-3}$$

Hence, setting

$$e^{-u/\gamma(\theta)} = 0.1$$

with $u = 10^{3.2} \theta^{-2.5}$ yields $\gamma(\theta) \approx 688\theta^{-2.5}$.

We choose for a simple example a "truncated" homogeneous model wherein we consider only the two interfering satellites on either side of the wanted one: these satellites, in the homogeneous model, account for about almost 80% of the interference. Further for simplicity we consider only the downlink. Thus we deal with

$$(C/I)^{-1} \approx \frac{1}{G_r} [G(\Delta \theta) + G(|-\Delta \theta|)]$$

which is the sum of two identically distributed exponential random variables. From [10] we can write the CDF for (C/I) as

$$Pr\,[(C/I) \geq u_o] = Pr\,[(C/I)^{-1} \leq u_o^{-1}]$$

$$= Pr \quad [G(\Delta \theta) + G(|-\Delta \theta|)] \leq u_o^{-1} \, G^r$$

$$= 1 - (x + 1)e^{-x}; \; x \triangleq u_o^{-1} \, G_r / 688 \,(\Delta \theta)^{-2.5} \qquad (C\text{-}4)$$

Now, in the traditional approach we compute for the (C/I)

$$(C/I)^{-1} = \frac{2 \cdot 10^{3.2}\,(\Delta \theta)^{-2.5}}{G_r} = \frac{3.17 \times 10^3\,(\Delta \theta)^{-2.5}}{G_r} \qquad (C\text{-}5)$$

If we identify the RHS of Eq. (C-5) as u_o^{-1}, then from Eq. (C-4) we have

$$Pr\,[(C/I) \geq u_o] = 1 - (5.605)e^{-4.605} = 0.944$$

Thus, we see that the probability of exceeding the standardly computed C/I is appreciably better than that associated with a single entry of equal power. This implies that for equal protection one could relax the spacing required in relation to that computed using the usual method, as in Eq. (C-5). However, before this advantage can be realized, it will be necessary to more fully verify the distribution which should be used.* Another obstacle to the application of this approach is that the improvement stems from the fact that it is unlikely for several interference entries to all encounter a sidelobe gain as high as the reference pattern; but as current practice has it, when interference calculations are required it is to establish the compability between two networks. Based on the reference pattern,

---

*A very recent unpublished study sponsored by INTELSAT has sought to statistically characterize the sidelobes of earth station antennas in the INTELSAT system. In all, the patterns at about 6 GHz of 335 antennas were examined. The conclusion obtained was that the Gamma density appeared to be a good fit to the sidelobe (power) gain, but for off-axis angles greater than 3° the exponential density also provided a reasonably good fit, in agreement with earlier results. We are indebted to INTELSAT for making this information available.

then, it would be necessary (to take advantage of the statistics) to increase the per-network interference allowance.

D
Statistical Analysis of the Effect of Station-Keeping Accuracy on Satellite Spacing

The results of Section 4.7, as stated, represent the worst-case combination of relative satellite positions. In orbit utilization studies a worst-case or nearly-worst case* deterministic assumption is typically made. In reality, the situation is probabilistic because the interfering satellites will assume, according to some probability law, random positions within their tolerance bands, independent of one another. It is interesting, therefore, to review the problem from a statistical standpoint [11]. This statistical approach is consistent with the manner in which performance specifications are typically stated, namely as fraction-of-time objectives. The information given here could ultimately be combined with other statistical variables to form, e.g., an overall distribution of carrier-to-interference ratio, much as in the approach taken by Bantin and Lyons [12] who considered other sources of variability.

We thus seek the distribution of the C/I ratio at a wanted earth station. Keeping to the co-channel co-polarized case, as in Section 4.7, this ratio is given by Eq. (4-54) and its statistical nature evidently depends on that of the ε_i. The behavior of the ε_i is farily complex, for reasons connected to the dynamical forces acting upon a spacecraft (see, e.g., [13]). It appears, however, that as a simplified abstraction, a uniform-*pdf* model is an adequate one. That is, the ε_i will be assumed uniformly distributed over the tolerance band. This model should be sufficient for the purpose of extracting the essential information from a statistical analysis. However, irrespective of this assumption, an exact formulation for the distribution of (I/C) is generally not tractable. But, with a reasonable assumption, an

---

*The nearly-worst case usually assumed amounts to having the wanted satellite at its nominal position but placing the other satellites at the worst extremes of their tolerance bands.

approximate analysis is possible and given next. It turns out that numerical results (without approximation) are readily available using Monte Carlo techniques, and results from a computer program written for this purpose will be shown.

D.1 Approximate Analytical Formulation

Taking (4-54) as the starting point, if we assume that the maximum longitudinal excursions are "small" compared to the nominal spacing $\Delta\theta$, the bracketed term in the summation can be approximated by the first two terms of its series expansion. The approximation improves, of course, with increasing index. Thus, (4-54) becomes

$$
\begin{aligned}
\frac{I}{C} &\approx (I/C)_o \sum_{-N}^{N} \frac{|i|^{-\beta}}{Z(N)} \left\{ 1 - \frac{\beta\,[\varepsilon_i - \varepsilon_o]\,sgn\,(i)}{|i|\Delta\theta} \right\} \\
&= (I/C)_o \sum_{-N}^{N} \frac{|i|^{-\beta}}{Z(N)} \left\{ 1 - \frac{\beta\,\varepsilon_i\,sgn\,(i)}{|i|\Delta\theta} \right\}
\end{aligned}
\tag{D-1}
$$

For simplicity, but without loss of generality, we assume $(I/C)_o = 1$. Let

$$x_i = a_i\,\varepsilon_i + b_i$$

where

$$a_i = -|i|^{-\beta-1}\,\beta\,sgn\,(i)/[\Delta\theta \cdot Z(N)]\,; b_i = |i|^{-\beta}/Z(N)$$

Let $f_i\,(\cdot)$ and $p_i\,(\cdot)$ be the pdf of ε_i and x_i, respectively. Then,

$$P_i\,(x_i) = \frac{1}{|a_i|}\,f_i\!\left(\frac{x_i - b_i}{a_i}\right)
\tag{D-2}$$

and since the G_i can reasonably be assumed to be independent, the pdf of (I/C), $p(x)$, is given by the convolution of the p_i, i.e.,

$$p(x) = p_{-N} * p_{-N+1} * \cdots * p_{-1} * p_1 * \cdots * p_{N-1} * p_N
\tag{D-3}$$

Equation (D-3) can be numerically evaluated, e.g., using the *FTT*, or in some instances, depending upon the nature of the f_i it can be manipulated analytically.

As an example of the latter instance, consider the case where the ε_i are identically distributed with a uniform *pdf*,

$$f_i\,(x) = \begin{cases} 1/2\,\varepsilon_{max}, & |\varepsilon| \le \varepsilon_{max} \\ 0, & elsewhere \end{cases}$$

Using (D-2), Figure D-1 illustrates p_1, p_2 and p_3 for the case $\varepsilon_{max} = 0.1$, $\Delta\theta = 1$, $N = 5$, $\beta = 2.5$. (Note in the situation considered, that $p_i = p_{-i}$.)

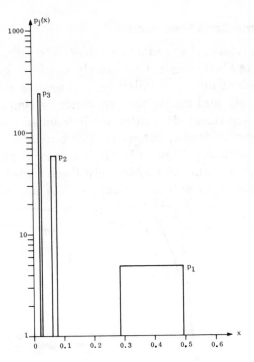

Figure D-1 Illustration of Probability Densities Involved in the
Evaluation of the Distribution of *C/I*

This figure suggests that for $|i| \geq 2$ we may approximate p_i by the
delta function $\delta(x\text{-}6_i)$ in comparison to the behavior of p_1. Thus, the
form of $p(x)$ is dominated by p_1 and p_{-1}, which convolve to a triangular distribution. Hence, we may finally write

$$p(x) \approx a^{-1} \, Tr \, [(x\text{-}1)/a] \qquad\qquad \text{(D-4)}$$

where $a = 2\beta\varepsilon_{max}/Z(N)$, and $Tr \, (\cdot)$ is the "unit" triangular density,
defined as an isosceles triangle centered at zero, with base 2 and
height 1. The distribution of I/C is then readily obtained and is
shown, for the case $\varepsilon_{max} = 0.1$, on Figure D-2 where it is compared with
that obtained by Monte Carlo methods. It can be seen that the
approximation is quite good. However, one can also ascertain that
for larger ε_{max} the approximation (D-4) begins to fail in the tails;
some improvement can be had from taking more terms $(i > 1)$ into the
convolution, but then one loses the desired simplicity and may as
well turn to the simulation.

D.2 Monte Carlo Simulation

The behavior of the random variable (I/C) can readily be determined by Monte Carlo trials. One merely needs to generate $L_1 + L_2 + 1$ independent uniformly distributed variates (the ε_i), compute I/C from (4-54), and repeat the experiment × times. The technique, of course, is not limited to uniformly distributed ε_i nor is it limited to the homogeneous model, but we restrict our numerical results to this case. Specifically, Figures D-2 to D-4 show the distribution of (I/C) for $\beta = 2.5$, 3.0, and 3.5 respectively. Each figure is parameterized in ε_r, the ratio of one-sided maximum error to nominal satellite spacing.

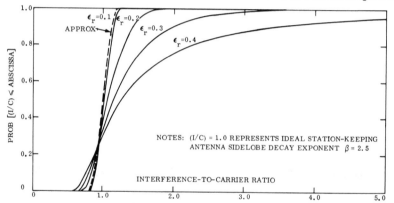

Figure D-2 Distribution Function of (C/I) for Antenna Sidelobe Decay Exponent Equal to 2.5 for Various Station-Keeping Accuracies

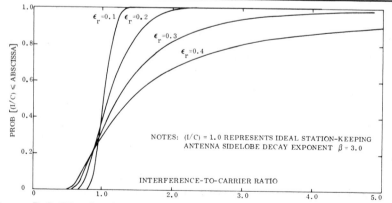

Figure D-3 Distribution Function of (C/I) for Antenna Sidelobe Decay Exponent Equal to 3.0 for Various Station-Keeping Accuracies

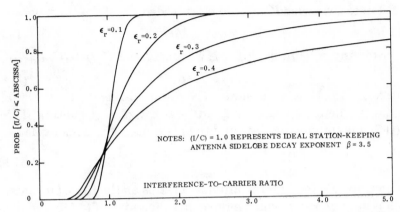

Figure D-4 Distribution Function of (*C/I*) for Antenna Sidelobe Decay
Exponent Equal to 3.5 for Various Station- Keeping
Accuracies

Ten interfering satellites were assumed. Each curve is the result of
100,000 trials, one trial consisting of a vector of 11 random positions
*(ε*ᵢ's). The curves are normalized so that unity on the abscissa corresponds to the ideal value *(I/C)*₀. However, for given antenna and
Δθ, the absolute value will be a function of β. Similarly, for given β,
Δθ, and antenna size, *(I/C)*₀ will be a function of frequency.

D.3 Discussion of Results

The figures shown may be used to ascertain the penalty, in terms of
the decrease of *C/I* for a specified percentage of time, given a
nominal spacing and station-keeping accuracy. Of perhaps greater
interest is the loss in orbit utilization (relative to ideal station-keeping) associated with requiring that a desired *C/I* be met or exceeded
for *P* percent of the time. Let this *C/I* be denoted *(C/I)*req; then we
note from (4-54) that with perfect station-keeping the required satellite separation Δθₚ is obtained from

$$(C/I)_{\text{req}} = K \ (\Delta\theta_{\text{p}})^{\beta} \tag{D-5}$$

Now, the curves in the previous figures display

$$Prob \ [(C/I) \geq \mu \ (C/I)_{\text{o}}] \tag{D-6}$$

and for any probability *P/100* satisfying (D-6) we can generate a
function $\mu = \mu(P, \ \varepsilon_{\text{r}}, \ \beta)$, for example by cross-plotting Figures D-2 to

D-4. Thus, if we wish $(C/I)_{req}$ to be exceeded $P\%$ of the time, we set

$$(C/I)_{req} = \mu(P,\ \varepsilon_r,\ \beta)\ (C/I)_o \qquad \text{(D-7)}$$

Since for any reasonable P, $\mu < 1$, we must have $(C/I)_o > (C/I)_{req}$, or,

$$(C/I)_o = K\ (\Delta\theta_o)^\beta \qquad \text{(D-8)}$$

with $\Delta\theta_o > \Delta\theta_p$. Now, we assume a fixed station-keeping accuracy $|\varepsilon_i| \leq \varepsilon_{max}$, and define $\varepsilon_{r,p} = \varepsilon_{max}/\Delta\theta_p$ and $\varepsilon_{r,o} = \varepsilon_{max}/\Delta\theta_o$. Equating (D-5) to (D-7), making use of (D-8) and the previous definitions leads to

$$\left(\frac{\Delta\theta_p}{\Delta\theta_o}\right)^\beta = \left(\frac{\varepsilon_{r,o}}{\varepsilon_{r,p}}\right)^\beta = \mu(P,\ \varepsilon_{r,o},\ \beta) \qquad \text{(D-9)}$$

which is to be solved for the unknown ratios on the LHS. Of course, (D-9) is simply a somewhat more general version of (4-56) and also needs to be solved iteratively. Equation (D-9) has been solved for $P = 90\%,\ 95\%,\ 99\%$ and 100% and for $\varepsilon_{r,p} = 0.1,\ 0.2,\ 0.3,$ and 0.4, and the resulting values of $\Delta\theta_o/\Delta\theta_p$ are shown in Table D-1. As before, it may be noted that the results are relatively insensitive to the antenna sidelobe exponent β. Also of interest is the fact that only modest improvements are available by asking for less than $P = 100\%$ protection on $(C/I)_{req}$.

TABLE D-1

Satellite spacing for *P* percent exceedance relative to spacing for ideal station-keeping

Percent of Time (C/I) Exceeded	Relative Station-keeping Error, $\varepsilon_r = 0.1$			Relative Station-keeping Error, $\varepsilon_r = 0.2$		
	$\beta=2.5$	$\beta=3.0$	$\beta=3.5$	$\beta=2.5$	$\beta=3.0$	$\beta=3.5$
P=90%	1.046	1.058	1.058	1.124	1.136	1.144
P=95%	1.064	1.064	1.067	1.143	1.156	1.17
P=99%	1.075	1.081	1.081	1.176	1.194	1.205
P=100%	1.11	1.11	1.117	1.235	1.258	1.266

Percent of Time (C/I) Exceeded	Relative Station-keeping Error, $\varepsilon_r = 0.3$			Relative Station-keeping Error, $\varepsilon_r = 0.4$		
	$\beta=2.5$	$\beta=3.0$	$\beta=3.5$	$\beta=2.5$	$\beta=3.0$	$\beta=3.5$
P=90%	1.212	1.217	1.253	1.307	1.344	1.371
P=95%	1.264	1.282	1.307	1.379	1.413	1.454
P=99%	1.302	1.338	1.357	1.454	1.498	1.509
P=100%	1.384	1.415	1.442	1.55	1.59	1.63

As an example, let $\varepsilon_{max} = 0.1$ degrees, which is the value currently mandated'for the fixed-satellite service by the *Radio Regulations* [14]. Suppose $\Delta \theta_p = 1.0$ degree, which is as small a nominal spacing as can be expected in the near future. Hence $\varepsilon_{rp} = 0.1$ and Table D-1 shows that the actual spacing required to meet $(C/I)_{req}$ for the four values of P are, respectively, $1.05°$, $1.06°$, $1.08°$, and $1.11°$. Thus, only relatively small increase in orbit utilization efficiency is possible by tightening the station-keeping accuracy to less than $0.1°$, though this might eventually become of significance in very congested arcs of the orbit.

E

Derivation of Optimum Orbit Capacity for Digital Signals

E.0 Introduction

In Section 4.13 of the text we presented some results on the orbit capacity for digital signals. In the first of these sections we related capacity to (relative) interference level, r, and obtained the optimum capacity with respect to r. In the first part of this Appendix we derive the relevant expressions. In the second section of the text mentioned, we showed pictorially the theoretically optimum capacity as a function of $(P/N_0 W)$, the carrier-to-noise ratio referred to the allocated bandwidth; this result is derived in the second part of this Appendix.

E.1 Orbit Capacity as a Function of Interference Level

As a starting point for our discussion we give for the orbit capacity

$$\dot{C} = \frac{\gamma B}{\Delta \theta} \; (bits/sec/deg) \tag{E-1}$$

which is obtained by combining (4-96) and (4-97).

Next, recall that
$$C/I = K \, (\Delta \theta)^{\beta}$$
which, in conjunction with Figure 4-31, implies that (in the homogeneous environment) there is a one-to-one relationship between C/I and C/N for given P_e. Hence there is a unique locus of pairs $(C/N, C/I)$ satisfying the P_e constraint. That relationship would have the

general appearance sketched in Figure E-1.* As shown, this relationship depends parametrically on the antenna decay slope, β. This is because the error probability depends not only on C/I but also on the distribution of amplitudes of the interferers. This distribution is governed by β. Functionally, Figure E-1 can be represented as

$$C/I = g\,(C/N;\,\beta) \tag{E-2}$$

but for simplicity we shall suppress the β dependence. The function g $(\cdot)$ is in general quite complex, although its behavior suggests that a reasonable approximation would be available from one of the two following forms

$$g\,(C/N) = \{(C/N)_T^{-1} - (C/N)^{-1}\}^{-\varepsilon_1} \tag{E-3a}$$

$$g\,(C/N) = \{\varepsilon_2\,[(C/N)_T^{-1} - (C/N)^{-1}]\}^{-1} \tag{E-3b}$$

where, in general,

$$\varepsilon_1 = \varepsilon_1\,(\beta,\,P_e);\ \varepsilon_2 = \varepsilon_2\,(\beta,\,P_e)$$

In the case where the interference is equivalent to thermal noise, $\varepsilon_1 = \varepsilon_2 = 1$. The implication of the preceding is that we can write

$$\Delta\theta = (1/K)^{1/\beta}\,\{g\,(C/N)\}^{1/\beta} \tag{E-4}$$

If we substitute (E-4) and (4-94) into (E-1) we obtain

$$\dot{C} = \frac{(P/N)_o}{C/N}\,\frac{\gamma\,K^{1/\beta}}{[g\,(C/N)]^{1/\beta}}, \ C/N \geq (C/N)_{\min} \tag{E-5}$$

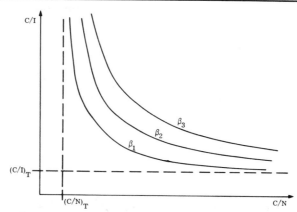

Figure E-1 Relationship Between *C/I* and *C/N* for Constant P_e

---

*Actual examples of these kinds of curves were presented in Figure 4-26 where the ordinate is $\Delta\theta$ rather than C/I.

One interesting formulation for the optimum orbit capacity is obtained by setting the derivative of (E-5) with respect to C/N equal to zero: $d\dot{C}/d\,(C/N) = 0$. This leads to the differential equation

$$(C/N)\,\dot{g}\,(C/N) + \beta g\,(C/N) = 0 \tag{E-6}$$

whose solution exists if it satisfies $(C/N) \geq (C/N)_{min}$.

Equation (E-6) does not explicitly reflect the role of the permissible interference level. In the text we adopted

$$r = (C/N)_T/(C/I) \tag{E-7}$$

as the definition of that notion. Different definitions are possible; for example,

$$\widetilde{r} = \frac{(C/I)^{-1}}{(C/I)^{-1} + (C/N)^{-1}} \tag{E-8}$$

The two are related by

$$r = \widetilde{r}\,\frac{(C/I)^{-1} + (C/N)^{-1}}{(C/N)_T^{-1}} \tag{E-9}$$

and the definitions coalesce when the interference is noise-like. However, (E-7) is more convenient for analytical purposes. In any event, results as a function of r can be converted to a function of $\widetilde{r}$ through (E-9), and furthermore there will probably be little numerical difference.

Using (E-7) and (E-3), a few manipulations lead us to

$$\dot{C} = \frac{\gamma\,K^{1/\beta}\,(P/N)_o}{(C/N)_T} \left\{ 1 - \frac{r^{1/\varepsilon_1}}{(C/N)_T^{1/\varepsilon_1 - 1}} \right\} \frac{r^{1/\beta}}{(C/N)_T^{1/\beta}} \tag{E-10a}$$

$$\dot{C} = \frac{\gamma\,K^{1/\beta}\,(P/N)_o}{(C/N)_T} \,(1 - r\,\varepsilon_2)\,\frac{r^{1/\beta}}{(C/N)_T^{1/\beta}} \tag{E-10b}$$

for the two forms of $g(\cdot)$, with constraint

$$r^{1/\varepsilon_1} \geq [(C/N)_T^{1/\varepsilon_1 - 1}]\,[1 - (C/N)_T/(C/N)_{min}] \tag{E-11a}$$

on the first equation, and constraint

$$(r/\varepsilon_2) \geq 1 - (C/N)_T/(C/N)_{min} \tag{E-11b}$$

on the second. Setting $d\dot{C}/dr = 0$, we get for the optimum solution of (E-10a) and (E-10b), respectively,

$$\hat{r} = \left\{ \frac{(C/N)_{\mathrm{T}}^{1/\varepsilon_1 - 1}}{1 + \beta/\varepsilon_1} \right\}^{\varepsilon_1} \qquad \text{(E-12a)}$$

$$\hat{r} = \frac{1}{\varepsilon_2 \, (\beta + 1)} \qquad \text{(E-12b)}$$

It is clear that when $\varepsilon_1 = \varepsilon_2 = 1$, the condition assumed in the test, Equations (E-10) both reduce to (4-98), Equations (E-11) both reduce to (4-99), and Equations (E-12) both reduce to (4-100).

E.2 Theoretically Optimum Orbit Capacity

Shannon's capacity formula [15],

$$C = W \log_2 (1 + \rho) \ (bits/sec) \qquad \text{(E-13)}$$

stipulates that C *bits/sec* can be transmitted with vanishing probability of error through a channel with bandwidth W and signal-to-noise ratio ρ. If we assume interference to be noise-like (Gaussian), we can apply (E-13) to the orbit utilization context by setting

$$\rho = [(P/N_o W)^{-1} + (K \Delta \theta^\beta)^{-1}]^{-1} \qquad \text{(E-14)}$$

and by so interpreting ρ, the orbit capacity is given by

$$\dot{C} = \frac{W}{\Delta \theta} \ \log \, (1 + \rho) \qquad \text{(E-15)}$$

In order to compare $\dot{C}$ computed from (E-15) to the results obtained for practical codes, we maintain the assumption that r is set at its optimum value for any given $(P/N_o W)$. Equivalently, we seek the value of $\Delta \theta$ which maximizes (E-15). Thus,

$$\frac{d\dot{C}}{d\Delta \theta} = \frac{W}{\ln 2} \left\{ \Delta \theta \frac{1}{1 + \rho} \frac{d\rho}{d\Delta \theta} - \ln \, (1 + \rho) \right\} \ (\Delta \theta)^{-2} = 0$$

establishes the necessary conditions. But

$$\frac{d\rho}{d\Delta \theta} = (\beta/K)\rho^2 \, (\Delta \theta)^{-(\beta + 1)}$$

so that the optimum $\Delta \theta$ is the solution to

$$(K \, \Delta \theta^\beta)^{-1} \, \beta \, \frac{\rho^2}{1 + \rho} - \ln \, (1 + \rho) = 0 \qquad \text{(E-16)}$$

For given $P/N_o W$, (E-16) actually finds $(K)^{1/\beta}\Delta\theta$, but in order to be compatible with Figures 4-29 to 4-31, we normalize by W as well. Thus, by assuming values of $(P/N_o W)$ we can find the corresponding optimum $\Delta\theta$, from which $\dot{C}/[K^{1/\beta} W]$ can be calculated as shown in Figure 4-32.

F

Relation Between Emission and Sensitivity Parameters and System Design Parameters

In Chapter 5 we mentioned that the emission and sensitivity parameters were intimately tied to the system design parameters, and therefore that constraints on the former implied certain constraints on the latter. This relationship was illustrated in Tables 5-2 and 5-3. Here we develop the equations which embody the relationship between the two sets of parameters, for the case of FDM/FM telephony and digital signalling. We also assume operation is in the 4/6 GHz frequency bands.

F.1 FDM/FM Telephony

F.1.1 Emission Parameters

The uplink and downlink carrier-to-noise ratios can be written in decibels as:

$$CNR_u = E_e - L_u + (G/T)_s - 10 \log (kB) \tag{F-1}$$

$$CNR_d = E_s - L_d + (G/T)_g - 10 \log (kB) \tag{F-1}$$

where

E_g, E_s = earth station and satellite eirp, respectively (dB);

L_u, L_d = uplink and downlink "space loss," respectively (dB);

$(G/T)_s$, $(G/T)_g$ = figures-of-merit $(dB/°K)$ for the satellite and earth station, respectively;

B = RF bandwidth.

Using the form of performance equation for telephony given in Equation (A-19) of Appendix A, and further invoking (A-16) as well as (F-1) and (F-2) leads to

$$10 \log N_b = 83.5 - E_g + L_u - (G/T)_s + 10 \log (kb) - m \qquad (F\text{-}3)$$

for the uplink, and to

$$10 \log N_b = 83.5 - E_s + L_d - (G/T)_g + 10 \log (kb) - m \qquad (F\text{-}4)$$

for the downlink, where

N_b = baseband noise in a telephone channel

$m = 20 \log M_{tt}$

b = channel bandwidth

and implicit in (F-3) and (F-4) are 4 dB pre-emphasis improvement and 2.5 dB psophometric weighting.

For 6 GHz uplink, $L_u \approx 201\ dB$ and for 4 GHz downlink $L_d \approx 197\ dB$. Assuming N_b is $5000p\,WOp$ for the downlink and $3000p\,WOp$ for the uplink, we get

$$E_g = 56 - (G/T)_s - m\ (dBW) \qquad (F\text{-}5)$$

$$E_s = 49.8 - (G/T)_g - m\ (dBW) \qquad (F\text{-}6)$$

If we write $E_g = P_g + G_g$ and assume the CCIR sidelobe antenna pattern $32 - 25 \log \theta$, then the uplink emission parameter defined in Section 5.3 can be expressed as

$$P_u = 88 - (G/T)_s - m - G_g - R - 25 \log \theta\ (dBW/4\ kHz) \qquad (F\text{-}7)$$

where R is the ratio of unmodulated carrier power to the maximum power in any 4 kHz.

The flux density at the earth associated with E_s is simply

$$P_d = -113.2 - (G/T)_g - m - R\ (dBW/m^2/4\ kHz) \qquad (F\text{-}8)$$

The parameters P_u and P_d are determined purely by performance requirements. Hence, constraining their values has implications on the parameters on the RHS of (F-7) and (F-8).

An appreciation of these implications is obtained by considering a set of FDM/FM telephony carriers described in Table F-1. For these carriers equations (F-7) and (F-8) have been evaluated and the results previously shown in Tables 5-2 and 5-3 of the text.

F.1.2 Sensitivity Parameters

While P_u and P_d are related to thermal noise objectives, the sensitivity parameters S_u and S_d are related to the allowable interference. The baseband noise N_{bj} induced in network j when interfered-with by network k is given by

$$10 \ log \ N_{bj} = 87.5 - CIR - R_{jk} \qquad (F\text{-}9)$$

where the RTC R_{jk} includes pre-emphasis but not weighting which is included in the constant. Assuming $N_b = 400 \ pWOp$, we find the required carrier-to-interference ratio

$$CIR_{req} = 61.5 - R_{jk} \ (dB) \qquad (F\text{-}10)$$

Further, we can write

$$(C/I) = E_j + G_{or} - G_{sr} - E_k$$

where

E_j = eirp of jth (wanted) carrier;

E_k = eirp of unwanted carrier in direction of wanted receiver;

G_{or} = on-axis gain of wanted receiver antenna;

G_{sl} = sidelobe gain of wanted receiver antenna at angle off boresight corresponding to direction of interfering signal.

Hence to satisfy the interference budget, the interfering e.i.r.p. is bounded by

$$E_{k,req} \leq E_j + (G_{or} \quad G_{sl}) - CIR_{req} \qquad (F\text{-}11)$$

Notice that the term $(G_{or} - G_{sl}) = (G_{or} - 32 + 25 \ log \ \theta$ for earth station antennas but cannot be further specified for the satellite antenna. For global beams, $G_{or} - G_{sl} \approx 0 \ dB$. For spot beams the same will be true if the interfering transmitters is within the beam coverage area. However, the narrower the beam the less likely this is to be so.

Calling upon (F-5) and (F-6) we thus get

$$S_u = -107 - (G/T)_s + (G_{or} - G_{sl})_s - m - CIR_{req} - R_k (dBW/m^2/4 \ kHz) (F\text{-}12)$$

$$S_d = 145.2 - M + T_g - CIR_{req} - R_k \ (dBW/m^2/4 \ kHz) \qquad (F\text{-}13)$$

where T_g is the earth station receiver noise temperature, and R_k is the factor earlier defined but applies here to the interfering signal. CIR_{req} must be calculated from (F-10) using the proper value of R_{jk}; the latter is developed in Part III.

For the telephony carriers in Table F-1 we have computed (F-12) and (F-13), assuming $(G_{or} - G_{sl})_s = 0$ dB. Also, where the bandwidth of the interfering signal is much smaller than that of the wanted signal, several interfering signals are assumed present in such a way as to occupy the full bandwidth of the wanted signal. Evidently, different R_{jk} apply for each interfering carrier, in this instance, because of a different carrier frequency offset. The results are given in Tables F-2 and F-3 for S_u and S_d, respectively.

In both this subsection and the preceding one we have made specific assumptions in order to present numerical results. It should be clear from the relevant equations how to modify these results if different assumptions are made.

TABLE F-1
CHARACTERISTICS OF SIGNALS CONSIDERED

CARRIER SIZE (NO. CHANNELS)	BANDWIDTH (MHz)	R (dB)	Sp (dB)
24	2.0	22.3	4.7
60	2.25	22.4	5.1
72	4.5	25.8	4.7
132	4.4	24.2	6.2
192	6.4	25.8	6.2
252	8.5	25.0	6.3
432	13.0	27.6	7.5
612	17.8	28.9	7.6
792	22.4	30	7.5
972	36.0	34.5	5.0
1092	36.0	32.2	7.3
PSK	--	$R_s/4000$	3.0

TABLE F-2
EXAMPLE CALCULATION OF UPLINK SENSITIVITY (S_u)

INTERFERING CARRIER

WANTED CARRIER	24	60	72	132	192	252	432	612	792	972	1092
24	-167.4	-167.5	-168.2	-167.7	-167.7	-168.0	-167.1	-167.3	-167.5	-169.0	-167.0
60	-166.6	-166.7	-168.1	-166.9	-167.5	-167.7	-167.5	-166.6	-166.9	-169.2	-167.2
72	-164.7	-164.8	-167.1	-165.8	-166.6	-167.1	-166.4	-166.3	-166.5	-169.0	-167.0
132	-164.5	-164.6	-166.7	-165.5	-166.4	-166.9	-166.3	-166.4	-166.8	-168.6	-166.6
192	-164.2	-164.3	-166.0	-164.5	-165.7	-166.2	-166.0	-166.3	-166.6	-169.4	-167.1
252	-164.1	-164.2	-165.2	-163.6	-164.8	-165.7	-165.5	-166.1	-166.4	-169.3	-167.0
432	-164.8	-164.9	-164.1	-162.5	-163.0	-164.1	-164.4	-165.0	-165.7	-168.6	-166.3
612	-164.5	-164.6	-164.9	-163.3	-162.3	-162.7	-163.1	-164.2	-165.1	-168.3	-165.7
792	-164.9	-165.0	-165.3	-163.7	-164.6	-162.0	-162.0	-163.3	-164.2	-168.5	-166.2
972	-164.3	-164.4	-164.8	-163.2	-163.2	-162.7	-161.5	-162.2	-163.2	-166.6	-164.2
1092	-165.1	-165.2	-165.4	-163.8	-163.4	-163.1	-161-6	-162.0	-163.0	-167.1	-164.8

Interfering Signal Density **(dBW/m²/4 kHz)**
at GSO Producing 400 **pWOp** *Interference*
Noise When Desired Signal Uplink
$(G/T)_s = 5 \, dB/°K.$
Entries Increase dB per dB as (G/T), **decreases.**

TABLE F-3
EXAMPLE CALCULATION OF DOWNLINK SENSITIVITY (S_u)

	INTERFERING CARRIER										
WANTED CARRIER	**24**	**60**	**72**	**132**	**192**	**252**	**432**	**612**	**792**	**972**	**1092**
24	-180.6	-180.7	-181.4	-180.9	-180.9	-181.2	-180.3	-180.5	-180.7	-182.2	-180.2
60	-179.8	-179.9	-181.3	-180.2	-180.7	-180.9	-180.7	-179.8	-180.1	-182.4	-180.4
72	-177.8	-178.1	-180.3	-179.	-179.8	-180.3	-179.6	-179.5	-179.7	-182.2	-180.2
132	-177.7	-177.8	-179.9	-178.7	-179.6	-180.1	-179.5	-179.6	-180.	-181.8	-179.8
192	-177.4	-177.5	-179.2	-177.7	-178.9	-179.4	-179.2	-179.5	-179.8	-182.6	-180.3
252	-177.3	-177.4	-178.4	-176.8	-178.	-178.9	-178.7	-179.3	-179.6	-182.5	-180.2
432	-178.0	-178.1	-177.3	-175.7	-176.2	-177.3	-177.6	-178.2	-178.9	-181.8	-179.5
612	-177.7	-177.8	-178.1	-176.5	-175.5	-175.9	-176.3	-177.4	-178.3	-181.5	-178.9
792	-178.1	-178.2	-178.5	-176.9	-177.8	-175.2	-175.2	-176.6	-177.4	-181.7	-179.4
972	-177.5	-177.6	-178.0	-176.4	-176.4	-175.9	-174.7	-175.4	-176.4	-179.8	-177.4
1092	-178.3	-178.4	-178.6	-177.0	-176.6	-176.3	-174.8	-175.2	-176.2	-180.3	-178.

Tabulation of Parameter $\bar{D}$ for $T_g = 100°K$
Downpath Sensitivity Parameter = $\bar{D}$ + 25 log θ
*Figures apply for 5000 **pWOp** internal noise = 400 **pWOp** interference noise*

F.2 Digital Telephony

F.2.1 Emission Parameters

The situation for digital signals is of course more complicated because (for a translating repeater) we cannot identify a separate BER for the uplink and downlink. However, there is a separate CNR for both links. This allows us to write for the uplink and downlink e.i.r.p., respectively, necessary to produce a specified BER:

$$E_g = CNR_t + d + L_u - (G/T)_s + 10 \log kB + p_u + m_u \qquad \text{(F-14)}$$

$$E_s = CNR_t + d + L_d - (G/T)_g + 10 \log kB + p_d + m_d \qquad \text{(F-15)}$$

where

CNR_t = required theoretical CNR to produce requisite BER;

d = degradation from theoretical for the given network;

p_u, p_d = ratio (in dB) of uplink and downlink CNR, respectively, to total CNR;

m_u, m_d = uplink and downlink margins required to meet BER in the presence of interference for specified availability objective.

To be specific assume coherent QPSK. Furthermore we take $B \approx 2R_s$ where R_s is the symbol rate so that the factor R in this case is $10 \log (R_s/4000)$. Let the $BER = 10^{-6}$ so that $CNR_t = 10.6 \, dB$, and suppose $d = 2$ dB, $p_d = 2 \, dB$, $p_u = 4.25 \, dB$, and $m_d = m_u = 1 \, dB$. We then obtain

$$P_u = 61.25 - (G/T)_s - G_g - 25 \log \theta \qquad \text{(F-16)}$$

$$P_d = -140 - (G/T)_g \qquad \text{(F-17)}$$

F.2.2 Sensitivity Parameters

We shall assume for present purposes that the allowable interference corresponds to a 4% increase in the equivalent system noise temperature, T_{eq}. We can write

$$I_d = (.04) \, k \, T_{eq} \, B \, (Watts)$$

$$I_u = (C_u/C_d) \, (.04) \, k \, T_{eq} \, B \, (Watts)$$

where

I_u, I_d = permissible uplink and downlink interference power, respectively;

C_u, C_d = uplink and downlink wanted carrier power, respectively, at the corresponding receiver terminals.

A few simple manipulations lead us to

$$S_u = -10 \ log \ (\lambda_u^2/4\pi) - G_s + 10 \ log \ (C_u/C_d)$$
$$+ \ 10 \ log \ S_p - 206.6 \ (dBW/m^2/4 \ kHz) \tag{F-18}$$

$$S_d = -10 \ log \ (\lambda_d^2/4\pi) + \ 10 \ log \ T_{eq} + 10 \ log \ S_p$$
$$- \ 238.6 + 25 \ log \ \theta \ (dBW/m^2/4 \ kHz) \tag{F-19}$$

where

$\lambda_u, \ \lambda_d$ = uplink, downlink wavelengths;

G_s = satellite antenna gain (dB);

S_p = spectral peaking factor, the ratio of power in the worst 4 kHz to the power which would reside in any 4 kHz if the interference power were uniformly distributed.

Values of S_p are given in Table F-1. Assuming $T_{eq} = 250°k$ and $C_u/C_d = 12$, we get for the 4/6 GHz band:

$$S_u = -130.3 - G_s + 10 \ log \ S_p \tag{F-20}$$

$$S_d = -181.1 + 10 \ log \ S_p + 25 \ log \ \theta \tag{F-21}$$

G

Analysis of Satellite Cluster Deployment

In this Appendix we develop the equations that describe the spacing relationships for the cluster alternating deployment (CAD) discussed in Section 5.4. The analysis applies also as a special case to the single alternating deployment (SAD) mode, which is simply the degenerate CAD with a cluster of one satellite. For simplicity we confine ourselves to the downlink only and we neglect path length differences.

We have adopted for this analysis the alternative earth station antenna sidelobe model described in Section 4.4, i.e.,

$$G \ (\theta) = \frac{G_0}{1 + (\theta/\Omega)^\beta}$$

$$\approx \frac{G_0}{(\alpha/\Omega)^\beta} \ , \ \alpha/\Omega \gg 1$$

which is somewhat more compact notationally than our standard model and is useful here for that reason because the equations tend to get somewhat unwieldy. Obviously, the conclusions will be unaltered by this choice.

The situation is depected in the sketch below, which is a somewhat expanded form of Figure 5-13.

G.1 Type a Satellites

For an earth station receving type a satellite the interference-to-carrier ratio, (I/C), due to other type a satellites is

$$\left(\frac{I}{C}\right)_{aa} = \sum_k G(k\,\phi'')/G_0 = \left(\frac{\Omega\,a}{\phi''}\right)^{\beta} 2\,\Sigma\,k^{-\beta}$$

$$= 2\,\xi\,(\rho)\left(\frac{\Omega\,a}{\phi''}\right)^{\beta} \tag{G-1}$$

where $\xi\,(\beta)\ \underline{\Delta}\ \Sigma\,k^{-\beta}$, and $2\,\Omega_a$ is the earth station 3 dB beamwidth.

The effective I/C due to type b satellites, where e.i.r.p. is $1/r$ that of type a satellites, is

$$\left(\frac{I}{C}\right)_{ba} = \frac{W_{ba}}{r}\,2\,\sum_{k=0}^{K}\,\sum_{j=0}^{n-1}\,G\,(\phi'' - \alpha - j\psi'' + k\phi'')/G_0 \tag{G-2}$$

$$= 2\,\frac{W_{ba}}{r}\,\frac{\Omega\,a}{\phi''}^{\beta}\,\sum_{k=0}^{K}\,\sum_{j=0}^{n-1}\,[k+1-(\alpha+j\,\psi'')/\phi'']^{-\beta} \tag{G-3}$$

where n is the number of type b satellites in a cluster and $K+1$ is the number of clusters considered on either side of the wanted type a satellite; for most practical cases $K = 0$ will be a good approximation thus obviating the need to sum on k.

Using the fact that $\alpha = \frac{1}{2}\,[\phi'' - (n-1)\,\psi'']$, the term $-(\alpha\,\mu\,j\,\psi'')/\phi''$ becomes

$$\frac{-\alpha+j\,\psi^4}{\phi''} = -\frac{1}{2} + \frac{1}{2}\,(n+1)\frac{\psi''}{\phi''} - j\frac{\psi''}{\phi''}$$

$$= -\frac{1}{2} + \frac{1}{2}\,[(n+1)-2j]\,z; \tag{G-4}$$

$$z\ \underline{\Delta}\ \psi''/\phi''$$

Substituting (G-4) in (G-3) and changing the inner index of summation from j to $j + 1$, we get

$$\left(\frac{I}{C}\right)_{ba} = \frac{W_{ba}}{r} \left(\frac{\Omega_a}{\phi''}\right)^\beta \sum_{k=0}^{K} \sum_{j=0}^{n} [k + \tfrac{1}{2} + \tfrac{1}{2}(n + 1 - 2j)z]^\beta \quad \text{(G-5)}$$

The total I/C for type $\underline{a}$ is simply $(I/C)_{aa} + (I/C)_{bb}$; i.e., Eq. (G-1) plus (G-5); thus,

$$\left(\frac{I}{C}\right)_a = 2\,\xi\,(\beta)\left(\frac{\Omega_a}{\phi''}\right)^\beta + \frac{2\,W_{ab}}{r}\left(\frac{\Omega_a}{\phi''}\right)^\beta$$
$$\sum_{k=0}^{K} \sum_{j=0}^{n} [k + \tfrac{1}{2} + \tfrac{1}{2}(n + 1 - 2j)z]^\beta \quad \text{(G-6)}$$

In a homogeneous environment, with spacing ϕ, we have

$$\left(\frac{I}{C}\right)_a = 2\,\xi\,(\beta)\left(\frac{\Omega_a}{\phi}\right)^\beta \quad \text{(G-7)}$$

and since we assume a given fixed level of interference, whatever the satellite deployment, we can equate (G-6) to (G-7) to get

$$\left(\frac{\phi''}{\phi}\right)^\beta = 1 + \frac{w_{ba}}{r\,\xi\,(\beta)} \sum_{k=0}^{K} \sum_{j=1}^{n} [k + \tfrac{1}{2} + \tfrac{1}{2}(n + 1 - 2\,j)z]^\beta$$
$$= 1 + \frac{w_{ba}\,2^\beta}{r\,\xi\,(\beta)} \sum_{k=0}^{K} \sum_{j=0}^{n} [2\,k + 1 + (n + 1 - 2j)z]^\beta \quad \text{(G-8)}$$

G.2 Type $\underline{b}$ Satellites

G.2.1 "Flank" Satellites

The interference arising from the type $\underline{a}$ satellites is

$$\left(\frac{I}{C}\right) = r\,w_{ab}\,\left\{(\alpha/\Omega_b)^{-\beta} + [(\phi'' - \alpha)/\Omega_b]^\beta + [(\phi'' + \alpha)/\Omega_b]^\beta\right.$$
$$\left. + [(2\,\phi'' - \alpha)/\Omega_b]^\beta + [(2\,\phi'' + \alpha)/\Omega_b]^\beta + \cdots\right\}$$
$$= r\,w_{ab}\,\left\{\sum_{j=0}^{J} [(\alpha + \phi'')/\Omega_b]^\beta + \sum_{k=0}^{K} [(k\,\phi'' - \alpha)/\Omega_b]^\beta\right\}$$
$$= r\,w_{ab}\,\left(\frac{\Omega_b}{\psi''}\right)^\beta\,\left\{\sum_{j=0}^{J} [\frac{1}{2z} - \tfrac{1}{2}(n - 1) + j/z]^\beta\right.$$
$$\left. + \sum_{k=0}^{K} [\tfrac{1}{2}(n - 1) - \frac{1}{2z} - \frac{k}{z}]^\beta\right\}$$

$$\left(\frac{I}{C}\right)_{ab} = r\, w_{ab} \;\; \left(\frac{\Omega_b}{\psi''}\right)^{\beta} 2^{\beta} \left\{ \sum_{j=0}^{J} \; [\frac{1}{z}\cdot(n-1)+2j/z]^{-\beta} \right.$$

$$\left. + \sum_{k=0}^{K} [n-1-\frac{1}{z}+2\,k/z]^{-\beta} \right\}$$

$$(G\text{-}9)$$

where $1/z = \phi''/\psi''$, $J+1$ is the number of type $\underline{a}$ satellites to the right of the wanted type $\underline{b}$ satellite, and K is the number of type $\underline{a}$ satellites to the left of the wanted type $\underline{b}$ satellite. (Note that in Eq. (G-9) the wanted type $\underline{b}$ flank satellite is the one at the right end of the cluster; however, the equation applies as well to the left flank satellite since the situation is symmetrical.)

The interference arising from other type $\underline{b}$ satellites is

$$\left(\frac{I}{C}\right)_{bb} = \sum_{k=0}^{n-1} (k\,\psi''/\Omega_b)^{-\beta} + \sum_{l=0}^{L} \sum_{j=0}^{n-1} [(l\,\phi''+j\,\psi'')/\Omega_b]^{-\beta}$$

$$+ \sum_{m=0}^{M} \sum_{i=0}^{n-1} [m\,\phi''+2\,\alpha+i\,\psi'')/\Omega_b]^{-\beta}$$

$$= \left(\frac{\Omega_b}{\psi''}\right)^{\beta} \left\{ \sum_{k=0}^{n-1} k^{-\beta} + \sum_{l=0}^{L} \sum_{j=0}^{n-1} [j+l/z]^{-\beta} \right.$$

$$\left. + \sum_{m=0}^{M} \sum_{i=0}^{n-1} [(m+1)/z+1-n+1]^{-\beta} \right\}$$

$$(G\text{-}10)$$

where L is the number of clusters considered to the left of the cluster containing the wanted satellite, and $M+1$ is the number of clusters considered to the right of the cluster containing the wanted satellite. The total (I/C) is simply $(I/C)_{bb} + (I/C)_{ab}$, namely the sum of Eq. (D-9) and (D-10). Now, in the homogeneous mode,

$$\left(\frac{I}{C}\right)_{b} = 2\,\xi\,(\beta)\left(\frac{\Omega_b}{\psi}\right)^{\beta}$$

and equating this to the $(I/C)_b$ in the cluster mode yields the condition for constant interference, namely

$$\left(\frac{\psi''}{\psi}\right)^{\beta} = \frac{r\,w_{ab}}{2\,\xi\,(\beta)}\,2^{\beta} \left\{ \sum_{j=0}^{J} \; [\frac{1}{z}\cdot(n-1)+2j/z]^{-\beta} \right.$$

$$\left. + \sum_{k=0}^{K} [n-1-\frac{1}{z}+2\,k/z]^{-\beta} \right\}$$

$$+ \frac{1}{2\,\xi\,(\beta)} \left\{ \sum_{k=1}^{n-1} k^{-\beta} + \sum_{l=1}^{L} \sum_{j=0}^{n-1} [j + l/z]^{-\beta} \right.$$

$$\left. + \sum_{m=0}^{M} \sum_{i=0}^{n-1} [(m+1)/z + i - n + 1]^{-\beta} \right\}$$

(G-11)

An alternate form of (G-11) which is also useful can be obtained as follows:

$$\left(\frac{\psi''}{\psi} \right)^{\beta} = \left(\frac{\phi''}{\phi} \right)^{\beta} \left(\frac{\psi''}{\phi''} \right)^{\beta} \left(\frac{\phi}{\psi} \right)^{\beta}$$

defining $\phi/\psi \triangleq w$ and since $\psi''/\psi'' = z$, we get, by multiplying both sides of (G-11) by $(zw)^{-\beta}$, the following:

$$\left(\frac{\phi''}{\phi} \right)^{\beta} = \frac{r\, w_{ab}\, 2^{\beta}\, \omega^{-\beta}}{2\,\xi\,(\beta)} \left\{ \sum_{j=0}^{J} [2j + 1 - (n-1)z]^{-\beta} \right.$$

$$\left. + \sum_{k=1}^{K} [2k + 1 + (n-1)z]^{-\beta} \right\}$$

$$+ \frac{\omega^{-\beta}}{2\,\xi\,(\beta)} \left\{ \sum_{k=1}^{n-1} (kz)^{-\beta} + \sum_{l=1}^{L} \sum_{j=0}^{n-1} [l + jz]^{-\beta} \right.$$

$$\left. + \sum_{m=0}^{M} \sum_{i=0}^{n-1} [(i - n + 1)z + m + 1]^{-\beta} \right\}$$

(G-12)

Equation (G-12) forms another independent condition, besides Eq. (G-8), for interrelating the various parameters.

G.2.2 Second Satellites

The interference arising from type $\underline{a}$ satellites is

$$\left(\frac{I}{C} \right) = r\,\omega_{ab} \left\{ [(\alpha + \psi'')/\Omega_b]^{-\beta} + [(\phi'' - \alpha - \psi'')/\Omega_b]^{-\beta} \right\}$$

$$+ [(\phi'' + \alpha + \psi'')/\Omega_b]^{-\beta} + [(2\phi'' - \alpha - \psi'')/\Omega_b]^{-\beta} + \cdots \}$$

$$= r\,\omega_{ab} \left\{ \sum_{j=0}^{J} {}_{\mathsf{L}}(j\,\phi'' + \alpha + \psi'')/\Omega_b]^{-\beta} + \sum_{k=0}^{K} {}_{\mathsf{L}}(k\,\phi'' - \alpha - \psi'')/\Omega_b]^{-\beta} \right\}$$

$$= r\,\omega_{ab} \left(\frac{\Omega_b}{\phi''} \right)^{\beta} 2^{\beta} \left\{ \sum_{j=0}^{J} [2j + - (n-3)z]^{-\beta} \sum_{k=1}^{K} [2k - 1 + (n-3)z]^{-\beta} \right.$$

(G-13)

where here $J + 1$ is the number of type a satellites to the right of the wanted type b *second* satellite, and K is the number of type a satellites to the left; J and K are the same numbers appearing in Eq. (G-9).

The interference arising from other type b satellites is

$$
\begin{aligned}
\left(\frac{I}{C}\right)_{bb} &= \sum_{k=1}^{n-2} (k\,\psi''/\Omega_b)^{-\beta} + \sum_{u=0}^{U} \sum_{i=0}^{n-1} [(u\phi'' + (n-2)\,\psi'' + 2\alpha + i\psi'')/\Omega_b]^{-\beta} \\
&\quad + (\psi''/\Omega_b)^{-\beta} + \sum_{m=0}^{M} \sum_{j=0}^{n-1} s\,[m\phi'' + \psi'' + 2\alpha + j\psi'')/\Omega_b]^{-\beta} \\
&= \left(\frac{\Omega_b}{\phi''}\right)^{\beta} \left\{ \sum_{k=1}^{n-2} (kz)^{-\beta} + \sum_{u=0}^{U} \sum_{i=0}^{n-1} [u + l + (i-1)z]^{-\beta} \right. \\
&\quad \left. + z^{-\beta} + \sum_{m=0}^{M} \sum_{j=0}^{n-1} [m + 1 + (j+1-n)z]^{-\beta} \right\}
\end{aligned}
$$

$$\text{(G-14)}$$

where $U+1$ is the number of clusters of type b satellites considered, to the left of the cluster containing the wanted satellite, and $M+1$ is the number of clusters to the right; M is the same quantity appearing in Eq. (G-10), while the latter the number $L = U + 1$.

Adding Eq. (G-13) and (G-14) and equating to the homogeneous case we get

$$
\begin{aligned}
\left(\frac{\phi''}{\phi}\right)^{\beta} &= \frac{r\,w_{ab}\,\omega^{\beta}\,2^{\beta}}{2\,\xi\,(\beta)} \left\{ \sum_{j=0}^{J} [2j + 1 - (n-3)\,z]^{-\beta} \right. \\
&\quad \left. + \sum_{k=1}^{K} [2k - 1 + (n-3)z]^{-\beta} \right\} \\
&\quad + \frac{\omega^{-\beta}}{2\,\xi\,(\beta)} \left\{ z^{-\beta} \left(1 + \sum_{k=1}^{n-2} k^{-\beta} \right) + \sum_{u=0}^{U} \sum_{i=0}^{n-1} [u + 1 + (i-1)z]^{-\beta} \right. \\
&\quad \left. + \sum_{m=0}^{M} \sum_{j=0}^{n-1} [m + 1 + (j+2-n)z]^{-\beta} \right\}
\end{aligned}
$$

$$\text{(G-15)}$$

G.3 Trade-Off Relations

If we take only the most "significant" terms in the previous equations we get

$$
\left(\frac{I}{C}\right)_{a} = \left(\frac{\Omega_a}{\phi''}\right)^{\beta} \left\{ 2\,\xi\,(\beta) + \frac{2\,w_{ba}\,2^{\beta}}{r} \sum_{j=1}^{n} [1 + (n+1-2j)z]^{-\beta} \right\}
$$

$$\text{(G-16)}$$

$$\left(\frac{I}{C}\right)_b (flank) = \left(\frac{\Omega_b}{\Phi''}\right)^\beta \left\{ z^{-\beta} \sum_{k=1}^{n-1} k^{-\beta} \right.$$

$$+ 2^\beta\, r\, w_{ab}\, [1 - (n-1)z]^{-\beta}$$

$$\left. + 2^\beta\, r\, w_{ab}\, [1 + (n-1)z]^{-\beta} \right\} \qquad (G\text{-}17)$$

$$\left(\frac{I}{C}\right)_b (second) = \left(\frac{\Omega_b}{\Phi''}\right)^\beta \left\{ z^{-\beta} \left(1 + \sum_{k=1}^{n-2} k^{-\beta} \right) \right.$$

$$+ 2^\beta\, r\, w_{ab}\, [1 - (n-3)z]^{-\beta}$$

$$\left. + 2^\beta\, r\, w_{ab}\, [1 + (n-3)z]^{-\beta} \right\} \qquad (G\text{-}18)$$

Eq. (G-8) reduces to

$$\left(\frac{\Phi''}{\phi}\right)^\beta \geq 1 + \frac{2^\beta\, w_{ba}}{r\,\xi\,(\beta)} \sum_{j=0}^{n} [1 + (n+1-2j)z]^{-\beta} \qquad (G\text{-}19)$$

and Eq. (G-15) reduces to

$$\left(\frac{\Phi''}{\phi\pi}\right)^\beta \geq \frac{\omega^{-\beta}}{2\,\xi\,(\beta)} \left\{ z^{-\beta} \left(1 + \sum_{k=1}^{n-2} k^{-\beta} \right) \right.$$

$$+ r\, w_{ab}\, 2^\beta (1 - (n-3)z)^{-\beta}$$

$$\left. + r\, w_{ab}\, 2^\beta (1 + (n-3)z)^{-\beta} \right\} \qquad (G\text{-}20)$$

Equations (G-19) and (G-20) form independent constraints. Note the inequalities; these are actually required because they need not necessarily be satisfied simultaneously with equality.

Setting (G-17) equal to (G-18) leads to a certain value of z (z_0 for example) for every value of n and rw_{ab}. Setting (G-19) equal to (G-20), with $z = z_0$, would lead to a relationship between w, w_{ba}, w_{ab}, r, and n, necessary to satisfy the equality. If we rewrite (G-19) as

$$\left(\frac{\Phi''}{\phi}\right)^\beta = 1 + \frac{2^\beta\, w_{ba}\, w_{ab}}{\xi\,(\beta) r\, w_{ab}} \sum_{j=1}^{n} [1 + (n+1-2j)z]^{-\beta} \qquad (G\text{-}21)$$

the number of independent variables that can be interrelated are: n, r w_{ab}, $w_{ab} \cdot w_{ab}$, and w. (Note that n and rw_{ab} together specify z_0.)

If we were to plot (G-20) and (G-21) as a function of $r\, w_{ab}$ we would get curves similar to the sketch shown in Figure G-1, and of course an infinite number of such figures could be drawn for different values of n and w.

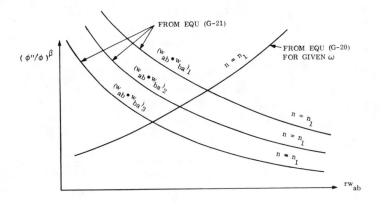

Figure G-1 Illustration of Constraint Behavior
(A) from Eq. (G-21)
(B) from Eq. (G-20) for given *W*

Any number of other curves could be derived from Figure G-1. For example, for a given rw_{ab} there is one value of $w_{ab} \cdot w_{ba}$ for which ϕ''/ϕ is the same as obtained from the two separate equations; in a sense these corresponding values are "optimum." These could be plotted as follows:

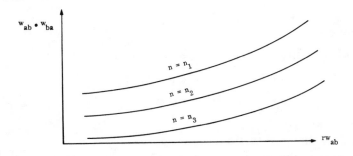

Another possible curve would be to plot the minimum ϕ''/ϕ as a function of n and $w_{ab} w_{ba}$. For a given $w_{ab} w_{ba}$ it can be seen that there is a minimum ϕ''/ϕ which occurs where the two types of curves intersect; furthermore this intersection will usually be a function of n. Thus, for a range of $w_{ab} \cdot w_{ba}$ the following types of curves could be drawn.

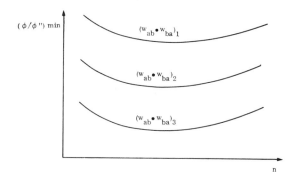

As can be seen there is no single solution. The spacing relationships depend on all the parameters involved as well as the constraints imposed. The preceding analysis also amply demonstrates the rapid increase in complexity that we encounter as we depart even slightly from the homogeneous model.

H

Constraints on EIRP Due to Flux-Density Limitations, for FDM/FM Telephony Signals

---

H.1 Introduction

FDM/FM telephony carrier can be expressed as:

$$S_c(t) = A_c cos \left[\omega_c t + \phi(t) \right] \tag{H-1}$$

where

A_c = carrier amplitude;
$f_c = \omega_c / 2\pi$ is the carrier frequency;
$\phi(t)$ = phase modulation on the carrier.

When the modulating signal, $x(t)$, is a multiplex of telephone channels, it can be modulated by a sample function of a zero-mean, stationary Gaussian random process (GRP) with a uniform spectral density [2], a representation which is implicit in all of our dealings with multichannel telephony. The transition between $x(t)$ and $\phi(t)$ is depicted in Figure H-1.

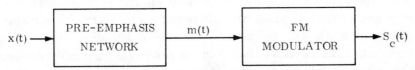

Figure H-1 Block Diagram of Transmitting End

As can be seen in the figure, the baseband signal is first processed (in general) through a pre-emphasis network (filter) with impulse response $h_p(t)$. The relationship between $x(t)$ and $m(t)$ is therefore given by:

$$m(t) = \int_{-\infty}^{\infty} x(\tau)\, h_p(t-\tau)\, d\tau \qquad \text{(H-2)}$$

The relationship between $m(t)$ and $\phi(t)$ is then:

$$\phi(t) = \int^{t} M(t')\, dt' \qquad \text{(H-3)}$$

The transitions from $x(t)$ to $m(t)$, and from $m(t)$ to $\phi(t)$, are linear, as shown by Equations (H-2) and (H-3). Hence, $m(t)$ and $\phi(t)$ are also stationary Gaussian random processes.

Let the power spectrum of $S_c(t)$ be represented by $G_c(f)$.*The power in some 4 kHz is then given by:

$$P_4 = \int_{\delta f} G_c(f)\, df \qquad \text{(H-4)}$$

where δf is the region of integration, and is such that is spans 4 kHz. Since the flux density limit applies to any 4 kHz band we are interested in the maximum value, $P_{4,m}$ of P_4 over all possible δf. This determination is critically dependent on the rms modulation index, M_0, defined as:

$$M_0 = \Delta f_{rms}/f_m \qquad \text{(H-5)}$$

where Δf_{rms} is the rms frequency deviation.

H.2 Evaluation of $P_{4,m}$ for Large Modulation Index

When M_0 is "large" the spectrum of $S_c(t)$ is relatively simple to obtain

---

*We shall use interchangeably the terms spectrum, power spectrum, or power spectral density (PSD).

and approaches a Ganssian shape independently of the spectrum of $m(t)$ and hence independently of the particular form of the pre-emphasis network. The definition of "large" modulation index is not precise but can be said to begin in the range $1 \lesssim M_0 \lesssim 2$. We assume that we are safely in the large M_0 range. Then, we have

$$G_c \ (f) = \frac{A_c^2/2}{\sqrt{2\pi}\ \Delta f_{\text{rms}}} e^{-(f-f_c)^2/2\Delta f_{\text{rms}}^2} \tag{H-6}$$

The peak of (H-6) is evidently at $f = f_c$, and assuming that 4 kHz is a small frequency increment compared to Δf_{rms}, we have

$$P_{4,m} \approx G_c \ (f_c) \times 4 \times 10^3 = \frac{(A_c^2/2)\ (4 \times 10^3)}{\sqrt{2\pi}\ \Delta f_{\text{rms}}}\ watts/4\ kHz$$

Substituting definition (H-5), and further assuming $f_m = \beta_n$, we have

$$P_{4,m} = \frac{(A_c^2/2)\ (4 \times 10^3)}{\sqrt{2\pi}\ \beta\ M_0\ n}$$

Using $\beta = 4200$, and realizing that $10\ log\ (A_c^2/2)$ is by definition the e.i.r.p. in dB, we get

$$10\ log\ P_{4,m} = eirp - 10\ log\ M_0 - 10\ log\ n - 4.2 \tag{H-7}$$

which, when substituted into (5-23) and (5-24) of the text, leads to the final result, (5-24).

H.3 Evaluation of Carrier Component ($P_{4,m}$) for Small Modulation Index

When M_0 attains moderate or small values, the spectrum, $G_c(f)$, has no longer a simple analytical form. However, under these conditions a significant component (spectral line) appears at the carrier frequency* and this component is such that it contains the predominant energy in any 4 kHz band. Since this carrier component is considerably simpler to calculate than the entire spectrum, it will be used to find an upper bound on the allowable e.i.r.p. when M_0 is small; this upper bound should also be a good approximation to the actual e.i.r.p. limit. This computation is carried out below.

---

*In general there is always a carrier component. For large M_{rms} however, its value is negligible.

As noted above, when $x(t)$ is modeled by a stationary Gaussian random process (GRP), then $\phi(t)$ is also a GRP. When a carrier is phase-modulated by a GRP it possesses a spectral component at the carrier frequency, $G_{cc}(f)$, given by [16]

$$G_{cc}(f) = \frac{A_c^2}{2} e^{-R_\phi(0)} \delta(f\text{-}f_c)$$ (H-8)

where

$$R_\phi(0) = mean\text{-}square\ value\ of\ \phi(t),\ (rad^2).$$

It is given by:

$$R_\phi(0) = \int_{-\infty}^{\infty} G_\phi(f)\ df$$ (H-9)

where $G_\phi(f)$ is the power spectral density of $\phi(t)$.

From Equation (H-3) it can be seen that

$$G_\phi(f) = \frac{1}{(2\pi f)^2}\ G_m(f)$$ (H-10)

where $G_m(f)$ is the spectral density of $m(t)$.

From Equation (H-2), we have the relationship

$$G_m(f) = G_x(f)\ |H_p(f)|^2$$ (H-11)

where

$G_x(f)$ = spectral density of $x(t)$;

$H_p(f)$ = transfer function of the pre-emphasis network.

Using Equations (H-9) to (H-11), we have:

$$R_\phi(0) = \int_{-\infty}^{\infty} \frac{G_x(f)\ |H_p(f)|^2}{(2\pi f)^2} df$$ (H-12)

To evaluate Equation (H-12) we need to know the properties of $x(t)$ and the form of the pre-emphasis network. As mentioned previously, $x(t)$ represents the multichannel baseband and is modeled by a GRP with uniform spectral density, $G_x(f)$. Specifically, the spectrum will be taken as shown in Figure H-2. In the figure, f_m and f_l represent, respectively, the upper and lower baseband frequency.

Before applying Equation (H-12), there is one more constraint that must be applied. This is the customary requirement that the mean-square frequency deviation with a pre-emphasis be the same as without pre-emphasis. It is easy to see that this constraint can be stated as:

$$\int_{-\infty}^{\infty} G_x(f)\, df = \int_{-\infty}^{\infty} G_m(f)\, df = (2\pi\,\Delta f_{rms})^2$$

or, using (H-11)

$$\int_{-\infty}^{\infty} G_x(f)\, df = \int_{-\infty}^{\infty} = G_x(f)\, |H_p(f)|^2\, df$$

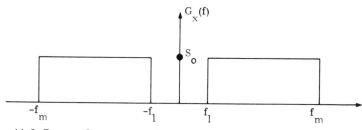

Figure H-2 Power Spectrum of $x(t)$

Equations (H-12) and (H-13) together determine the strength of the carrier component which, as can be seen, depends on the pre-emphasis network. We now evaluate $R_\phi(0)$ for two types of pre-emphasis networks.

H.3.1 CCIR Pre-Emphasis

The standard CCIR pre-emphasis network has the following form [17]

$$|H_p(f)|^2 = A\ \frac{7.9 + 1.52\, g^2\, (f)}{1 + 1.52\, g^2\, (f)} \tag{H-14}$$

where

A = a normalizing constant, determined by Equation (H-13)

$G(f) = f/f_R - f_R/f$

$f_R = 1.25 f_m$

Substituting for g(f), Equation (H-14) takes the form

$$|H_p(f)|^2 = A \; \frac{f_0^4 + 3.2 f_0^2 + 1}{f_0^4 - 1.34 f_0^2 + 1} \tag{H-15}$$

where $f_0 = f/f_R$. Equation (H-15) is plotted in Figure H-3.

Equation (H-15) is somewhat unwieldy to work with. Consequently, we have made a two-point fit of this equation, as follows:

$$|H_p(f)|^2 \cong A \; \{1 + 4.5 f_0^2 + 6.75 f_0^4\} \tag{H-16}$$

Equation (H-16) is also plotted on Figure H-3 and, as can be seen, the fit is excellent for $f_0 \leq 0.8$ or $f \leq f_m$, which is the region of interest. Consequently, we shall adopt Equation (H-16) as the pre-emphasis characteristic.

To evaluate A, we use Equation (H-13) and Figure H-2 as follows:

$$2 S_o \int_{f_l}^{f_m} df = 2 S_o \int^{f_m} A \; (1 + 4.5 f_0^2 + 6.75 f_0^4) \; df$$

$$(f_m - f_l) = f_R \int_{0.8 \, f_l/f_m} A \; (1 + 4.5 f_0^2 + 6.75 f_0^4) \; df_0$$

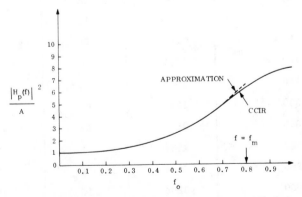

Figure H-3 CCIR Pre-emphasis Characteristic and Polynominal Fit

Recalling that $f_R = 1.25 f_m$ and defining $f_l/f_m = \varepsilon$, we have

$$(1 - \varepsilon) = 1.25 A \int_{0.8\varepsilon}^{0.8} (1 + 4.5 f_o^2 + 6.75 f_o^4) \, df_o$$

Performing the indicated integration yields:

$$(1 - \varepsilon) = 1.25 A \{0.8 (1 - \varepsilon) + 1.5 (0.8)^3 (1 - \varepsilon^3) + 1.35 (0.8)^5 (1 - \varepsilon^5)\}$$

After further manipulation, we get:

$$A = \{1 + 0.96 (1 + \varepsilon + \varepsilon^2) + 0.555 (1 + \varepsilon + \varepsilon^2 + \varepsilon^3 + \varepsilon^4)\}^{-1} \quad \text{(H-17)}$$

Numerical evaluation of Equation (H-17) evidently depends on ε which, in turn, is a function of the number of channels and the particular baseband arrangement. Some typical values of ε, as given by the CCIR [18] are presented in Table H-1.

TABLE H-1

SOME TYPICAL VALUES OF $f_m, f_l,$ and f_m/f_l

Number of Channels	f_m (kHz)	f_l (kHz)	$f_m/f_l = 1/\varepsilon$
12	60	12	5
24	108	12	9
60	300	60	5
120	552	60	9.2
240	1052	60	17.5
300	1296	64	20.2
600	2660	64	41.5
960	4188	316	13.2
1260	5564	316	17.6
1800	8204	316	25.9

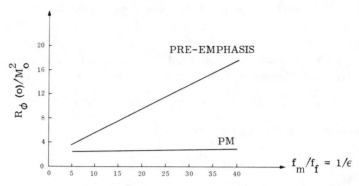

Figure H-4 Comparison of Factors Influencing Carrier Component

H.3.3 Evaluation of $P_{4,m}$ for CCIR Pre-Emphasis

Using Equations (H-4) and (H-8), we have:

$$P_{4,m} = \frac{A_c^2}{2} \ e^{-R_\phi(0)} \quad \delta \ (f\text{-}f)_c) \ df = \frac{A_c^2}{2} \ e^{-R_\phi(0)} \tag{H-26}$$

It should be recalled that Equation (H-26) is really a lower bound since only the carrier component is assumed within the 4 kHz band around the carrier. Taking logs and using (H-21) we find

$$10 \ log \ P_{4,m} = eirp - 4.34 \ M_0^2 \ \{0.4/\varepsilon + 1.15 + 0.368 \ (1 + \varepsilon + \varepsilon^2)\}$$

which, when used in conjunction with (5-23) and (5-24) yields the final answer (5-26).

------------------------------ **I** ------------------------------

Constraints on EIRP Due to Flux-Density Limitations for Multilevel PSK Signals

I.1 Introduction

As with FDM/FM telephony, the quantity which determines the power in a given bandwidth for digital signals is, of course, the power spectrum. The spectrum is a function of the shape of the phase modulating pulses and of the statistics of the input symbols. To simply matters somewhat we shall assume a rectangular signalling waveform.

As can be seen from the table we have, roughly, $0.025 < \varepsilon < 0.2$. This yields, using Equation (H-17), $0.347 < A < 0.391$. There is, therefore, relatively little variation in A as ε varies. As an approximation, then, we pick for A the constant value 0.4, which is the number usually used. Hence,

$$|H_p(f)|^2 = (1 + 4.5\, f_0^2 + 6.75\, f_0^4)/2.5 \tag{H-18}$$

Using (H-18) in (H-12), we have:

$$R_\phi(0) = \frac{2\, S_o}{2.5} \int_{f_l}^{f_m} \frac{(1 + 4.5\, f_0^2 + 6.75\, f_0^4)}{(2\pi)^2}\, df$$

$$= \frac{S_o}{(2\pi)^2} \frac{1}{(1.25)^2\, f_m} \int_{0.8}^{0.8} \frac{(1 + 4.5\, f_0^2 + 6.75\, f_0^4)}{f_0^4}\, df_o$$

Performing the indicated integration, we get:

$$R_\phi(0) = \frac{2\, S_o}{(2\pi)^2} \frac{1}{(1.25)^2\, f_m} \{1.25\, \frac{(1-\varepsilon)}{\varepsilon} + 3.6\, (1-\varepsilon) + 1.15\, (1-\varepsilon^3)\} \tag{H-19}$$

Equation (H-19) can be put into a more convenient form by recognizing that:

$$(2\pi\Delta f_{rms})^2 = \int_{-\infty}^{\infty} G_x(f)\, df = 2\, S_o\, f_m\, (1 - \varepsilon) \tag{H-20}$$

Combining Equations (H-19) and (H-20), and using the fact that $M_o = \Delta f_{rms}/f_m$, we finally get:

$$R_\phi(0) = M_0^2\, \{0.4/\varepsilon + 1.15 + 0.368\, (1 + \varepsilon + \varepsilon^2)\} \tag{H-21}$$

Equation (H-21) is plotted in Figure H-14 as a function of $1/\varepsilon$. The quantity $P_{4,m}$, implied by Equation (H-21) is considered in Section H.3.3.

H.3.2 Ideal Pre-Emphasis (Phase Modulation)

For the sake of comparison, it is of interest to study the "ideal" pre-emphasis characteristic:

$$|H_p(f)|^2 = Bf^2 \tag{H-22}$$

where B is normalizing constant determined from Equation (H-13). This characteristic is "ideal" in the sense that the corresponding de-emphasis characteristic (i.e., the inverse) is such that the parabolic noise output from an ideal FM receiver will be "flattened" or equalized over all channels. It is also evident that the combination of the pre-emphasis characteristic of Equation (H-22) and the FM modulator is completely equivalent to phase modulation of the carrier by $x(t)$.

To determine B we use Equation (H-13), which gives:

$$2 \, S_o \, f_m \, (1 - \varepsilon) = 2 \, S_o \int_{f_l}^{f_m} B f^2 df = \frac{2}{3} \, S_o \, B f_m^3 \, (1 - \varepsilon^3) \qquad \text{(H-23)}$$

whence

$$B = 3 \, (1 - \varepsilon) / f_m^2 \, (1 - \varepsilon^3) \qquad \text{(H-24)}$$

Using Equation (H-12) we have

$$R_\phi (0) = \frac{B}{(2\pi)^2} \, 2 \, S_o \int_{f_l}^{f_m} df = 2 \, S_o \, B f_m \, (1 - \varepsilon) / (2\pi)^2$$

However, we see from (H-23), that:

$$(2\pi \Delta f_{rms})^2 = 2 \, S_o \, f_m \, (1 - \varepsilon) = \frac{2}{3} \, S_o \, B f_m^3 \, (1 - \varepsilon^3)$$

Hence, from the definition of M_o, we have:

$$R_\phi (0) = 3 \, M_{rms}^2 / (1 + \varepsilon + \varepsilon^2) \qquad \text{(H-25)}$$

Equation (H-25) is also plotted in Figure H-4. It can be seen, particularly at small values of ε, that there can be a substantial difference between phase modulation and CCIR pre-emphasis.* For example, if $\varepsilon = 0.1$ and $M_{rms} = 0.5$, the carrier component in PM is $exp(-0.675) = 0.51$, while for CCIR pre-emphasis the carrier component is $exp(-1.39) = 0.25$. Thus, in the former case, the carrier component is more than 3 dB greater.

---

*We point this out specifically because it is frequently assumed that FM with pre-emphasis is more or less equivalent to PM. This may be the case for signal-to-noise calculations but not necessarily so, as shown, for other purposes.

I.2 Evaluation of $P_{4,m}$

Under the above assumption the phase waveform $\alpha(t)$ is constant over each symbol interval, T. Thus,

$$\alpha_k = \frac{2\pi k}{M} \quad k = 0, 1, \ldots, M\text{-}1$$

specifies the m possible phase values. With each α_k is associated a probability of occurence p_k. With these stipulations it can be shown [19] that, with total power P_c, the power spectral density is given by

$$G_c(f) = P_c T \left\{ 1 - \left| \sum_{k=1}^{M} p_k e^{j\alpha_k} \right|^2 \right\} \left\{ \frac{\sin \pi (f\text{-}f_c) T}{\pi (f\text{-}f_c) T} \right\}^2$$

$$+ P_c \left| \sum_{k=1}^{M} p_k e^{j\alpha_k} \right|^2 \sigma (f\text{-}f_c) \tag{I-1}$$

The first term is the continuous part of the spectrum and the second is a spectral line which appears at f_c.

Equation (I-1) cannot be further reduced without specific assumptions. To illustrate the method of evaluation we take as an example the case $M = 4$ and we further assume that three of the four symbols occur with equal probability while the fourth occurs with a different probability. Symbolically

$$p_0 = p; \; p_k = \frac{1}{3}(1 - p), \; k = 1, 2, 3 \tag{I-2}$$

Under the assumption of (I-2) the summation in (I-1) reduces easily to the following:

$$\left| \sum_{k=1}^{M} p_k e^{j\alpha_k} \right|^2 = \left(\frac{4p-1}{3} \right)^2 \tag{I-3}$$

Let $\widetilde{G}_c(f)$ and $G_{cc}(f)$ represent the continuous and carrier components, respectively. We assume that 4 kHz is sufficiently small with respect to $(1/T)$ that due to the first alone,

$$P_{4,m} \approx \widetilde{G}_c(f_c) \times 4 \times 10^3.$$

Hence we find

$$P_{4,m} = P_c \, (4000/R) \left\{ 1 - \left(\frac{4\cdot p - 1}{3} \right)^2 \right\} \tag{I-4}$$

where $R = 1/T$ is the symbol rate. Taking logs,

$$10 \log P_{4,m} = eirp + 36 - 10 \log R + 10 \log \left\{ 1 - \left(\frac{4\cdot p - 1}{3} \right)^2 \right\}$$

which leads to Equation (5-27) of the text.

The carrier component is immediately obtained from (I-3) and (I-1). For this component taken by itself:

$$P_{4,m} = G_c(f) \, P_c \left(\frac{4p - 1}{3} \right)^2$$

or,

$$10 \log P_{4,m} = eirp + 10 \log \left(\frac{4p - 1}{3} \right)^2$$

from which Equation (5-28) is obtained.

J

Interference NPR and Receiver Transfer Characteristic for High-Index Angle-Modulated Telephony Signals

---

Our purpose here is to derive (6-28) of the text and the related Equation (6-26), which are, respectively, the NPR and RTC for high-index FM (or angle-modulated) telephony carriers interfered-with by signals not necessarily identical but of the same nature. Although restricted in applicability, this case is in itself of some importance and it is instructive as an example of the procedure used to obtain specific results.

J.1 Baseband Interference Spectrum

From Equation (6-20), assuming wanted and interfering signals given by (6-21) and (6-22), respectively, we have for the baseband PSD due to interference:

$$S_\lambda(f) = \frac{r^2}{4} \left\{ S_i(f\text{-}f_d) * S_w(f) + S_i(\text{-}f\text{-}f_d) * S_w(\text{-}f) \right\} \tag{J-1}$$

where now both $S_w(f)$ and $S_i(f)$ are the normalized equivalent low-pass spectra. Under the high-index assumption these spectra are given by

$$S_w(f) \cong \frac{e^{-f^2/2\Delta f_1^2}}{\sqrt{2\pi} \, \Delta f_1}; \, S_i(f) \cong \frac{e^{-f^2/2\Delta f_2^2}}{\sqrt{2\pi} \, \Delta f_2}$$

where $\Delta f_{1,1}$ is the rms frequency deviation of wanted and interfering signals, respectively. Therefore, (J-1) becomes explicitly

$$S_\lambda (f) = \frac{r^2}{4} \int_{-\infty}^{\infty} \left\{ \frac{e^{-(\alpha-f_d)^2/2\Delta f_2^2}}{\sqrt{2\pi}\,\Delta f_2} + \frac{e^{-(\alpha+f_d)^2/2\Delta f_2^2}}{\sqrt{2\pi}\,\Delta f_2} \right\} \frac{e^{-(f_d-\alpha)^2/2\Delta f_1^2}}{\sqrt{2\pi}\,\Delta f_1} d\alpha$$

Performing the indicated integration results in

$$S_\lambda (f) = \frac{r^2/4}{\sqrt{2\pi}\,\sqrt{\Delta f_1^2 + \Delta f_2^2}} \; exp \; \frac{-(f+f_d)^2}{2\,(\Delta f_1^2 + \Delta f_2^2)} + exp \; \frac{-(f-f_d)^2}{2\,(\Delta f_1^2 + \Delta f_2^2)} \qquad \text{(J-1)}$$

Defining

$M_1 = \Delta f_1/f_{m_1}$
$M_2 = \Delta f_2/f_{m_2}$
$M_2^1 = M_2 f_{m_2}/f_{m_1}$
$M^2 = M_1^2 + M_2^{1^2}$
$u = f/f_{m_1}$
$\upsilon = f/f_d$

Equation (J-2) can be re-written as

$$S_\lambda (f) = \frac{r^2/4}{\sqrt{2\pi}\,f_{m_1}\,M} \left\{ exp\,[-(u+\upsilon)^2/2M^2] + exp\,[-(u-\upsilon)^2/2M^2] \right\} \qquad \text{(J-3)}$$

At the output of an FM discriminator the PSD is $(2\pi f)^2$ time that of (J-3). Further, if the pre-emphasis network is given by $G_p(f)$, the de-emphasis network is of the form $G_d(f) = K/G_p(f)$. Hence the baseband interference spectrum $G_i(f)$ is given by

$$G_i(f) = \frac{r^2(2\pi f)\,G_d\,(f)}{4\,\sqrt{2\pi}\,f_{m_1}\,M} \left\{ exp\,[-(u+\upsilon)^2/2M^2] + exp\,[-(u-\upsilon)^2/2M^2] \right\} \qquad \text{(J-4)}$$

Equation (J-4) is applicable to pure PM if $G_p(f)\,\alpha\,f^2$ and to the FM case when $G_p(f)$ is constant. If a weighted spectrum is desired, then the weighting function simply multiplies $G_i(f)$.

J.2 Output Signal Power

A baseband multichannel telephony signal is customarily modeled as a white Gaussian process with the spectrum shown in Figure J-1.

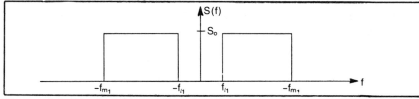

Figure J-1. Wanted Signal Baseband Spectrum

At the receiver, the output baseband spectrum is exactly that of Figure J-1 except for the de-emphasis filter constant. Thus,

$$S_o(f) = KS(f), f_{l_1} \leq |f| \leq f_{m_1}$$

The output "signal" power in any small band, b, is then:

$$P_o = 2KS_o b \tag{J-5}$$

The rms frequency deviation, by convention, is maintained the same with or without pre-emphasis. Therefore,

$$2S_o(f_{m_1} - f_{l_1}) = (2\pi \Delta f_1)^2$$

or defining $\varepsilon_1 = f_{l_1}/f_{m_1}$,

$$2S_o f_m (1 - \varepsilon_1) = (2\pi \Delta f_1)^2$$

and substituting into Equation (J-5), we have

$$P_o = Kb (2\pi \Delta f_1)^2 / f_{m_1} (1 - \varepsilon_1)$$
$$= Kb (2\pi)^2 M_1{}^2 f_{m_1} (1 - \varepsilon_1) \tag{J-6}$$

Equation (J-6) gives the output power in the signal-simulating white noise load, in a channel bandwidth. We are often concerned, however, with the power in a standard test-tone in the same channel. This is obtained as follows:

First we note from Appendix A that

$$10 \log (2\pi \Delta f_1)^2 = -15 + 10 \log n_1, (dBm\ 0), n_1 \geq 240$$
$$= -1 + 4 \log n_1, (dBm\ 0), 12 < n_1 < 240$$

where n_1 is the number of channels in the baseband. Let $(2\pi \delta f)$ be the test-tone rms radian frequency deviation; since, by definition, test-tone power is $0\ dBm\ 0$, we have

$$10 \log (\delta f / \Delta f_1)^2 = 15 - 10 \log n_1, n_1 \geq 240$$
$$= -1 - 4 \log n_1, 12 \leq n_1 \leq 240$$

or

$$(\delta f / \Delta f_1)^2 = 31.6 / n_1, n_1 \geq 240$$
$$= -1.26 / n_1{}^{0.4}, 12 \leq n_1 < 240 \tag{J-7}$$

As indicated in Appendix A we may write

$$Bn_1 = f_{m_1} (1 - \varepsilon_1) \tag{J-8}$$

where, usually, $4000 \leq B \leq 4500$, for most channel arrangements.

The test-tone power is given by:

$$P_{tt} = K (2\pi \delta f)^2$$

which, using Equations (J-7) and (J-8) becomes:

$$P_{tt} = K (2\pi)^2 \Delta f_1^2 \, 31.6/n_1$$

$$= K (2\pi)^2 \Delta f_1^2 \left(\frac{31.6}{n_1} \frac{b}{f_{m_1} (1 - \varepsilon_1)}\right) \left(\frac{\beta \, n_1}{b}\right)$$

$$= \frac{Kb \, (2\pi\Delta f_v)^2}{f_{m_1} (1 - \varepsilon_1)} 36.6 \, (B/b)$$

$$= P_o \, 31.6 \, (B/b), \, n_1 \geq 240$$

(J-9)

and similarly,

$$P_{tt} = P_o \, 1.26 n_1^{0.6} \, (\beta/b), \, 12 \leq n_1 < 240 \qquad \text{(J-10)}$$

J.3 NPR and Receiver Transfer Characteristic

As discussed in Appendix A the NPR is the ratio of the signal-simulating noise load power to the noise power in a small band, b, usually a voice channel. Thus,*

$$NPR(f) = P_o/2bg_i(f)$$

which, upon using Equations (J-4) and (J-6), becomes

$$NPR(f) = \frac{2 \sqrt{2\pi} \, M_1^2 \, M \, G_p(f)}{(1 - \varepsilon_1) \, r^2 \, u^2 \{exp - [(u + v)^2] + exp - [(u - v)^2/2M^2]\}} \qquad \text{(J-11)}$$

which is (6-28) of the text.

The test-tone-to-noise ratio, $(S/N)_{tt}$, is simply related to the NPR through Equations (J-9) and (J-10). Thus,

$$\left(\frac{S}{N}\right)_{tt} = NPR(f) \, 31.6 \, (B/b), \, n_1 \geq 240 \qquad \text{(J-13)}$$

$$\left(\frac{S}{N}\right)_{tt} = NPR(f) \, 1.26 \, n_1^{0.6} \, (B/b), \, 12 \leq n_1 < 240 \qquad \text{(J-13)}$$

To evaluate the constants in (J-12) and (J-13) we take the commonly used values $B = 4.2 \times 10^3$, $b = 3.1 \times 10^3$, and for psophometric weighting the factor 1.78, so that

---

*NPR is defined for positive frequencies, hence the factor of 2 is necessary since $G_i(f)$ is a two-sided PSD.

$$\left(\frac{S}{N}\right)_{tt} = \begin{cases} 42.8 \ NPR(f), \ n \geq 240 & \textit{without} \\ & \textit{psophometric} \\ 1.71 \ n^{0.6} \ NPR(f), \ 12 \leq n \leq 240 & \textit{weighting} \\ \\ 76.1 \ NPR(f), \ n \geq 240 & \textit{with} \\ & \textit{psophometric} \\ 3.04 \ n^{0.6} \ NPR(f), \ 12 \leq n \leq 240 & \textit{weighting} \end{cases}$$

By definition, the receiver transfer characteristic, R_x, is given by:

$$\left(\frac{S}{N}\right)_{tt} = \frac{1}{r^2} \ R_x, \tag{J-14}$$

since $1/r^2$ is the carrier-to-interference ratio. Therefore, combining (J-12), (J-13), and (J-14) we may write

$$R_x = NPR(f) \ (B/b) \ L(n_1) \ r^2 \tag{J-15}$$

with an obvious identification of terms, which is exactly Eq. (6-26) of the text. Both (J-11) and (J-15) are unweighted, but of course a psophometric weighting factor can be applied to either.

For high-index identical co-channel carriers, Eq. (J-11) becomes for the top channel $(u = 1)$,

$$NPR(f_m) = \frac{2 \sqrt{\pi} \ M_0^3 \ G_p \ (f_m)}{(1 - \varepsilon) \ r^2 \ exp \ (-1/2 \ M_0^2)}$$

where M_0 is the common multi-channel rms modulation index. Assuming $G_p = 4 \ dB$, this reduces to

$$NPR(f_m) = \frac{8.9 \ M_0^3}{(1 - \varepsilon) \ r^2 \ exp \ (-1/M_0^2)}$$

since the terms in the denominator (other than r^2) are slightly less than one, we "round-off" the constant to 9.5, so that

$$NPR(f_m) \approx 1/r^2) \ 9.5 \ M_0^3$$

which for large M_0 is the form to which (6-31) reduces.

J.4 NPR and RTC Behavior as a Function of Frequency

The NPR ro RTC is a function of the position of the channel in the baseband, as can be seen from (J-11). We are usually interested in the worst channel since, for given minimum performance, this channel will set the parameter constraints required to meet this performance.

It has been implied that the worst channel is the top channel; that this is the case for practical purposes, at least for certain pre-emphasis characteristics, will now be shown.

Retaining only those terms in (J-11) which are a function of baseband frequency, we are led to a consideration of the following expression:

$$B(u) = u^2 \, G_d(u) \, \{exp \text{-} [(u+v)^2/2M^2] + exp \text{-} [(u-v)^2/2M]\} \quad \text{(J-16)}$$

which, except for positive constants, is proportional to the reciprocal of NPR; the term $G_d(u)$ is the de-emphasis characteristic written as a function of $u = f/f_{m_1}$.

Since $B(u)$ evidently depends on the specific pre-emphasis characteristic, it is convenient to consider the different possibilities separately.

J.4.1 Frequency Modulation (No Pre-emphasis)

For the case of pure frequency modulation, i.e., no pre-emphasis, we may take $G_d(u) = 1$, so that (J-16) reduces to:

$$B(u) = u^2 \, \{exp \text{-} [(u+v)^2/2M^2] + exp \text{-} [(u-v)^2/2M]\} \quad \text{(J-17)}$$

The behavior of (J-17) is not immediately obvious since the two terms may have opposing tendencies. It turns out to be convenient to consider the derivative which, after some manipulations, can be written as:

$$B'(u) = 2 \, u \, exp \text{-} [(u^2 + v^2)/2M^2]$$

$$\left\{ \left(\frac{1 - u^2}{2 \, M^2}\right) [exp \, (uv/M^2) + exp \, (\text{-}uv/M^2)] \right. \quad \text{(J-18)}$$

$$\left. + \frac{uv}{2M^2} [exp \, (uv/M^2) \text{-} exp \, (\text{-}uv/M^2)] \right\}$$

Note that the second term within the brackets is never negative.* Therefore, a sufficient condition for (J-18) to be non-negative is that the first term in the brackets also be non-negative, which is satisfied if:

$$1 \text{-} u^2/2M^2 \geq 0$$

or, since $u \leq 1$, if $M \geq 1\sqrt{2}$ \quad (J-19)

Condition (J-19) is sufficient to ensuer that $B'(u)$ is non-negative which means that B(u) is non-decreasing and, hence, that NPR is a non-increasing function of baseband position. Recall, however, that the original assumptions for which (J-16) is valid to begin with, included the high-index condition, implying at least $M \geq 1$. Since (J-19) includes this condition, we conclude that for pure, wide-index FM, the top channel is the worst channel.

---

*Note that v is defined positively and, of course, $\epsilon \leq u \leq 1$.

J.4.2 Phase Modulation ("Ideal" Pre-emphasis)

Phase modulation represents "ideal" pre-emphasis in the sense that it perfectly equalizes or "flattens out" the theoretically triangular output thermal noise in all channels. It is not ideal in that sense, in the present case, since the baseband interference noise is not triangular. Nevertheless, PM is frequently studied as a limiting case of more practical forms of pre-emphasis. The de-emphasis characteristic for PM may be written, to within a constant, as $G_d(u) = 1/u^2$.

Therefore, Eq. (J-16) becomes

$$B(u) = \{exp - [(u + v)^2/2M^2] + exp - [(u - v)^2/2M]\}$$

It is evident that when $v = 0$ (co-channel case), $B(u)$ is a decreasing function of u and hence, contrary to the FM case, the lowest channel is the worst. When $v \neq 0$, the top or the bottom channel may be the worst, depending upon the specific values of v and M.

J.4.3 CCIR Pre-emphasis

The CCIR de-emphasis characteristic, to within a constant, may be written as

$$G_d(u) = 2.5 \ \frac{1 - 0.86\,u^2 + 0.41\,u^4}{1 + 2.05\,u^2 + 0.41\,u^4} \tag{J-20}$$

When Eq. (J-20) is substituted into Eq. (J-16), the behavior of the resultant is not apparent, and a study of the derivative is quite cumbersome. It is simpler instead, to plot several cases and note the resulting trends. Figures (J-2) and (J-3) show plots of $B(u)$ for various values of v and M. It can be seen that for small values of v and M, the worst channel is somewhere in the middle of the baseband. The

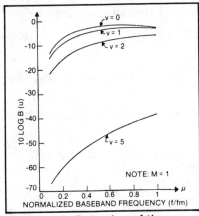

Figure J-2. Behavior of the Function B [u]

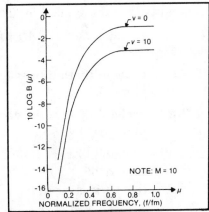

Figure J-3. Behavior of the Function B [u]

maximum difference, however, between the worst channel and the top channel is only about 1 dB so that, in such cases, the top channel can still be considered the worst with little error. For large values of v/M, the top channel, is indeed the worst, while for large values of M, the top channel is not necessarily the worst but the top third of the baseband performs virtually the same, varying only by hundredths of a dB. Thus, for practical purposes, the top channel can be considered the worst one. Since CCIR pre-emphasis represents perhaps the most common situation, we are justified, in the high-index case, in considering only the top channel.

K

Bounding Relations for M-ary CPSK

In Chapter 7 we noted that the error probability for M-ary CPSK, P_M, can be bounded by $P_M \leq P_M \leq P_M$ where P_M is given by (7-38) in a distortionless system corrupted by noise and interference. The utility of using the upper bound as an approximation depends on the closeness of P_M to P_M. Here we demonstrate that under fairly general conditions P_M and P_M are sufficiently close so that P_M is adequately tight for most purposes. Two sets of relationships are derived under differing levels of generality.

To obtain the first relationship, refer to Figure K-1, in which the decision process is graphically represented in terms of the familiar decision cones, illustrated for $M = 8$, and assuming without loss of generality that the symbol for 0° is sent. From the figure we have

$$\hat{P}_M = P(AA') + P(BB') - P(A'B') \tag{K-1}$$

where

P(AA') = probability that the vector terminus lies in the upper half-plane AA'

P(BB') = probability that the vector terminus lies in the lower half-plane BB'

P(A'B') = probability that the vector terminus lies in the hatched cone $A'B'$.

From Eq. (K-1) it is clear that

$$\hat{P}_M \leq P_M = P(AA') + P(BB') \tag{K-2}$$

On the other hand, from the figure we have the obvious inequality

$$P(A'B') \leq P_M$$

which along with Eq. (K-1) is easily shown to yield

$$2\, P_M \leq P(AA') + P(BB') \tag{K-3}$$

Combining (K-2) and (K-3) gives

$$0.5\, \hat{P}_M \leq P_M < P_M \tag{K-4}$$

Considering the simplicity of the argument, (K-4) gives surprisingly tight bounds. Note that no assumptions concerning the statistics of the processes involved needed to be made. Observe also from the figure that the bound applies for a more general class of channels than the one in whose context the initial discussion was made.

A tighter but slightly less general bound can be obtained by making a very weak assumption that can always be expected to be satisfied under normal operating conditions. In Figure K-1 the angle β_0, defined by Eq. (7-29) is identified. The error probability can be formulated in terms of the probability density of β_0, $p(\beta_0)$, which is typically as sketched in Figure K-2, and in which the positions of the lines AA', BB', are shown. Then, P_M is the area under the curve, exclusive of the central portion from B to A. Our only assumption is that $p(\beta_0)$, be non-increasing away from 0; this is clearly very weak, and actually can be phrased even less restrictively. From this assumption, and realizing that the abscissa of Figure K-2 can be visualized as M contiguous segments of phase, one can write immediately

$$P(A'B') \leq P_M/(M-1)$$

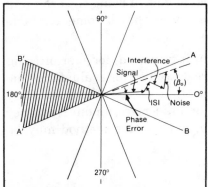
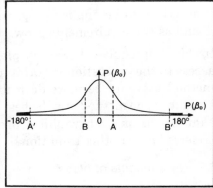

Figure K-1 Illustration of Decision Cones for M-8

Figure K-2 typical behaviour of P (β_0)

which, introduced into (K-1), yields

$$P(A'B') \leq \frac{1}{M}[P(AA') + P(BB')] \qquad (K-5)$$

Inserting (J-5) into (J-1) to obtain a lower bound, and using (J-2) again for the upper, we obtain

$$\left(\frac{M-1}{M}\right) PM \leq PM \leq PM$$

which for $M \geq 4$ is quite tight for most applications.

———————————————— **L** ————————————————

CCIR	Doc. IWP 4/1-981-E
Interim Working Party 4/1	27 May 1981
Geneva, May 1981	Original: English

Working Group 4/1-C
CHAPTER 6: EQUITABLE ACCESS TO THE ORBIT AND SPECTRUM

6.1 *Introduction*

The potential capacity of the geostationary satellite orbit and the radio frequency spectrum for the various space radiocommunications services is large compared with the demands expected to be placed on it. However, if all administrations are to have access to it when they need it and to a sufficient degree, it is necessary:

for efficient methods to be used to plan and coordinate the use of orbit and spectrum, and;

for improvements in the technical means of using the orbit and spectrum to be applied progressively as they become available and as the requirements grow.

In this chapter, five alternative planning methods for guaranteeing access to the geostationary orbit are described for evaluation. One method may not be more desirable than all the others for every situation: one method may be best for one service, for one frequency band or for one geographical area, but another method may be preferred under other conditions.

6.2 *Key elements of plans*

The key elements of a plan for space services are descriptions or specifications of: (1) the telecommunications service requirements to be satisfied, (2) the range of technical parameters of the networks and their associated interference criteria, (3) the structure of the plan and the resulting allotments, and (4) the methods for subsequent modifications to accommodate changing service requirements and technology. Each of the key elements will be discussed in turn.

6.2.1 *Service requirements*

Service requirements are prepared and/or coordinated by individual administrations, groups of administrations, international consortia, and by specialized international agencies to reflect domestic, regional and international telecommunication service requirements. The specification of these requirements could range from a single statement of the total bandwidth, preferred orbital position and associated service area to a complete delineation of the types and number of circuits (telephony, data and television) and their connectivity within the service area.

Service requirements are closely coupled to the time element associated with the particular approach to planning. On the one hand, the service requirements may be an estimate of the requirements for the first two to three years of operation of the network: the plan of the network operator/owner being to expand the network based on actual requirements derived from network operating experience. On the other hand, the service requirements could be fully specified (types and number of circuits and connectivity) over a given period of time; 20 years for example.

6.2.2 *Technical parameters and criteria*

As a range of options, each Administration or network owner would be given the flexibility to design a system which uniquely satisfied the service requirements according to some criteria, minimum cost for example, determined by the network owner. Inter-system interference would be negotiated in the same way appropriate to the planning used. The technical parameters and intersystem interference allowance of the network could change with time, in response to changing service requirements, technology, and cost.

However, on the other hand, the technical parameters and intersystem interference allowance would be fully specified for all networks for a long period of time (say 20 years).

6.2.3 *Structure and allotments*

Under the present *Radio Regulations* specific band segments of the radio frequency spectrum are allocated in the various ITU regions to the space services. These bands are available to all administrations to satisfy their telecommunications requirements. Assignments are recorded in the *Master Register* after successful coordination with other networks. Similarly, any orbital position is available to an administration for its network subject to successful coordination. These assignments in the Master Register constitute one possible structure of a plan. Other structures and allotment methods are possible. For example, segments of the orbital arc could be allocated exclusively to international networks, to regional networks, and to domestic networks. Similarly, sub-allocations within bands allocated to the space services could be made for specific types of transmission parameters such as those associated with single-channel-per-carrier (SCPC), multichannel frequency division multiplex FM, and TV/FM carriers. Using still another allotment method, band segments and orbital locations would be allocated to each administration or network owner for perhaps a 20 year period of time.

6.2.4 *Modification*

The final major element in planning is the accommodation of changes in service requirements, technical parameters, criteria, and technology, and consequential changes in the allotment. Over a period of time, service requirements are likely to change due to a number of factors. Different service categories will respond differently to political, demographic, economic and technological change. New applications, such as data communications for business, may develop and experience a demand for service which was not foreseen. The ability and the degree to which changing requirements may be accommodated within a plan depends on the approach to planning and that portion of the total capacity of the geostationary orbit-spectrum resource not already alloted in the plan. The capacity is a function of the technical parameters used to develop the plan and the position along the geostationary orbit. Under the present Radio Regulations and the practices currently employed in the international coordination process, especially for the fixed-satellite service, changes could be made relatively easily. During international coordination of networks, every effort would be made to preserve as much flexibility as possible to accommodate future requirements for each network.

6.3 Planning methods [1-5]

6.3.1 World or Regional Detailed Long-Term [10-20 years] a priori Allotment Plan – METHOD 1

A long-term world or regional *a priori* frequency/orbit allotment plan with a procedure for the revision of requirements that is similar to Article 4 of Appendix 30 (the 1977 *Broadcasting Satellite Plan*). Under this procedure new requirements may be accommodated only if they do not cause unacceptable interference to those networks within the Plan.

6.3.1.1 Service requirements

Radiocommunication service requirements and their associated service areas expected to be implemented during the period of the plan (10-20 years) would be submitted to the conference by each administration.

6.3.1.2 Technical parameters and criteria

Technical parameters and interference criteria necessary to provide the basis for the plan would be adopted by the conference.

6.3.1.3 Structure and allotments

Based on the service requirements, technical criteria, and available orbit/spectrum capacity, allotments would be made to each administration taking into account existing systems. The results of this allotment agreement would be contained in the Final Acts of the conference.

6.3.1.4 Modifications

Changes in service requirements, service areas, technical parameters, or the introduction of new systems would only be possible if such modifications did not introduce unacceptable interference to the systems operating in accordance with the plan.

6.3.2 Periodically Revised [3-5 years] World or Regional Detailed Allotment Plan – METHOD 2

Conferences would be convened periodically (3-5 years) to revise the technical parameters. At each conference it is understood that all of the existing networks and all of the new or modified requirements would be accomplished. During the interval between conferences, new requirements would be accommodated to the extent that they did not cause unacceptable interference to networks in the plan.

6.3.2.1 *Service requirements*

Radiocommunication service requirements and their associated service areas expected to be implemented during the period of the plan (3-5 years) would be submitted to the conference by each administration. At succeeding conferences, requirements that had not been implemented in the interval between conferences would be considered on an equal basis with new requirements.

6.3.2.2 *Technical parameters and criteria*

Technical parameters and interference criteria necessary to provide the basis for the plan would be adopted by the conference. Succeeding conferences would review them to ensure the efficient use of the orbit/spectrum resource based upon the most recent technological advances.

6.3.2.3 *Structure and allotments*

Based on the service requirements, technical criteria, and available orbit/spectrum capacity, allotments would be made to each administration taking into account existing systems. The results of this allotment agreement would be contained in the Final Acts of the conference.

6.3.2.4 *Modifications*

Modifications could be made in two ways. First, the technical parameters, interference criteria and regulatory procedures would be reconsidered and modified at future conferences; and secondly, modifications during the interim period would be those that did not cause unacceptable interference to systems within the plan.

6.3.3 *World, Regional or Sub-Regional Allotment Plan with Guaranteed Access – METHOD 3*

Conferences would be convened from time to time as required (at intervals of 10 years or less) to revise the overall technical parameters and regulatory procedures. At these conferences, all existing networks and new requirements would be accommodated in the plan. Between conferences, there would be guaranteed access for new requirements. Access would be guaranteed by such mechanisms as reserving spectrum/orbit capacity for future requirements unforeseen at the time of the conference or by the subsequent convening of a special meeting.

6.3.3.1 *Service requirements*

Radiocommunication service requirements and their associated service areas expected to be implemented during the period of the plan (10 years or less) would be submitted to the conference by administrations (world-wide, regional, or sub-regional, as appropriate).

6.3.3.2 *Technical parameters and criteria*

The technical parameters and criteria necessary to accommodate the service requirements and to create an orbit/spectrum capacity reserve would be adopted by the conference. In the case of regional or sub-regional planning, these technical parameters and criteria would be chosen to result in relative independence between the regions or sub-regions, taking into account those parameters and criteria for inter-regional systems.

6.3.3.3 *Structure and allotments*

The structure and allotments in the plan could specify orbital locations, channel bandwidths, transmission parameters and other technical characteristics.

Alternatively, the allotments could be made to provide flexibility in implementing systems, e.g., using such techniques as block frequency allotments.

6.3.3.4 *Modifications*

Modifications to accommodate new requirements which arise between planning conferences could be implemented in two ways; first, from the reserve capacity established at the prior conference, and secondly, in the event there is saturation within the plan or because of constraints resulting from sharing with another service of equal status, then regulatory procedures would be invoked to guarantee access.

Modifications to the plan are also made at the periodic conferences. These conferences would also modify the technical parameters and interference criteria to accommodate new requirements and to re-establish the reserve capacity.

6.3.4 *Guaranteed access by means of multilateral coordination – METHOD 4*

The conference would not establish a formal plan but would establish procedures for guaranteed frequency/orbit access for new requirements. Normally, frequency/orbit access would be coordinated in accordance with the procedures contained in Method 5. When a new requirement could not readily be accommodated a special meeting would be called of those administrations which might be affected and a means be found to accommodate the new requirement.

6.3.4.2 *Service Requirements*

Service requirements are those for which the appliant has designed a system and which he is prepared to implement within a given period of time.

6.3.4.2 *Technical parameters and criteria*

Technical parameters and interference criteria would comply with the most recent design, operating and coordination guidelines for systems operating in that portion of the geostationary arc.

6.3.4.3 *Structure and allotments*

The decisions reached in multilateral coordination represent a re-allotment of the orbit/spectrum resource among the new system operator and the existing and previously notified systems.

6.3.4.4 *Modifications*

Modifications to the technical and operating guidelines for particular segments of the geostationary orbital arc are made at the multilateral coordination meetings. Assignments to operating and previously notified systems may also be modified if required to accommodate a new entrant.

6.3.5 *Co-ordination Procedures and Technical Factors which are Revised Periodically – METHOD 5*

This approach to planning is a phased revision of the existing regulatory procedures, regulations and CCIR *Recommendations* as well as the development of new procedures, regulations and *Recommendations* (simplified to the extent possible) leading to more efficient use of the geostationary satellite orbit/spectrum resource.

6.3.5.1 *Service requirements*

The near term requirements of administrations would be implicit in the filings submitted to the IFRB at the time of advance publication.

6.3.5.2 *Technical parameters and criteria*

As in the case of the other planning methods, it is implicit that the system technical characteristics would conform to the Radio Regulations and CCIR Recommendations effective at the time of advance publication.

6.3.5.3 *Structure and allotments*

Assignments in the Master Register would continue to be the basic instrument of the plan.

6.3.5.4 *Modifications*

Modifications are incorporated on the basis of an expressed need for orbit/spectrum capacity for individual systems. The CCIR has the responsibility to upgrade the technical standards for the particular service as technology permits, and would continue to do so on a formal basis in order to develop orbit/spectrum capacity in excess of the need. Consequently, modifications to the operational and regulatory provisions relating to the particular service will continue to be improved leading to a simplification in the use of these provisions, and enhancing access to the geostationary orbit by all administrations.

6.4 *Suggested criteria for study of the application of planning methods*

Objective criteria could be used in determining which of the various possible methods of planning the use of orbit and spectrum will be most satisfactory in any specific set of circumstances. The following criteria are suggested taking into account the relative magnitudes of system and administrative costs:

(A) Access criteria

1. Equitable Access

Does the method guarantee in practice for all countries equitable access to the geostationary satellite orbit and the frequency bands allocated to space services?

2. Service requirements

(a) Is it possible to establish realistic forecasts of traffic requirements to be used as the basis for allotments in the plan?

(b) Can allotments be defined in the plan that will accommodate the likely variety of requirements?

3. Accommodation of unforeseen new networks or changes in traffic requirements

Is there a procedure for accommodating unforeseen new networks or increasing or decreasing (e.g., when traffic is transferred to new satellites in other frequency bands) traffic requirements?

4. Accommodation of existing networks

Does the method ensure equitable treatment and minimize the dislocation of existing operational networks during hte implementation and operation of the method?

5. Access for multi-administration satellite networks

Does the method cater for the introduction or expansion of multi-Administration satellite networks; global, regional or sub-regional?

6. Establishment and modification of technical parameters and interference criteria

Can technical parameters and interference criteria be established and maintained for the life of the plan that will accommodate changing technology and service requirements? Is there a provision for modifying the technical parameters and interference criteria of the plan to take advantage of the technical developments that are more efficient and/or less costly?

(B) Technical criteria

7. Restrictions due to sharing with terrestrial services

Does the method impose any additional sharing constraints on space or terrestrial services due to sharing the same frequency allocation?

8. Restrictions due to sharing with an unplanned space service

Does this method impose any additional sharing constraints either on the elaboration of the plan or on the unplanned space service due to sharing a frequency allocation?

9. Efficient use of the orbit/spectrum

(a) Does the method make efficient use of the orbit/spectrum resource?

(b) Is there incentive to use optimum technical standards?

(C) Cost criteria

10. Impact on satellite system costs

Are there features of the plan that, over the life of the plan, are likely to force administrations to utilize more costly satellite systems?

11. Administrative costs

Does the administrative implementation and operation of the plan involve substantial work for administrative and technical staff, taking into account the relative magnitude of system and administrative costs?

REFERENCES
for the Appendices

1. Reinhart, E. E., "Radio-Relay System Performance in an Interference Environment," RAND Corp. Memo. RM-5786-NASA, Oct. 1968.

2. CCIR: XIVth Plenary Assembly, Kyoto, Japan, Vol. IV, *Rec. 353-3* (Allowable noise power in the hypothetical reference circuit for frequency-division multiplex telephony in the fixed satellite service), Geneva, 1978.

3. CCIR: XIVth Plenary Assembly, Kyoto, Japan, Vol. IX, *Rec. 405-1,* (Pre-emphasis characteristics for frequency modulation radio-relay systems for television), Geneva, 1978.

4. Campanella, S. J., H. G. Snyderhood, and M. Wachs, "Frequency Modulation and Variable-Slope Delta Modulation in SCPC Satellite Transmission," *Proc. IEEE,* Vol. 65, No. 3, March 1977, pp. 419-434.

5. Johannsen, K., G. Tustison, and S. Egh, "Notes on SCPC-FM System Design and Performance," *IEEE Trans. Aerosp. & Electronic Sys.,* Vol. 16, No. 5, Sept. 1980, pp. 683-711.

6. CCIR: XIVth Plenary Assembly, Kyoto, Japan, Vol. IV, *Report 391-3* (Radiation diagrams of antenna for earth stations in the fixed satellite service for use in interference studies and the determination of a design objective), Geneva, 1978.

7. CCIR; XIVth Plenary Assembly, Kyoto, Japan, Vol. IV, *Rec. 465-1* (Reference earth station radiation pattern for use in coordination and interference assessment in the frequency range from 2 to about 10 GHz) Geneva, 1978.

8. Jeruchim, M. C., "A Statistical Approach to Satellite Interference Levels," *Conf. Rcd.,* Vol. 3, pp. 35.3.1-35.3.4 Intern. Conf. Commun. (ICC '78) Toronto, June 1978.

9. CCIR: Study Groups Doc. 4/32. "Proposed Amendment to Report 391-2: Some Statistical Properties of Antenna Side Lobes," 19 Jan, 1976.

10.   Karmel, P. R., "Statistical Properties of Antenna Sidelobes," *COMSAT Tech. Rev.*, Vol. 9, No. 1, Spring 1979, pp. 91-120.

11.   Jeruchim, M. C., and J. H. Moore, "The Effect of Station-Keeping Error on the Distribution of Carrier-to-Interference Ratio," *IEEE Trans. Commun.*, (Part II), Vol. COM-30, No. 7, pp. 1786-1791, July 1982.

12.   Bantin, C. C., and R. G. Lyons, "The Evaluation of Satellite Link Availability," *IEEE Trans. Commun.*, Vol. COM-26, No. 6, June 1978, pp. 847-853.

13.   CCIR: XIVth Plenary Assembly, Kyoto, Japan, Vol. IV, *Report 556-1* (Factors Affecting station-keeping of geostationary satellites of the fixed satellite service), Geneva, 1978.

14.   ITU, Final Acts of the 1979 World Administrative Radio Conference, Geneva, Switzerland, 1979.

15.   Gallager, R. G., "Information Theory and Reliable Communication," New York, John Wiley, 1968.

16.   Abramson, N., "Bandwidth and spectra of phase and frequency-modulated waves," *IEEE Trans. Commun. Syst.*, Vol. CS-11, pp. 407-414, Dec. 1963.

17.   CCIR: XIVth Plenary Assembly, Kyoto, Japan, Vol. IV, *Report 464* (Pre-emphasis characteristics for frequency-modulation systems for frequency-division multiplex telephony in the fixed satellite service), Geneva, 1978.

18.   CCIR: XIVth Plenary Assembly, Kyoto, Japan, Vol. IV, *Report 380-3* (Interconnection at baseband frequencies of radio-relay systems for telephony using frequency-division multiplex), Geneva, 1978.

19.   Glance, B., "Power Spectra and Multilevel Digital Phase-Modulated Signals," *B. S. T. J.*, Vol. 50, No. 9, pp. 2857-2878, Nov. 1971.

SELECT BIBLIOGRAPHY

Listed here are works generally applicable to the subject matter of *Communication Satellites in the Geostationary Orbit* which have not been cited elsewhere in the book.

During the preparation of this book, the latest edition of the CCIR "Green Books" was the 1978 edition. In the meantime, the XVth Plenary Assembly was held in 1982, and the reader consulting CCIR *Recommendations* and *Reports* cited in the book should refer to the latest edition.

Baghdady, E.J., *Lectures on Communication System Theory*, New York: McGraw Hill, 1961, pp. 483-508.

Blanchard, A., "Interference in phase-locked loops," *IEEE Trans. Aerosp. Electron. Syst.*, Vol. AES-10, pp. 686-697 (corrections in Vol. AES-11, pp. 285-286), Sept. 1974.

Bradley, W.E., "Communications stategy of geostationary orbit," *Astronaut. Aeronaut.*, Vol. 6, No. 4, pp. 35-41, Apr. 1968.

Brown, R.J., M.L. Guha, R.A. Hedinger, and M.L. Hoover," Companded Single Sideband Satellite Transmission," *GLOBECOM 1982 Record,* Miami, Florida, 1982.

Cahn, C.R., "Worst interference for coherent binary channel," *IEEE Trans. Inform. Theory*, Vol. IT-17, No. 2, pp. 209-210, Mar. 1971.

CCIR: XIVth Plenary Assembly, Kyoto, Japan, Vol. IV, *Rec. 446-2* (Carrier energy dispersal for systems employing modulation by analogue signals or digital modulation in the fixed satellite service), Geneva, 1978.

CCIR: XIVth Plenary Assembly, Kyoto, Japan, Vol. IV, *Rec. 483-1* (Maximum permissible level of interference in a television channel of a geostationary satellite network in the fixed satellite service employing frequency modulation caused by other networks of this service), Geneva, 1978.

CCIR: XIVth Plenary Assembly, Kyoto, Japan, Vol. XI, *Rec. 500* (Method for the subjective assessment of the quality of television pictures), Geneva, 1978.

CCIR: XIVth Plenary Assembly, Kyoto, Japan, Vol. IV, *Rep. 213-3* (Use of pre-emphasis in frequency modulation systems for frequency division multiplex telephony and television in the fixed satellite service), Geneva, 1978.

CCIR: XIVthe Plenary Assembly, Kyoto, Japan, Vol. IV, Rep. 559 (The effect of modulation characteristics on the efficiency of use of the geostationary-satellite orbit in the fixed satellite service), Geneva, 1978.

CCIR: Canadian Contribution to IWP 4/1 Meeting, "Pairing of frequency bands used by the fixed-satellite service," 3-8 June 1975.

CCIR: Canadian Contribution to IWP 4/1 Meeting, "Space diversity as a possible means of improving the efficiency of utilization of the geostationary satellite orbit at frequencies above 10 GHz," 3-8 June 1975.

CCIR: Draft New Report, "A Survey of Interference Cancellers for Application in the Fixed Satellite Service," Doc. USSG 4/7 (USA), 12 March 1981.

CCIR: Japanese Contribution to IWP 4/1 Meeting, "Efficiency of orbit/spectrum utilization by use of orthogonal polarization," 3-8 June 1975.

CCIR: Study Group Doc. IWP 4/1-4 (USA), "The effects of geography on the use of the geostationary orbit," 3 April 1980.

Chayavadhanangkur, C., and J.H. Park, Jr., "Analysis of FM systems with co-channel interference using a click model," *IEEE Trans. Commun.*, Vol. COM-24, No. 8, pp. 903-910, Aug. 1976.

Childs, W.H., P.A. Carlton, R.G. Egri, C.E. Mahle, and A.E. Williams, "A 120-MBITS/S 14 GHz Regenerative Receiver for Spacecraft Applications," *COMSAT Tech. Rev.*, Vol. 11, No. 1, pp. 103-130, Spring 1981.

Colavito, C., and M. Sant'Agostino, "Multiple co-channel and interchannel interference effects in binary and quaternary PSK radio system," presented at Int. Conf. Communications, Philadelphia, PA, June 19-21, 1972.

Colavito, C., and M. Sant'Agostino, "Binary and quaternary PSK radio systems in a multiple-interference environment," *IEEE Trans. Commun.*, Vol. COM-21, pp. 1056-1067, Sept. 1973.

Corrington, M.S., "Frequency modulation distortion caused by common- and adjacent-channel interference," *RCA Review*, Vol. 7, No. 4, pp. 522-560, Dec. 1946.

Curtis, H.E., "Radio frequency interference considerations in the TD-2 radio relay system," *Bell Syst. Tech. J.*, Vol. 39, pp. 369-387, Mar. 1960.

Curtis, H.E., "Interference between satellite communication systems and common carrier surface systems," *Bell Syst. Tech. J.*, Vol. 41, pp. 921-943, May 1962.

Evans, H.W., "Technical background, AT&T domestic satellite proposal," presented at AIAA 2nd Communications Satellite Systems Conf., San Francisco, CA, 8-10 Apr. 1968, AIAA Paper 68-411.

Fuenzalida, J.C., "A comparative study of the utilization of the geostationary orbit," *Proc. INTELSAT/IEE Conf. Digital Satellite Communication*, pp. 213-225, Nov. 1969.

George, D.A., and J.S. Bird, "The use of the Fourier-Bessel series in calculating error probability for digital communication systems," *IEEE Trans. Commun.*, Vol. COM-29, No. 9, pp. 1357-1365, Sept. 1981.

Gould, R.G., "Regulatory Aspects of Digital Communications (Radio Frequency Considerations)," *IEEE Trans. Commun.*, Vol. COM-24, pp. 118-131, Jan. 1976.

Granlund, J., "Interference in frequency-modulation reception," M.I.T. Research Lab. Electronics, Cambridge, MA, *Tech. Rep. 42*, 20 Jan. 1949.

Horstein, M., "Satellite adjacent-channel interference due to multicarrier transponder operation," presented at Int. Conf. Communications, Philadelphia, PA, 19-21 June 1972.

Hult, J.L., et al., "The technology potentials for satellite spacing and frequency sharing," RAND Corp. Memo. RM-5785-NASA, Santa Monica, CA, Oct. 1968.

Hult, J.L., and E.E. Reinhart, "Satellite spacing and frequency sharing for communication and broadcast services," *Proc. IEEE*, Vol. 59, pp. 118-128, Feb. 1971.

IEEE: Special Issue on the 1979 World Administrative Radio Conference (WARC 79), *IEEE Trans. Commun.*, Vol. COM-29, No. 8, Aug. 1981.

Jansky, D.M., and M.C. Jeruchim, "A technical basis for communication satellites to share the geostationary orbit," presented at AIAA 3rd Communications Systems Conf., Los Angeles, CA, Apr. 1970, AIAA Paper 70-441.

Jansky, D.M., and M.C. Jeruchim, "Technical factors and criteria affecting geostationary orbit utilization," *Communications Satellite Systems for the 70's: Systems*, N.E. Feldman and C.N. Kelly, Ed., Cambridge, MA: M.I.T. Press, 1971.

Jeruchim, M.C., "Orbit utilization," paper presented at Electronics and Aerospace Systems Conv., Washington, DC, Oct. 1971.

Jeruchim, M.C., D.A. Kane, F.C. Moore, and T.C. Sayer, "Orbit/Spectrum Utilization Study," General Electric Co., Space Systems Organization, Valley Forge Space Center, 1969-1970. Vol. I (Interim Report), G.E. Co. Doc. No. 69SD4270 (NTIS: PB-194-781), 15 May 1969. Vol. II, G.E. Co. Doc. No. 69SD4348 (NTIS: PB-194-780), 12 Sept. 1969. Vol. III, G.E. Co. Doc. No. 70SD4246 (NTIS: PB-194-782), 30 June 1970. Vol. IV, G.E. Co. Doc. No. 70SD4293 (NTIS: PB-203-389), 31 Dec. 1970.

Jeruchim, M.C., H.Ng, and D.M. Jansky, "Regulatory and technical factors in geostationary orbit utilization," *IEEE Trans. Commun.*, Vol. COM-27, No. 10, pp. 1544-1550, Oct. 1979.

Jowett, J.K.S., and A.K. Jefferis, "Ultimate communications capacity of the geostationary-satellite," *Proc. Inst. Elec. Eng.*, Vol. 116, pp. 1304-1310, Aug. 1969.

Kane, D.A., and M.C. Jeruchim, "Orbital and frequency sharing between broadcasting-satellite service and fixed-satellite service," paper presented at Int. Conf. Communications, Philadelphia, PA, 19-21 June 1972.

Kostic, I.M., "Error rate of phase-modulated signals in multilink system with interference," *Electronic Letters*, Vol. 16, pp. 599-600.

Krishnamurthy, J., and G.L. Paradeep, "Co-channel interference in partially coherent communication," *IEEE Trans. Aerosp. Electron. Syst.*, Vol. AES-13, No. 2, pp. 120-126, Mar. 1977.

Kurjan, D.J., and M.R. Wachs, "TV co-channel interference on a PCM-PSK SCPC system," *COMSAT Tech. Rev.*, Vol. 6, No. 2, pp. 413-424, Fall 1976.

Lester, R.M., "The Introduction of New Satellites to an Operating System,' Intern. Conf. Commun. (ICC), pp. 58.2.1-58.2.4, Denver, CO, June 1981.

Lyons, R.G., "Signal and interference output of a bandpass nonlinearity," *IEEE Trans. Commun.*, Vol. COM-27, pp. 888-891, June 1979.

Maciejko, R., "Digital modulation in Rayleigh fading in the presence of co-channel interference and noise," *IEEE Trans. Commun.*, Vol. COM-29, No. 9, pp. 1379-1386, Sept. 1981.

Matsumoto, M., and G.R. Cooper, "Multiple narrow-band interferers in an FH-DPSK spread-spectrum communication system," *IEEE Trans. Vehicular Technol.*, Vol. VT-30, Feb. 1981.

Park, J.H., and C. Chayavadhanangkur, "Effect of fading on FM reception with cochannel interference," *IEEE Trans. Aerosp. Electron. Syst.*, Vol. AES-13, No. 2, pp. 127-132, Mar. 1977.

Sawitz, P.H., "Spectrum-orbit utilization: An overview," paper presented at National Telecommunications Conf., New Orleans, LA, pp. 43-1, 43-7, Dec. 1-3, 1975.

"SSB interference earmarked as most damaging to digital communications systems," *Commun. Design*, Oct. 1972.

Stavroulakis, P., and S.C. Moorthy, "A statistical approach to the interference reduction of a class of satellite transmissions," *Conf. Rd.*, Vol. 3, pp. 52.3.1-52.3.5, Nat'l Telecommun. Conf. (NTC 79), Washington, DC, Dec. 1979.

Stavroulakis, P., ed., *Interference Analysis of Communication Systems*, New York: IEEE Press, 1980.

Stumpers, F.L.H.M., "Interference problems in frequency modulation," *Philips Research Reports*, Vol. 2, No. 2, pp. 136-160, Apr. 1947.

Withers, D.J., "Effective utilization of the geostationary orbit for satellite communications," *Proc. IEEE*, Vol. 65, No. 3, pp. 308-317, Mar. 1977.

Yue, O., "Saddle-point approximation for M-ary phase-shift-keying with adjacent satellite interference," *B.S.T.J.*, Vol. 59, No. 7, pp. 1139-1152, Sept. 1980.

INDEX